T0292401

A PARADIGM SHIFT TO PREVENT AND TREAT ALZHEIMER'S DISEASE

A PARADIGM SHIFT TO PREVENT AND TREAT ALZHEIMER'S DISEASE

From Monotargeting Pharmaceuticals to Pleiotropic Plant Polyphenols

HOWARD FRIEL

and

SALLY FRAUTSCHY

ACADEMIC PRESS

An imprint of Elsevier

Academic Press is an imprint of Elsevier
125 London Wall, London EC2Y 5AS, United Kingdom
525 B Street, Suite 1800, San Diego, CA 92101-4495, United States
50 Hampshire Street, 5th Floor, Cambridge, MA 02139, United States
The Boulevard, Langford Lane, Kidlington, Oxford OX5 1GB, United Kingdom

Notices
Knowledge and best practice in this field are constantly changing. As new research and experience broaden our understanding, changes in research methods, professional practices, or medical treatment may become necessary.

Practitioners and researchers must always rely on their own experience and knowledge in evaluating and using any information, methods, compounds, or experiments described herein. In using such information or methods they should be mindful of their own safety and the safety of others, including parties for whom they have a professional responsibility.

To the fullest extent of the law, neither the Publisher nor the authors, contributors, or editors, assume any liability for any injury and/or damage to persons or property as a matter of products liability, negligence or otherwise, or from any use or operation of any methods, products, instructions, or ideas contained in the material herein.

Library of Congress Cataloging-in-Publication Data
A catalog record for this book is available from the Library of Congress

British Library Cataloguing-in-Publication Data
A catalogue record for this book is available from the British Library

ISBN: 978-0-12-812259-4

For information on all Academic Press publications visit our website at
https://www.elsevier.com/books-and-journals

Working together
to grow libraries in
developing countries

www.elsevier.com • www.bookaid.org

Publisher: Mica Haley
Acquisitions Editor: Kristine Jones/Erin Hill-Parks
Editorial Project Manager: Molly Mclaughlin/Tracy Tufaga
Production Project Manager: Chris Wortley
Designer: Matthew Limbert

Typeset by TNQ Books and Journals
Howard Friel is the copyright holder of the tables that appear in this book.

Picasso always smiles when you look the right way.
Lutz Engelke, "Picasso's Smile," Ithaca, New York, 1985

PRAISE FOR *A PARADIGM SHIFT TO PREVENT AND TREAT ALZHEIMER'S DISEASE*

"This is a wonderful book for people who are looking for ways by which we will overcome Alzheimer's disease, the most common dementing disorder affecting middle-aged and old people.

On understanding of the pathophysiology of Alzheimer's disease, the authors clearly explain and discuss multiple ('pleiotropic') effects of plant polyphenols on Alzheimer's disease from multiple viewpoints, including epidemiology, sequential events in the brain that lead to disease initiation and progression, and current status of human trials with polyphenols.

Plant polyphenols that have effects on the multiple pathomechanisms of Alzheimer's disease are expected to prevent or delay the disease onset and progression safely, effectively, and inexpensively. We need further studies to establish this."

Masahito Yamada is Professor and Chair, Department of Neurology and Neurobiology of Aging, Kanazawa University Graduate School of Medical Sciences, Japan

"This timely book by Howard Friel and Sally Frautschy provides an in-depth discussion of an approach to the treatment of Alzheimer's disease (AD) that is very distinct from the one that has been taken by the pharmaceutical industry. Since none of the drug candidates promoted by the pharmaceutical industry for AD treatment over the past 20 years have succeeded in the clinic, alternative approaches are desperately needed. This book should be of great interest to clinicians, health care workers and research scientists as well as the informed layperson. For clinicians/health care workers, the book provides a potential road map for the treatment of patients ranging from those with only very mild cognitive problems to cases of more severe memory impairment.

Since there are currently no effective treatments for AD, this road map could give new hope to patients and their care providers. For the research scientist outside the AD field, the book provides an excellent, thoroughly referenced introduction to AD. And, for the AD scientist, the book suggests new avenues of research. Finally, for the informed lay person, the book identifies safe approaches for the prevention of this devastating disease. Thus, this volume makes a significant and much needed contribution to the AD field and should be highly appreciated by many."

Pamela Maher is a Senior Staff Scientist in Cellular Neurobiology at the Salk Institute for Biological Studies in La Jolla, California, United States

"This book from Friel and Frautschy rigorously supports the thesis that multi-mechanistic, non-toxic, polyphenolic compounds found in the ancient foods of tea, turmeric, olive oil, and red wine are superior to mono-targeting pharmaceuticals for the prevention and treatment of numerous chronic diseases in humans, in this case Alzheimer's disease."

Navindra P. Seeram is an Associate Professor in the Department of Biomedical and Pharmaceutical Sciences, Botanical Research Laboratory, College of Pharmacy, University of Rhode Island, United States

CONTENTS

PART III: RATIONAL BASIS VERSUS STRICT SCRUTINY

PREFACE BY FRIEL

In the summer of 2013, on a bayside beach at Cape Cod, Massachusetts, as the sun was quickening its dusky pivot down toward Provincetown, my wife, who occupied the beach chair next to mine, handed me a copy of *The New Yorker* and said, "You should read this." The magazine was open to "Before Night Falls" by Jerome Groopman. I asked, "Is this about what to do at Cape Cod before the sun sets?" "It's about Alzheimer's disease," she said. I read the article and have been reading about Alzheimer's disease (AD) ever since. Chapter 1 begins with a reference to Groopman's article as a salute to my wife's farsighted instincts.

While reading the published literature on AD, I occasionally encountered the words "monotargeting" and "monotherapy" and their synonyms (references to the limited pharmaceutical targeting of the biomarkers of AD and other chronic diseases) and "monoclonal antibodies" (the likely next generation of FDA-approved drugs for AD). When I read "Why Pleiotropic Interventions Are Needed for Alzheimer's Disease" by UCLA neuroscientists Sally Frautschy and Greg Cole,[1] wherein pleiotropism ("multi-targeting") could be viewed as (1) an upgrade over monotargeting and (2) the antidote to the intellectual monocentrism of pharmaceutical research, "I had a feeling there was a book in me."[2]

What occurred to me was that pleiotropism is not a theory of pathogenesis; rather, it is an approach to treatment that could accommodate various theories of pathogenesis. In other words, a promising treatment for AD may not have to wait for "the one right theory" about AD pathogenesis to emerge. Even assuming that any such hypothesis might come forward as a consensus view or scientifically provable, the current "one compound for one target within the right hypothesis" approach to treatment probably would still fall short, given the enduring fact of the pathophysiological complexity of AD. In contrast, given what we currently know about AD pathophysiology, much of which is sketched in this volume, it is unlikely that any pleiotropic treatment model that might emerge from any newly hatched consensus framework of AD would look much different to the one presented here, which, while favoring the "amyloid hypothesis" of AD, nevertheless accommodates numerous other credible ideas about the AD-related disease process. This "accommodationist pleiotropism" toward prevention and treatment, in contrast with "rejectionist monotargeting," which excludes potentially viable targets beyond the one target that is targeted, is herein grounded in the pleiotropic capabilities of the plant-based polyphenolic compounds featured in this volume, our best expression of which is Table 6.4 on page 208. It also seems plausible that, in addition to AD, these observations apply to other neurological diseases and major chronic diseases beyond those.

A somewhat self-evident feature in the scientific literature on AD is the disparity between the exceptionally high level of research into the biomechanics of AD pathophysiology and the disappointing efforts to apply that research to safely and effectively prevent and treat the disease. The discussions about this dichotomy seemed strangely unsatisfactory, and for the most part appeared to avoid any focus on the fundamental imbalance involved in monotargeting a complex, multidimensional disease process. Because the Frautschy–Cole piece on "pleiotropism" and AD also served as a gateway to a large cache of applied research featuring plant polyphenol multitargeting of AD, the structural flaws of the monotargeting pharmaceutical model loomed even larger.

In short, the basic science of AD has established that AD involves a highly complex disease process with multiple dynamic targets and feedback systems, which renders it largely resistant to the pharmaceutical-based "one compound, one target" model of treatment; yet, "one compound, one target" is still the dominant paradigm in applied research on AD. However, the dominance of monotargeting as applied to AD is viewed in growing segments in the scientific literature as an inherently flawed approach to treating the disease.

The alternative, pharmacologically, and as increasingly recognized, is the pleiotropism of certain food-based plant polyphenols, which possess not only multiple-targeting characteristics, but may be capable of nontoxically and comprehensively targeting all of the major pathogenic hallmarks of AD; this is regardless of when and whether they proceed linearly (upstream–downstream), in parallel, in destructive cycles, or as a complex of each of these, or whether the "amyloid hypothesis" holds or not.

Part of the narrative framework in this volume is the textual display of many of the references used. This includes the title of the science or medical journal in which a cited work was published, the journal's country of origin, date of publication of the cited study or review (which for the most part are organized chronologically to reflect any progression in the scientific thinking), and the medical centers or university-based labs in which the research was conducted. This "display" method of referencing clearly exhibits the formidable nature of the research that has been published over the past 10–15 years as the counternarrative to monotargeting. Since most counternarratives bear a higher burden of evidence, hopefully this method of referencing sources and content more than meets that rigorous standard.

I would like to thank Sally Frautschy for her contributions, constructive criticisms, and indispensable scientific insights. Sally and I also would like to thank Elsevier's superb senior acquisitions editor for Biomedicine and Pharmacology, Kristine Jones, for her ceaseless diligence and work on behalf of this project. We also would like to thank many other members of Elsevier's excellent staff, especially Erin Hill-Parks, Chris Wortley, Molly McLaughlin, and Rajaganapathy Essakipandyan. I most especially would like to thank my wife, Michelle Joffroy, without whom this project would not have begun, continued, or been completed, and who spent many hours reading and commenting on each chapter of this book. I also thank my late parents, Anna and Howard, both of whom to this day serve as my nearly constant counselors and companions.

Howard Friel
October 15, 2016

REFERENCES

1. Frautschy, Cole. Why pleiotropic interventions are needed for Alzheimer's disease. *Molecular Neurobiology* June 2010;**41**(2–3):392–409.
2. This invokes the words of Norman Mailer about his Pulitzer–Prize winning, National–Book Award winning, historical novel of 1968, The Armies of the Night. See, YouTube, William F. Buckley Interviews Norman Mailer on Firing Line Part 1, as of October 15, 2016, at https://www.youtube.com/watch?v=O3vTu99Vpd8.

PREFACE BY FRAUTSCHY

When Howard Friel contacted me, expressing interest in a book on the use of plant polyphenols to treat or prevent Alzheimer's disease (AD), I was at first skeptical, but quickly became intrigued when I saw that he had written books on many disciplines, including climate change,[1] which my father and Roger Revelle had worked on at the Scripps Institution of Oceanography. As the advantages of Howard's interdisciplinarity and detailed familiarity with AD soon became obvious, our discussion advanced to several substantive matters, including the slow progress on finding ways to treat AD. The four drugs that had been approved by the U.S. Food and Drug Administration, and that are currently in use to treat AD, are grounded in decades-old theories about the disease—cholinergic loss and/or excitotoxicity. These drugs, at best, have meager impacts on AD. We both viewed the combination of the ineffectiveness of medical treatment and the abundance of recent research on plant polyphenols as an opportunity to contribute a scholarly monograph about a different approach to looking at and treating AD. So, our partnership began.

I am frequently bombarded with questions from family members and others suffering from AD or other intractable diseases who are eager to learn more about the science behind botanical compounds. This also led me to believe that there was an urgent need for the type of book that Howard was proposing. People outside the field of AD research needed to understand more about the impasse in the scientific arena about preventing and treating AD. And a book of the kind that we were considering also could spark a perceptual shift within the profession toward better ways of preventing and treating AD and improving the way we test phytochemical approaches through clinical trials.

Numerous epidemiological and some clinical studies show that diets high in fruits and vegetables are associated with a reduced risk of AD, due perhaps to their high content of polyphenols. Further cellular and animal models were known to show potent phytochemical effects, as good as any drug, on pathways implicated in AD. The scientific community remained skeptical, and not just because without patents and support from the pharmaceutical industry no one would pay for it. Many expressed disbelief that animal models could translate to humans, even though the amyloid vaccine developed in animal models that cleared amyloid from the brain did the same in human trials. In humans and mice, vaccines prevented cognitive deficits, but did not do much if given at late stages. It seemed obvious that reducing amyloid would not be sufficient to suppress its downstream pathogenic pathways and cure AD, just as reducing mutagen exposure by quitting smoking would not be sufficient to cure lung cancer. Without doubt, people with certain lifestyles appear to delay or prevent AD. Certainly the "impossible" is possible, because we are already doing it.

Interest by me and my husband, Greg Cole, in phytochemicals began after reports that the first AD prevention trials failed with vitamin E, which blocks amyloid-associated oxidative damage in the cell culture dish. The results were negative, but led to the erroneous conclusion that oxidative damage was not contributing to the disease and ignored the fact that vitamin E poorly penetrates the brain. Clinical trialists and NIH dropped strategies against oxidative damage altogether, while we suspected that attenuating oxidative damage would be an insufficient treatment as evidence mounted that (1) people on nonsteroidal antiinflammatory drugs (like ibuprofen) for at least 2 years had a much lower risk for AD and (2) several inflammatory pathways were dysregulated in AD brains preceding cognitive deficits. Now, much later, researchers have found that half of the many genes that increase risk of AD control innate immune function and inflammation. So now the need for immunomodulators is driven by compelling new genetic data.

Greg and I decided to test molecules with multiple activities ("pleiotropic") in animal models that could both correct inflammatory dysregulation as well as reduce oxidative damage. The polyphenol curcumin was one of the pleiotropic molecules we screened, and it worked the best. Curcumin was also an amyloid binding molecule, and later turned out to stimulate processes inside and outside the cells to clear the misfolded proteins characteristic of nearly all the neurodegenerative diseases as well as traumatic brain injury. In a nutshell, curcumin is a pleiotropic molecule that can target and correct dysregulation in at least five pathways critical for AD pathogenesis. We need to see what it can do in the clinic.

My personal mission to help heal the injured brain started in the late 1970s, when I woke up after days in the hospital to learn that I had suffered from a traumatic brain injury in a bicycle accident. I was an undergraduate at the University of California at Davis, and confident that I would quickly recover as they withdrew the feeding tube out of my nose. I had no idea that I would spend much of the next decade struggling to overcome the obstacles associated with the phases of posttraumatic injury. There was no magic bullet to overcome these obstacles. That personal experience and my background in physiology studying paracrine and endocrine feedback systems and homeostatic mechanisms taught me beyond any doubt about the futility of a one-target molecular approach for a disease so much more complex than posttraumatic brain injury.

Sally Frautschy
October 29, 2016

REFERENCE

1. Friel H. *The lomborg deception: setting the record straight about global warming.* New Haven: Yale University Press; 2010.

PART I

Pleiotropism

CHAPTER 1

Monotargeting Versus Multitargeting

Contents

In summer 2013, Jerome Groopman, a physician and widely published commentator on medical issues, published an article in *The New Yorker* about Alzheimer's disease (AD).[1] Because Groopman identified a number of key characteristics and conundrums of the disease that are best addressed from the start, his article functions here as a suitable way to begin.

As Groopman observed, an unsurprising but little-known feature of AD is the lengthy interval between onset of the AD-related disease process and the emergence of AD-related symptoms. Citing the work of neurologists at Massachusetts General Hospital in Boston, Groopman wrote "structural changes in the brain's memory networks can begin a decade before symptoms become obvious."[1] This turned out to be a conservative estimate. Referring to an early stage of the AD-related disease process, the Mayo Clinic reports "amyloid deposition appears to begin as early as 20 years prior to symptom development."[2] A detailed 2013 study published in *The Lancet Neurology* reported: "Similar to most chronic diseases, Alzheimer's disease develops slowly from a preclinical phase into a fully expressed clinical syndrome," and that "Aβ [amyloid-beta] deposition is slow and protracted" and "likely to extend for more than two decades."[3] A 2014 report in *Expert Review of Neurotherapeutics* also observed that "pathophysiological brain alterations occur decades before clinical signs and symptoms of cognitive decline can be diagnosed."[4] A 2015 study on preclinical (presymptomatic) AD reported in *JAMA Psychiatry* that "Alzheimer disease is now known to have a long preclinical phase in which pathophysiologic processes develop many years, even decades, before the onset of clinical symptoms."[5] And in May 2015, the *New York Times*, citing the *Journal of the American Medical Association*, reported that "amyloid [a protein involved in AD pathogenesis] can appear 20 to 30 years before symptoms of dementia." Samuel Gandy, a prominent neurologist at Mount Sinai Medical Center in New York, and an expert on amyloid, noted in the *Times* that "amyloid could appear as early as age 30." For Gandy, this meant that any medications to prevent the disease must be both safe and effective, since "we'll be exposing people to a drug for decades when they are healthy."[6] We refer hereinafter to this statement by Samuel Gandy as the "Gandy test," which we interpret as a standard of

A Paradigm Shift to Prevent and Treat Alzheimer's Disease
ISBN 978-0-12-812259-4
http://dx.doi.org/10.1016/B978-0-12-812259-4.00001-1

safety for the long-term intake of any biologically active agent to prevent onset of the AD-related disease process or its symptoms. Thus, while the multidecadal prelude to clinical AD provides an opportunity to impede the symptomatic bloom of the disease, the requirement that any preventive agents must be both safe and effective for long-term use represents a significant obstacle that has not been overcome to date.

The primary prevention of AD—which refers to prevention of onset of the AD-related disease process as opposed to postonset prevention to AD dementia—presents additional challenges. For example, lengthy human trials, perhaps lasting 10 years or longer, likely would be required before the U.S. Food and Drug Administration would approve a pharmaceutical product for the primary prevention of AD. However, no such drugs with any likelihood of passing the Gandy test appear to be within reach.

Furthermore, the cost of an FDA-approved drug to treat AD would likely be high, given the prices of pharmaceutical drugs for roughly comparable diseases. This includes multiple sclerosis, about which the *New York Times*, in an editorial called "Runaway Drug Prices," reported in May 2015 that "there are no multiple sclerosis drugs available in the United States with a list price below $50,000 a year."[7] In 2013, in an article about the cancer drug Gleevec titled, "Doctors Blast Ethics of $100,000 Cancer Drugs," CNN reported that "an annual course of Gleevec now wholesales for more than $76,000 in the U.S.," that "the retail price that patients or their insurers pay [for Gleevec] is typically much higher," and that "drug prices for all sorts of conditions are far out of line with any economic basis."[8] Perhaps a drug for the secondary prevention of AD, for example, one that safely cleared amyloid plaque from the brains of cognitively normal individuals, would not be quite as expensive, but it likely would not be cheap.

These questions flow from an arguably optimistic scenario, since they assume that the pharmaceutical industry had designed and developed a drug that could safely prevent and treat AD. However, there are significant scientific challenges to a drug-based remedy for AD. To begin, a fact sheet issued by the National Institute on Aging (NIA) of the U.S. National Institutes of Health, current as of September 2015, reported: "Scientists don't yet fully understand what causes Alzheimer's disease in most people."[9] Another NIA fact sheet stipulates: "Alzheimer's disease is complex, and it is unlikely that any one drug or other intervention can successfully treat it."[10]

The difficulty of developing a drug that solves the highly complex AD-related disease process is reflected in the status of treatment efforts after 30 years of AD research. In his *New Yorker* piece, and addressing the issue of the degree to which pharmaceutical drugs have produced the desired results to date, Groopman wrote that "three decades of Alzheimer's research has done little to change the course of the disease," that "although several initially promising agents have been developed to reverse or at least slow the decline of cognitive function, successive experimental trials have failed," and "some drug therapies come with unpleasant side effects, including headaches and bleeding and swelling of the brain."[1] With regard to drug-related adverse events, an NIA fact sheet on the four FDA-approved medications for AD lists "common side effects" for the three FDA-approved cholinesterase inhibitors as "nausea, vomiting, and diarrhea," in addition to "loss of appetite" for two of the three cholinesterase inhibitors, and "muscle weakness" for one of them. It also lists "common side effects" for the fourth FDA-approved drug for AD (an NMDA inhibitor) as "dizziness, headache, constipation, confusion."[11]

Groopman also observed that an experimental AD drug, "solanezumab," which was the focus of much of his article, was aimed at a single pathogenic target, "a protein called beta-amyloid" (Abeta).[1] Yet, as the Mayo Clinic (in addition to the NIA) has observed—concurrently with the clinical trials on solanezumab and other humanized monoclonal antibodies that target Abeta—"AD is now recognized as a complex disease for which a single therapeutic target may not be sufficient."[12] And with regard to solanezumab in particular, in January 2014, 7 months after the publication date of Groopman's *New Yorker* article, the *New England Journal of Medicine* reported that "solanezumab, a humanized monoclonal antibody that binds amyloid, failed to improve cognition or functional ability" in two large Phase III trials on humans with mild to moderate AD.[13] According to a report issued in 2012 by Pharmaceutical Research and Manufacturers of America, "there have been more than 100 failed attempts to develop a treatment for the neurodegenerative disease [AD] since 1998."[14] Given the complexity of the disease and its unknown and perhaps unknowable pathophysiological quarters, it seems imprudent to assume that a single, safe, AD-treatment product might emerge any time soon from the pharmaceutical and biotech industries.

Prior to the 2014 results on solanezumab, George Perry, an AD scientist at the University of Texas at San Antonio, and editor of the *Journal of Alzheimer's Disease*, as Groopman reported, doubted that singularly targeting amyloid-beta to treat AD would succeed: "These clinical trials," involving monoclonal antibodies, including solanezumab, which are designed to bind to amyloid in the brain, "are unlikely to have a direct therapeutic benefit." Perry continued: "They are extremely naïve and reflect a simplistic view of the [Alzheimer's] disorder." He also observed with frustration that "the beta-amyloid field has dominated both N.I.H. funding and pharmaceutical funding for twenty years." Another AD scientist, Peter Davies, director of the Feinstein Institute of the North Shore-Long Island Jewish Health System, told Groopman that by focusing exclusively on amyloid, the research community has overlooked other potential causes and approaches to prevention and treatment. On this count, Davies said: "It's depressing, because I watch a lot of young people who come up with clever ideas, new ways of thinking, and they are just destroyed." Groopman wrote that Davies "despairs that the string of failed trials targeting beta amyloid has led researchers only to a belief that the [amyloid] protein needs to be targeted earlier, instead of raising the possibility that it might be the wrong target altogether."[1]

Although these criticisms are possibly or partially correct, it is also possible that a number of failed human trials targeting amyloid were the product of faulty research design and flawed experimental drugs, as Harvard's Dennis Selkoe convincingly argued in 2011.[15]

This includes narrowly targeting amyloid too far into the disease process when it is likely best targeted at the earliest stages of AD, in addition to the indiscriminate action of a key experimental drug (semagacestat) on substrates other than amyloid, which was the protein expressly targeted (as amyloid precursor protein). On the other hand, while the amyloid hypothesis may largely explain the pathogenic course of the disease, it does not necessarily follow that narrowly targeting amyloid is the best approach to the treatment of AD dementia. Overall, however, the comments by Perry and Davies reflect a key theoretical plank of this volume, which asserts that aiming at a single pathogenic target with a single pharmaceutical drug with respect to AD may be a dated approach.

For example, in 2011 scientists from the Salk Institute for Biological Studies, the Research Institute of Pharmaceutical Sciences at Musashino University in Japan, the Molecular Neurosciences Department at the Scripps Research Institute, and the Laboratorium für Physikalische Chemie in Switzerland reported in the science journal, *PLoS One*: "[T]he currently most widely used approach to drug discovery is to identify a single molecular target and then to make a drug that alters this target. Unfortunately, drugs for AD that were developed through this approach have all failed in clinical trials, perhaps because one target is not sufficient or because the targets are also critical for normal brain function so that their inactivation results in toxicity."[16] In 2014, a review of AD-related drug design reported in *Current Neuropharmacology*: "Alzheimer's disease is a complex neurodegenerative disorder with a multi-faceted pathogenesis. So far, the therapeutic paradigm 'one-compound-one-target' has failed and despite enormous efforts to elucidate the pathophysiology of AD, the disease is still incurable."[17] And in 2015, researchers reporting in the *British Journal of Pharmacology* wrote: "Alzheimer's disease is accepted nowadays as a complex neurodegenerative disorder with multifaceted cerebral pathologies... This may explain why it is currently widely accepted that a more effective therapy for AD would result from multifunctional drugs, affecting more than one brain target involved in the disease pathology."[18]

We would add that the narrow targeting of AD-related pathogenic elements descends in part from the limited-targeting capabilities of the pharmaceutical industry, and that the therapeutic paradigm of "one-compound-one-target" appears to wag the dog of applied research on AD, despite its apparent limitations. Although clinical AD emerges from a complex pathophysiology with a number of potential pathogenic targets, the research objective of locating a single "right target" is mandated paradigmatically by the well-stocked stable of mostly one–trick pony pharmaceutical drugs.

These include the following drug classes, the designations of which literally describe a singularly targeted biochemical action: HMG-CoA reductase inhibitors to lower LDL cholesterol; selective serotonin reuptake inhibitors for depression; proton pump inhibitors to block stomach acid production; beta-blockers, which block the action of epinephrine and norepinephrine on beta receptors of the sympathetic nervous system; ACE inhibitors, which, to lower blood pressure, inhibit angiotensin-converting-enzyme; and so on. Add to this the only FDA-approved drugs to treat AD to date: three cholinesterase inhibitors and one NMDA receptor inhibitor. And to those the likely next generation of AD-related drugs—the monoclonal antibodies—that target amyloid with modest clinical benefits to date. While these drugs are often effective in achieving their narrowly targeted biochemical endpoints (with varying degrees of therapeutic benefit), a problem arises upon applying the monotargeting paradigm of pharmaceuticals to the treatment of a full-blown chronic disease with a billowing pathogenic course over two or three decades. This may help explain why the therapeutic mission of the Abeta-targeting monoclonal antibodies has been largely scaled back from arresting or reversing AD dementia to reducing the speed limit to it.

To be clear, the point here is not that Abeta is a "wrong target." Substantial evidence indicates that amyloid is an essential target with respect to efforts aimed at the primary prevention of AD (see Chapter 3). If there is an unsettled view of the precise scientific standing of the "amyloid hypothesis" of AD pathophysiology, it may be because the hypothesis has seldom if ever been tested within a primary prevention setting (as defined in this volume), within which inhibiting amyloid production and enhancing amyloid clearance would almost certainly be key points of focus. However, targeting Abeta alone to treat AD dementia seems theoretically deficient with little real-world support. In his article, Groopman correctly points out that the amyloid-targeting trials involving solanezumab "don't even aim at helping current patients" with AD dementia. As things stand, the AD-related disease process is so complex it has easily evaded the monotargeting, safety-challenged drugs that have been tested to date to treat AD. In contrast, multitargeting treatment agents that could address the multidimensionality of the clinical stages of AD offer a much stronger theoretical rationale.

In response, many AD researchers, working in medical centers and universities around the world, have shifted the focus of applied research from pharmacentric monotargeting to plant–polyphenol-based pleiotropism, in part because epidemiological evidence compelled such research, including one such report, which found that "dietary, lifestyle, and sociocultural interventions may be protective against dementia."[19] This was in addition to an earlier study, published in *Neurology*, which found that "frequent consumption of fruits and vegetables, fish, and omega-3 rich oils may decrease the risk of dementia and Alzheimer disease."[20]

Pleiotropism (ply-o-trōpism), for our purposes, refers to the multitargeting therapeutic capabilities of the plant polyphenols featured in this volume, including as applied to AD. Thus, the converse of monotargeting, in addition to being its potential successor, is pleiotropic targeting for the prevention and treatment of AD.

There are many references in the research literature to the multiple targeting capabilities of several nontoxic plant polyphenols, which are naturally occurring compounds found in fruits, vegetables, cereals, dry legumes, coffee, tea, red wine, olive oil, and cocoa.[a] For example, in 2015, in a multicenter research project titled, "Broad Targeting of Angiogenesis for Cancer Prevention and Therapy," over 30 cancer researchers assessed the pleiotropic prevention and treatment attributes of a number of plant polyphenols

[a] Fruits like grapes, apples, pears, cherries, and berries contain up to 200–300 mg of polyphenols per 100 g of fresh weight. Typically, a glass of red wine or a cup of green or black tea contains about 100 mg of polyphenols.[20a]

"Broad Targeting of Angiogenesis for Cancer Prevention and Therapy," *Seminars in Cancer Biology* **(2015).**

- Department of Urology, Massachusetts General Hospital, Harvard Medical School, Boston, Massachusetts.
- Department of Oncology and the Department of Clinical and Experimental Medicine, Linköping University, Linköping, Sweden.
- Medicine and Research Services, Veterans Affairs San Diego Healthcare System and the University of California at San Diego.
- Lineberger Comprehensive Cancer Center, University of North Carolina at Chapel Hill.
- Molecular Therapy and Pharmacogenomics Unit, AO Isituti Ospitalieri di Cremona, Cremona, Italy.
- Department of Hematology and Medical Oncology, Emory University in Atlanta, Georgia.
- Department of Basic Medical Sciences, Neurosciences, and Sensory Organs, University of Bari Medical School, Bari, Italy.
- National Cancer Institute Giovanni Paolo II, Bari, Italy.
- Department of Biology, Alderson Broaddus University, Philippi, West Virginia.
- Department of Orthopedic Surgery, Arthroplasty, and Regenerative Medicine, Nara Medical University, Nara, Japan.
- Physics Department, School of Applied Mathematics and Physical Sciences, National Technical University of Athens, Athens, Greece.
- Mayo Graduate School, Mayo Clinic College of Medicine, Rochester, Minnesota.
- Department of Experimental and Clinical Medicine, University of Florence, Italy.
- Department of Biology, College of Science, United Arab Emirate University, United Arab Emirates.
- Faculty of Science, Cairo University, Cairo, Egypt.
- Department of Chemistry, College of Science, United Arab Emirate University, United Arab Emirates.
- University of Illinois at Urbana, Illinois.
- School of Chemical and Biotechnology, SASTRA University, Thanjavur, India.
- Department of Biology, University of Rome "Tor Vergata," Rome, Italy.
- Ovarian and Prostate Cancer Research Trust Laboratory, Guilford, Surrey, United Kingdom.
- New York Medical College, New York City.
- Department of Comparative Pathobiology, Purdue University Center for Cancer Research, West Lafayette, Indiana.
- School of Medicine, Wayne State University, Detroit, Michigan.
- Institute of Cancer Sciences, University of Glasgow, United Kingdom.
- Department of Medical and Health Sciences, Linköping University, Linköping, Sweden.
- Department of Microbiology, Tumor and Cell Biology, Karolinska Institutet, Stockholm, Sweden.

against tumor angiogenesis. The institutional affiliations of the researchers involved in the study are listed in the adjacent sidebar.

This extensive investigation of plant polyphenols for the treatment of cancer was prompted by epidemiology and the inadequacy of "targeted theory" as applied to the complex pathophysiology of tumor angiogenesis, as the researchers reported:

> [T]he clinically approved anti-angiogenic drugs in use today are only effective in a subset of the patients, and many who initially respond develop resistance over time. Also, some of the anti-angiogenic drugs are toxic and it would be of great importance to identify alternative compounds which could overcome these drawbacks and limitations of the currently available therapy. Finding "the most important target" may, however, prove a very challenging approach as the tumor environment is highly diverse, consisting of many different cell types, all of which may contribute to tumor angiogenesis. Furthermore, the tumor cells themselves are genetically unstable, leading to a progressive increase in the number of different angiogenic factors produced as the cancer progresses to advanced stages. As an alternative approach to targeted therapy, options to broadly interfere with angiogenic signals by a mixture of non-toxic natural compounds with pleiotropic actions were viewed by this team as an opportunity to develop a complementary anti-angiogenesis treatment option.[21] (Emphasis on pleiotropic added here and below.)

Similar to the research methodology and analytic framework that has been applied to this volume on AD, the cancer research team "identified 10 important aspects of tumor angiogenesis and the pathological tumor vasculature which would be well suited as targets for anti-angiogenic therapy."[22] The team then "scrutinized the available literature on broadly acting anti-angiogenic natural products, with a focus on finding qualitative information on phytochemicals which could inhibit these targets and came up with 10 prototypical phytochemical compounds."[23] They subsequently identified 10 such compounds, including curcumin, resveratrol, and EGCG (epigallocatechin-gallate, a green tea extract).[24]

The researchers also stipulated: "We suggest that these plant-derived compounds could be combined to constitute a broader acting and more effective inhibitory cocktail at doses that would not be likely to cause excessive toxicity." They concluded: "The aim is that this discussion could lead to the selection of combinations of such anti-angiogenic compounds which could be used in potent anti-tumor cocktails, for enhanced therapeutic efficacy, reduced toxicity, and circumvention of single-agent anti-angiogenic resistance, as well as for possible use in primary or secondary cancer prevention strategies."[23]

In short, this multicenter report represents, not a precedent, but a coexisting paradigm for the application in this volume of pleiotropic plant polyphenols to the complex, multidimensional pathophysiology of AD.

Many research reports and expert reviews, specifically citing the pleiotropic attributes of plant polyphenols, have been published at an accelerated rate in recent years. For example, about curcumin, researchers from Texas Southern University in Houston, in *Advances in Experimental Medicine and Biology* (2007), reported: "Curcumin is the active ingredient of turmeric that has been consumed as a dietary spice for ages. Turmeric is widely used in traditional Indian medicine to cure biliary disorders, anorexia, cough, diabetic wounds, hepatic disorders, rheumatism, and sinusitis. Extensive investigation over the last five decades has indicated that curcumin reduces blood cholesterol, prevents low-density lipoprotein oxidation, inhibits platelet aggregation, suppresses thrombosis and myocardial infarction, suppresses symptoms associated with type II diabetes, rheumatoid arthritis, multiple sclerosis, and Alzheimer's disease, inhibits HIV replication, enhances wound healing, protects from liver injury, increases bile secretion, protects from cataract formation, and protects from pulmonary toxicity and fibrosis. Evidence indicates that the divergent effects of curcumin are dependent on its *pleiotropic* molecular effects."[26]

Researchers from the University of California at Los Angeles, also in *Advances in Experimental Medicine and Biology* (2007), reported: "Curcumin has an outstanding safety profile and a number of *pleiotropic* actions with potential for neuroprotective efficacy, including anti-inflammatory, antioxidant, and anti-protein-aggregate activities... [C]urcumin has at least 10 known neuroprotective actions and many of these might be realized in vivo."[27]

Also with respect to curcumin, researchers from the School of Medicine at Wake Forest University, in *Cellular and Molecular Life Sciences* (2008), reported: "Curcumin has a surprisingly wide range of beneficial properties, including anti-inflammatory, antioxidant, chemopreventive and chemotherapeutic activity. The *pleiotropic* activities of curcumin derive from its complex chemistry as well as its ability to influence multiple signaling pathways."[28]

Researchers from the Department of Medicine at the University of British Columbia likewise reported in *Scientifica* (2012): "Unquestionably, the natural food additive curcumin, derived from the colorful spice turmeric used in many Asian cuisines, possesses a diverse array of biological activities. These range from its anti-inflammatory, antineoplastic, and metabolic modifying properties to surprising roles in disorders ranging from Alzheimer's disease to cystic fibrosis. Its effects on growth factor receptors, signaling molecules, and transcription factors, together with its epigenetic effects are widely considered to be extraordinary. These *pleiotropic* attributes, coupled with its safety even when used orally at well over 10 g/day, are unparalleled amongst pharmacological agents."[29]

With respect to resveratrol, researchers from the Department of Biological Sciences at the University of L'Aquila in Italy, in *Current Drug Metabolism* (2009), reported: "Resveratrol, a naturally occurring polyphenol, shows *pleiotropic* health beneficial effects, including antioxidant, anti-inflammatory, anti-aging, cardioprotective and neuroprotective activities. Due to the several protective effects, and since this compound is widely distributed in the plant kingdom, resveratrol can be envisaged as a chemo-preventive/curative agent... In addition to these important functions, resveratrol is reported to exhibit several other biological/biochemical protective effects on heart, circulation, brain, and age-related diseases."[30]

Researchers from the Department of Biochemistry and Biophysics at Second University of Naples in Italy, in *Nutrition, Metabolism, and Cardiovascular Diseases* (2010), also reported: "[R]esveratrol is endowed with significant positive activities by protecting against cardiovascular diseases and preventing the development and progression of atherosclerosis... The [resveratrol] molecule shows a *pleiotropic* mode of action. Particularly, its cellular targets are crucial for cell proliferation and differentiation, apoptosis, antioxidant defence and mitochondrial energy production. The complexity of resveratrol activities might account for its effectiveness in ameliorating multifactorial processes, including the onset and/or progression of several degenerative diseases such as myocardial infarction, atherosclerosis and type 2 diabetes."[31]

In another summary of the pleiotropic abilities of plant polyphenols, scientists from the Department of Biology at the University of Alabama at Birmingham reported in *Current Aging Science* (2010): "Age-associated changes within an individual are inherently complex and occur at multiple levels of organismal function. The overall decline in function of various tissues is known to play a key role in both aging and the complex etiology of certain age-associated diseases such as Alzheimer's disease and cancer... Numerous lines of evidence suggest that dietary polyphenols such as resveratrol, (-)-epigallocatechin-3-gallate (EGCG), and curcumin have the capacity to mitigate age-associated cellular damage induced via metabolic production of reactive oxygen species. However, recently acquired evidence also demonstrates a likely role for these polyphenols as anticancer agents capable of preventing formation of new vasculature in neoplastic tissues. Polyphenols have also been shown to possess other anticancer properties, such as specific cell-signaling actions that may stimulate the activity of the regulatory protein SIRT1... These increasingly well-documented results have begun to provide a basis for considering the use of polyphenols in the development of novel therapies for certain human diseases."[32]

And researchers from a number of universities and medical centers in Spain reported in *Aging* (2014): "Aging is associated with common conditions, including cancer, diabetes, cardiovascular disease, and Alzheimer's disease. The type of multi-targeted pharmacological approach necessary to address a complex multifaceted disease such as aging might take advantage of *pleiotropic* natural polyphenols affecting a wide variety of biological processes."[33]

In addition to the broader application of plant polyphenols to chronic disease, as noted above, there is a substantial body of scientific literature on the application of pleiotropic plant polyphenols in particular to the prevention and treatment of AD.

In "Why Pleiotropic Interventions Are Needed for Alzheimer's Disease," published in *Molecular Neurobiology* (2010), Sally Frautschy and Greg Cole, both neuroscientists at UCLA, presented an early multitargeted approach to AD: "By the MCI [mild cognitive impairment] stage, we stand a greater chance of success [in preventing progression to fully symptomatic AD] by considering *pleiotropic* drugs or cocktails that can independently limit the parallel steps of the AD cascade at all stages, but that do not completely inhibit the constitutive normal functions of these pathways."[34]

Several expert reviews in effect have concurred with this approach. In *Oxidative Medicine and Cellular Longevity* (2012), in a review titled, "Pleiotropic Protective Effects of Phytochemicals in Alzheimer's Disease," researchers at the Clinical Biochemistry and

Clinical Molecular Biology Laboratory at the University of Molise in Italy noted that "considering the heterogeneity of AD, therapeutic agents acting on multiple levels of the pathology are needed."[35]

In *Medicinal Research Reviews* (2013), researchers in the Department of Chemistry at the University of Cambridge in England observed: "With 27 million cases worldwide documented in 2006, Alzheimer's disease constitutes an overwhelming health, social, economic, and political problem to nations. Unless a new medicine capable to delay disease progression is found, the number of cases will reach 107 million in 2050. So far, the therapeutic paradigm one-compound-one-target has failed."[36]

In *Phytotherapy Research* (2014), researchers at the Atta-ur-Rahman School of Applied Biosciences at the National University of Sciences and Technology in Islamabad, Pakistan, referenced the "multiple medicinal uses" of curcumin and other curcuminoids as applied to AD as follows:

Alzheimer's disease (AD) is the most common form of dementia. There is limited choice in modern therapeutics, and the drugs available have limited success with multiple side effects in addition to high cost. Hence, newer and alternate treatment options are being explored for effective and safer therapeutic targets to address AD. Turmeric possesses multiple medicinal uses, including treatment for AD. Curcuminoids, a mixture of curcumin, demethoxycurcumin, and bisdemethoxycurcumin, are vital constituents of turmeric. It is generally believed that curcumin is the most important constituent of the curcuminoid mixture that contributes to the pharmacological profile of parent curcuminoid mixture or turmeric. A careful literature study reveals that the other two constituents of the curcuminoid mixture also contribute significantly to the effectiveness of curcuminoids in AD.

They concluded: "The progress in understanding the [AD] disease etiology demands a multiple-site-targeted therapy, and the curcuminoid mixture of all components, each with different merits, makes this mixture more promising in combating the challenging disease."[37]

In *Frontiers in Aging Neuroscience* (2014), researchers at the Icahn School of Medicine at Mount Sinai in New York, the James J. Peters Veterans Affairs Medical Center in the Bronx, and the Department of Food Sciences at Purdue University reported: "[Alzheimer's disease] onset and progression are influenced by multiple factors. There is growing consensus that successful treatment will rely on simultaneously targeting multiple pathological features of AD." The researchers concluded: "Our studies provide experimental evidence that application of polyphenols targeting multiple disease mechanisms may yield a greater likelihood of therapeutic efficacy."[38]

And in *Biochimica et Biophysica Acta* (2015), researchers at the Icahn School of Medicine at Mount Sinai and the James J. Peters Veterans Affairs Medical Center in the Bronx reported: "There is mounting evidence that dietary polyphenols, including resveratrol, may beneficially influence Alzheimer's disease... Several research groups worldwide with expertise in AD, plant biology, nutritional sciences, and botanical sciences have reported very high quality studies that ultimately provided the necessary information showing that polyphenols and their metabolites, which come from several dietary sources, including grapes, cocoa etc., are capable of preventing AD. The ultimate goal of these studies was to provide novel strategies to prevent the disease even before the onset of clinical symptoms."[39]

The model of applying the nontoxic, pleiotropic actions of plant polyphenols to the prevention and treatment of AD, as presented in this volume, is influenced by these and numerous other AD-related studies. However, one point of departure in response to the appeal for "novel strategies to prevent [AD] even before the onset of clinical symptoms," as advocated by many AD scientists, is to note that the vital focus on the prevention of AD-related clinical symptoms needs to be supplemented with an emphasis on preventing onset of the AD-related disease process in the first place, as this book illustrates in the chapters ahead.

REFERENCES

1. Before night falls. *The New Yorker* June 24, 2013.
2. Preclinical Alzheimer's Disease: For Medical Professionals, Mayo Clinic. At: http://www.mayoclinic.org/medical-professionals/clinical-updates/neurosciences/preclinical-alzheimers-disease.
3. Villemagne, Burnham, Bourgeat, et al. Amyloid β deposition, neurodegeneration, and cognitive decline in sporadic Alzheimer's disease: a prospective cohort study. *Lancet Neurology* April 2013;**12**(4):357–67. The researchers added: "Our projections suggest a prolonged preclinical phase of AD in which Aβ [amyloid-beta] deposition reaches our threshold of [AD] positivity at 17.0 (14.9 to 19.9) years, hippocampal atrophy at 4.2 (3.6 to 5.1) years, and memory impairment at 3.3 (2.5 to 4.5) years before the onset of dementia."
4. Hampel, Schneider, Giacobini, et al. Advances in the therapy of Alzheimer's disease: targeting amyloid beta and tau and perspectives for the future. *Expert Review of Neurotherapeutics* January 2015;**15**(1):83–105.
5. Pietrzak, Lim, Neumeister, et al. Amyloid-β, anxiety, and cognitive decline in preclinical Alzheimer disease: a multicenter, prospective cohort study. *JAMA Psychiatry* March 2015;**72**(3):284–91.
6. Studies confirm brain plaque can help predict Alzheimer's. *New York Times* May 19, 2015.
7. Runaway drug prices. *The New York Times* May 5, 2015.
8. Doctors Blast Ethics of $100,000 Cancer Drugs, CNN, April 23, 2013. At: http://money.cnn.com/2013/04/25/news/economy/cancer-drug-cost/.
9. About Alzheimer's Disease: Causes, Alzheimer's Disease Education and Referral Center, National Institute on Aging. At: https://www.nia.nih.gov/alzheimers/topics/causes.
10. About Alzheimer's Disease: Treatment, Alzheimer's Disease Education and Referral Center, National Institute on Aging. At: https://www.nia.nih.gov/alzheimers/topics/treatment.
11. Alzheimer's Disease Medications: Fact Sheet, National Institute on Aging. At: http://www.nia.nih.gov/alzheimers/publication/alzheimers-disease-medications-fact-sheet.
12. Preclinical Alzheimer's Disease: For Medical Professionals, Clinical Updates, Mayo Clinic. At: http://www.mayoclinic.org/medical-professionals/clinical-updates/neurosciences/preclinical-alzheimers-disease.
13. Doody, Thomas, Farlo, et al. Phase 3 trials of solanezumab for mild-to-moderate Alzheimer's disease. *New England Journal of Medicine* January 23, 2014;**370**(4):311–21.
14. Newest hope for Alzheimer's tempered by earlier failures. *Bloomberg Business* December 3, 2014.
15. Selkoe. Resolving controversies on the path to Alzheimer's therapeutics. *Nature Medicine* September 7, 2011;**17**(9):1060–5.
16. Chen, Prior, Dargusch, et al. A novel neurotrophic drug for cognitive enhancement and Alzheimer's disease. *PLoS One* 2011;**6**(12):e27865.

17. Dias, Viegas. Multi-target directed drugs: a modern approach for design of new drugs for the treatment of Alzheimer's disease. *Current Neuropharmacology* May 2014;**12**(3):239–55.

18. Weinreb, Amit, Bar-Am, Youdim. Neuroprotective effects of multifaceted hybrid agents targeting MAO, cholinesterase, iron and β-amyloid in aging and Alzheimer's disease. *British Journal of Pharmacology* July 2016; **173**(13): 2080–94.

19. Tripathi, Vibha, Gupta, et al. Risk factors of dementia in North India: a case-control study. *Aging and Mental Health* 2012;**16**(2):228–35. This study reported the results as follows: "Diabetes, depression, hyperhomocysteinemia, hyperlipidemia, APOE ε4 gene, BMI, use of saturated fatty acids, pickles in diet, urban living, and lack of exercise were associated with independent risk of dementia. Various dietary factors and sociocultural factors, like cognitively stimulating activities, active socialization, living in joint families, increased intake of polyunsaturated fats, fruits, and salads conferred protection against dementia. CONCLUSIONS: Dietary, lifestyle, and sociocultural interventions may be protective against dementia."

20. Barberger-Gateau, Raffaitin, Letenneur, et al. Dietary patterns and risk of dementia: the three-city cohort study. *Neurology* November 13, 2007;**69**(20):1921–30. Results were summarized as follows: "Daily consumption of fruits and vegetables was associated with a decreased risk of all cause dementia (hazard ratio [HR] 0.72, 95% CI 0.53 to 0.97) in fully adjusted models. Weekly consumption of fish was associated with a reduced risk of AD (HR 0.65, 95% CI 0.43 to 0.994) and all cause dementia but only among ApoE epsilon 4 noncarriers (HR 0.60, 95% CI 0.40 to 0.90). Regular use of omega-3 rich oils was associated with a decreased risk of borderline significance for all cause dementia (HR 0.46, 95% CI 0.19 to 1.11). Regular consumption of omega-6 rich oils not compensated by consumption of omega-3 rich oils or fish was associated with an increased risk of dementia (HR 2.12, 95% CI 1.30 to 3.46) among ApoE epsilon 4 noncarriers."

20a. Pandey KB, Rizvi SI. Plant polyphenols as dietary antioxidants in human health and disease. *Oxid Med Cell Longev* November–December 2009;**2**(5):270–8.

21. Wang, Debrosin, Yin, et al. Broad targeting of angiogenesis for cancer prevention and therapy. *Seminars in Cancer Biology* January 16, 2015:S224–43. [Epub ahead of print].

22. The research team listed the ten important aspects of tumor angiogenesis and the pathological tumor vasculature which would be well suited as targets for anti-angiogenic therapy that inhibit blood vessel growth-supporting tumors as follows: (1) endothelial cell migration/tip cell formation, (2) structural abnormalities of tumor vessels, (3) hypoxia, (4) lymphangiogenesis, (5) elevated interstitial fluid pressure, (6) poor perfusion, (7) disrupted circadian rhythms, (8) tumor promoting inflammation, (9) tumor promoting fibroblasts and (10) tumor cell metabolism/acidosis.

23. Wang, Broad Targeting of Angiogenesis for Cancer Prevention and Therapy.

24. The ten phytochemical compounds were listed as follows: (1) oleic acid, (2) tripterine, (3) silibinin, (4) curcumin, (5) epigallocatechin-gallate, (6) kaempferol, (7) melatonin, (8) enterolactone, (9) withaferin A and (10) resveratrol.

25. Deleted in review.

26. Modulation of transcription factors by curcumin. *Advances in Experimental Medicine and Biology* 2007;**595**:127–48.

27. Cole, Teter, Frautschy. Neuroprotective effects of curcumin. *Advances in Experimental Medicine and Biology* 2007;**595**:197–212.

28. Hatcher, Planalp, Cho, et al. Curcumin: from ancient medicine to current clinical trials. *Cellular and Molecular Life Sciences* June 2008;**65**(11):1631–52.

29. Soni, Salh. A neutraceutical by design: the clinical application of curcumin in colonic inflammation and cancer. *Scientifica* 2012;**2012**:757890.

30. Brisdelli, D'Andrea, Bozzi. Resveratrol: a natural polyphenol with multiple chemopreventive properties. *Current Drug Metabolism* July 2009;**10**(6):530–46.

31. Borriello, Cucciolla, Della Ragione, Galletti. Dietary polyphenols: focus on resveratrol, a promising agent in the prevention of cardiovascular diseases and control of glucose homeostasis. *Nutrition, Metabolism, and Cardiovascular Diseases* October 2010;**20**(8):618–25.

32. Queen, Tollefsbol. Polyphenols and aging. *Current Aging Science* February 2010;**3**(1):34–42.

33. Corominas-Faja, Santangelo, Cuyàs, et al. Computer-aided discovery of biological activity spectra for anti-aging and anti-cancer olive oil oleuropeins. *Aging* September 2014;**6**(9):731–41.

34. Frautschy, Cole. Why pleiotropic interventions are needed for Alzheimer's disease. *Molecular Neurobiology* June 2010;**41**(2–3):392–409.
35. Davinelli, Sapere, Zella, et al. Pleiotropic protective effects of phytochemicals in Alzheimer's disease. *Oxidative Medicine and Cellular Longevity* 2012;**2012**:386527.
36. León, Garcia, Marco-Contelles. Recent advances in the multitarget-directed ligands approach for the treatment of Alzheimer's disease. *Medicinal Research Reviews* January 2013;**33**(1):139–89.
37. Ahmed, Gilani. Therapeutic potential of turmeric in Alzheimer's disease: curcumin or curcuminoids? *Phytotherapy Research* April 2014;**28**(4):517–25.
38. Wang, Bi, Cheng, et al. Targeting multiple pathogenic mechanisms with polyphenols for the treatment of Alzheimer's disease—experimental approach and therapeutic implications. *Frontiers in Aging Neuroscience* March 14, 2014;**6**:42.
39. Pasinetti, Wang, Ho, et al. Roles of resveratrol and other grape-derived polyphenols in Alzheimer's disease prevention and treatment. *Biochimica et Biophysica Acta* June 2015;**1852**(6):1202–8.

CHAPTER 2

The Pleiotropic Pharmacology of Plant Polyphenols

Contents

The health-providing, bioactive natural compounds in fruits and vegetables are broadly categorized as plant polyphenols. In recent years, a number of plant-based polyphenols have been studied for their effects on neurodegenerative diseases, including Alzheimer's disease (AD). For example, in 2006 Charles Ramassamy of the Institute of Nutraceuticals and Functional Foods at Laval University in Quebec, Canada, wrote in the *European Journal of Pharmacology*: "Increasing number[s] of studies [have] demonstrated the efficacy of polyphenolic antioxidants from fruits and vegetables to reduce or to block neuronal death occurring in the pathophysiology of these disorders. These studies revealed that other mechanisms than the antioxidant activities could be involved in the neuroprotective effect of these phenolic compounds." Ramassamy also noted that "particular emphasis" in his review would be given "to polyphenolic compounds from green tea, *Ginkgo biloba* extract EGb 761, blueberry extracts, wine components and curcumin."[1]

Similarly, in 2011 scientists at the Eve Topf Center for Neurodegenerative Disease Research and the Department of Molecular Pharmacology at the Technion medical center in Haifa, Israel, reported in the *Journal of Alzheimer's Disease*: "Natural plant polyphenols (flavonoids and non-flavonoids) are the most abundant antioxidants in the diet and, as such, are ideal nutraceuticals for neutralizing stress-induced free radicals and inflammation. Human epidemiological and new animal data suggest that green tea drinking (enriched in a class of flavonoids named catechins) may help protect the aging brain and reduce the incidence of dementia, AD, and PD [Parkinson's disease]." They reported that "mechanistic studies on the neuroprotective/neuroregenerative effects of green tea catechins" have been published, demonstrating that "these dietary compounds are receiving significant attention as therapeutic multifunctional cytoprotective agents that simultaneously manipulate various brain targets."[2]

A Paradigm Shift to Prevent and Treat Alzheimer's Disease
ISBN 978-0-12-812259-4
http://dx.doi.org/10.1016/B978-0-12-812259-4.00002-3

In 2014, scientists at the Department of Biotechnology at Alagappa University in Tamil Nadu, India, likewise reported in *Current Pharmaceutical Biotechnology* that "polyphenols are the most abundant components of our daily food, occupying the major portion of naturally occurring phytochemicals in plants," that these "polyphenols have received special attention from the scientific community" in recent years, and that "a large number of polyphenolic compounds showing promising results against AD pathologies have been identified and described in the past decade."[3]

This brief survey of food-based polyphenols, their multimechanistic biological effects, and their ability to favorably influence AD-related pathogenic pathways, in addition to the international character of the research itself, functions here as a brief study of the interrelated themes reported in finer detail and on a much larger scale in the pages ahead, all the while reflecting an extensive but growing body of published research on plant polyphenols and chronic disease, including AD.

In this chapter, we review the evidence in epidemiological and population studies with respect to the health benefits of a few ancient foods and medicinal extracts—green tea, red wine and other grape products, olive oil, turmeric, and *Ginkgo biloba* extracts— that contain well-studied polyphenolic compounds. In subsequent chapters, we review the "mechanistic studies" of the polyphenols themselves as applied directly to the known cellular and molecular mechanisms involved in the pathogenic course of AD. The results of these exercises point toward the possibility that certain ancient foods, in addition to the standardized polyphenolic compounds extracted from such foods, possess an ability to enhance or inhibit key biological processes as needed to prevent and treat AD.

GREEN TEA (*Camellia sinensis*)

After water, tea is the most widely consumed beverage in the world.[4] Or, as the *Cambridge World History of Food* reports: "Tea, a drink made from the leaves and buds of the shrub *Camellia sinensis*, is the most culturally and economically significant nonalcoholic beverage in the world." By all accounts, tea originated in China and "had spread to surrounding nations before European contact, after which it was made a commodity of world importance by the British and Dutch East India companies."[5]

Camellia sinensis is a small evergreen shrub that is native to mountainous regions of China, Japan, and India. Today it is cultivated around the world in countries with tropical and subtropical climates and is used extensively in the traditional medicine systems of China, Hong Kong, Japan, and Korea. In China, the use of tea as a beverage dates to 2700 BCE.[6] According to one of the classic works on food history, Reay Tannahill's *Food and History*: "Tea has been a popular drink in China since the Tang dynasty and was generally believed to have medicinal value and to contribute to longevity."[7]

Depending on the level of fermentation, tea can be categorized into three types: green (unfermented), oolong (partially fermented), and black (highly to fully fermented).

According to *The Cambridge World History of Food*, which was first published in 2000, "about 98 percent of the tea that enters the world is black tea,"[5] although that degree of dominance may have receded in recent years with increased demand for green tea, including in the United States.[8]

The beneficial effects of green tea have been attributed to a class of polyphenolic compounds called catechins. Of the catechins found in green tea—namely epicatechin-3-gallate, epigallocatechin, epicatechin, and epigallocatechin-3-gallate (EGCG)—EGCG is the most abundant and powerful.[9] In general, green tea is likely superior to black and oolong teas with regard to its health-promoting properties due to the higher content of catechins, including EGCG.[4] In "Beneficial Effects of Green Tea—A Review," the *Journal of the American College of Nutrition* (2006) reported: "Green tea is a 'non-fermented' tea, and contains more catechins than black tea or oolong tea. Catechins are in vitro and in vivo strong antioxidants. In addition, its content of certain minerals and vitamins increases the antioxidant potential of this type of tea. Since ancient times, green tea has been considered by the traditional Chinese medicine as a healthful beverage."[10]

Today, significant scientific evidence exists supporting the long-term safety and extensive health benefits of green tea and its biologically active catechin compounds. In "Medicinal Benefits of Green Tea: Part I. Review of Noncancer Health Benefits," the *Journal of Alternative and Complementary Medicine* (2005) reported: "Tea, in the form of green or black tea, is one of the most widely consumed beverages in the world… Green tea contains a unique set of catechins that possess biological activity in antioxidant, antiangiogenesis, and antiproliferative assays potentially relevant to the prevention and treatment of various forms of cancer."[11]

In "Green Tea Consumption and Mortality Due to Cardiovascular Disease, Cancer, and All Causes in Japan: The Ohsaki Study," which was "a population-based, prospective cohort study initiated in 1994 among 40,530 Japanese adults aged 40 to 79 years without history of stroke, coronary heart disease, or cancer at baseline," and where "participants were followed up for up to 11 years (1995-2005) for all-cause mortality and for up to 7 years (1995-2001) for cause-specific mortality," the *Journal of the American Medical Association* (2006) reported: "Green tea consumption is associated with reduced mortality due to all causes and due to cardiovascular disease but not with reduced mortality due to cancer."[12]

In "Green Tea Consumption and Mortality among Japanese Elderly People: The Prospective Shizuoka Elderly Cohort," which "investigate[d] the association between green tea consumption and mortality from all causes, cancer, and cardiovascular disease (CVD)" among 14,001 elderly people ages 64–84 years and followed for 6 years, *Annals of Epidemiology* (2009) concluded: "Green tea consumption is associated with reduced mortality from all causes and CVD. This study also suggests that green tea could have protective effects against colorectal cancer."[13]

And in "Association of Green Tea Consumption with Mortality Due to All Causes and Major Causes of Death in a Japanese Population: The Japan Public Health Center-Based Prospective Study," which "examined the association between green tea consumption and mortality due to all causes, cancer, heart disease, cerebrovascular disease, respiratory disease, injuries, and other causes of death" among 90,914 Japanese between ages 40 and 69 years and followed for 18.7 years, *Annals of Epidemiology* (2015) concluded: "This prospective study suggests that the consumption of green tea may reduce the risk of all-cause mortality and the three leading causes of death in Japan."[14]

Scientists at Technion's Faculty of Medicine in Israel are among the top researchers in the world on green tea and neurodegenerative diseases. In 2004, two scientists from that group, Silvia Mandel and Moussa Youdim, in *Free Radical Biology and Medicine*, and in an early exposition of the potential therapeutic advantages of a multitargeted (pleiotropic) approach to neurodegenerative diseases, reported: "Neurodegeneration in Parkinson's, Alzheimer's, and other neurodegenerative diseases seems to be multifactorial, in that a complex set of toxic reactions including inflammation, glutamatergic neurotoxicity, increases in iron and nitric oxide, depletion of endogenous antioxidants, reduced expression of trophic factors, dysfunction of the ubiquitin-proteasome system, and expression of pro-apoptotic proteins leads to the demise of neurons... This has led to the current notion that drugs directed against a single target will be ineffective and rather a single drug or cocktail of drugs with pluripharmacological properties may be more suitable." With respect to the catechins contained in green tea, the scientists wrote: "Green tea catechin polyphenols, formerly thought to be simple radical scavengers, are now considered to invoke a spectrum of cellular mechanisms of action related to their neuroprotective activity. These include pharmacological activities like iron chelation, scavenging of radicals, activation of survival genes and cell signaling pathways, and regulation of mitochondrial function and possibly of the ubiquitin-proteasome system."[15]

In 2005, in "Multifunctional Activities of Green Tea Catechins in Neuroprotection," and thus again with respect to the pleiotropic pharmacology of green tea extracts, Technion researchers reported in *Neurosignals*: "Tea flavonoids (catechins) have been reported to possess potent iron-chelating, radical-scavenging and anti-inflammatory activities and to protect neuronal death in a wide array of cellular and animal models of neurological diseases. Recent studies have indicated that in addition to the known antioxidant activity of catechins, other mechanisms such as modulation of signal transduction pathways, cell survival/death genes and mitochondrial function, contribute significantly to the induction of cell viability."[16]

In 2006, in *Molecular Nutrition and Food Research*, scientists at the Technion group reported: "Neurodegeneration in Parkinson's, Alzheimer's, or other neurodegenerative diseases appears to be multifactorial, where a complex set of toxic reactions, including oxidative stress, inflammation, reduced expression of trophic factors, and accumulation of protein aggregates, lead to the demise of neurons... Tea flavonoids (catechins) have been reported to possess divalent metal chelating, antioxidant, and anti-inflammatory activities, to penetrate the brain barrier and to protect neuronal death in a wide array of cellular and animal models of neurological diseases. This review aims to shed light on the multipharmacological neuroprotective activities of green tea catechins with special emphasis on their brain-permeable, nontoxic, transitional metal (iron and copper)-chelatable/radical scavenger properties."[17]

In 2007, in *Progress in Neurobiology*, Technion scientists observed: "Considering the multi-etiological character of Alzheimer's disease (AD), the current pharmacological approaches using drugs oriented towards a single molecular target possess limited ability to modify the course of the disease and thus offer a partial benefit to the patient. In line with this concept, novel strategies include the use of a cocktail of several drugs and/or the development of a single molecule, possessing two or more active neuroprotective-neurorescue moieties that simultaneously manipulate multiple targets involved in AD pathology... This review will discuss two separate scenarios concerning multiple therapy targets in AD, sharing in common the implementation of iron chelation activity: (i) novel multimodal brain-permeable iron-chelating drugs, possessing neuroprotective-neurorescue and amyloid precursor protein-processing regulatory activities; (ii) natural plant polyphenols (flavonoids), such as green tea epigallocatechin gallate (EGCG) and curcumin, reported to have access to the brain and to possess multifunctional activities, such as metal chelation-radical scavenging, anti-inflammation and neuroprotection."[18]

In 2008, in *Genes & Nutrition*, Technion researchers reported: "Vast epidemiology data indicate a correlation between occurrence of neurodegenerative disorders, such as Parkinson's and Alzheimer's diseases, and green tea consumption. In particular, recent literature strengthens the perception that diverse molecular signaling pathways, participating in the neuroprotective activity of the major green tea polyphenol, (-)-epigallocatechin-3-gallate (EGCG), renders this natural compound as a potential agent to reduce the risk of various neurodegenerative diseases."[19]

In 2011, in *Current Alzheimer Research*, researchers in the Department of Psychology at McGill University in Montreal reported: "Epidemiological studies have reported that elderly people have a lower risk (up to 50%) of developing dementia if they regularly eat fruits and vegetables and drink tea and red wine (in moderation). Numerous studies indicate that polyphenols derived from these foods and beverages account for the observed neuroprotective effects," including, "in particular... polyphenols extracted from green tea (i.e., epigallocatechin gallate or EGCG) and red wine (i.e., resveratrol)."[20]

And in 2012, in *Recent Patents on CNS Drug Discovery*, Mandel and Youdim reported: "To address the etiological complexity of aging and age-associated conditions, a new paradigm gaining increasing acceptance considers the use of multi-targeted ligands or combination of drugs to modulate several targets at once... The catechin polyphenol constituents of green tea, which were for a long time regarded merely as dietary antioxidants, have caught our and other scientists' attention because of their diverse pharmacological activities, which have been allied to a possible beneficial action on brain health."[21]

RED WINE AND GRAPE PRODUCTS

According to a 2014 history of food published by the University of California Press and edited by the food studies scholar, Paul Freedman: "The earliest archaeological evidence of cultivated grapes for winemaking dates to 5000-7000 BCE in what is now the Eurasian state of Georgia. The ancient Babylonian and Egyptian civilizations made wine, but it was the ancient Greek and Roman civilizations that embraced viniculture, spreading grapes and winemaking knowledge to their far-flung colonies."[22] In another classic work on food history, Maguelonne Toussaint-Samat wrote in *A History of Food*: "We do not know the precise geographical origin of the grapevine, *Vitis vinifera*, 'the vine that bears wine,' or rather bears the grapes from which wine can be made. It is generally thought to have come from the southern Caucasus, situated between Turkey, Armenia and Iran. This is more or less where Noah, famous as the first of all drunkards, is supposed to have landed his ark after the Flood."[23] In another colorful depiction of the beginnings of winemaking, Reay Tannahill wrote: "There are many picturesque tales about the origins of wine, but what probably happened was that at some time during the Neolithic era a container of *vinifera* grapes was left neglected in a corner. The juice would run, and in the right conditions, ferment and then settle; someone (tradition often makes it a woman) had the courage to taste the result and found it congenial. This suggests that the grapes in question may originally have been dried for keeping, but not quite enough. If they had been fresh, the accidental 'wine' would probably have been very rough, whereas fermentation of dried grapes would produce something sweeter, less alcoholic, and more palatable." Tannahill continued: "The wild vine flourished in the Caucasus, and it may have been there that it was brought under cultivation."[24] Amy Trubek at the Nutrition and Food Science department at the University of Vermont rounds out the history: "The Greeks are known to have brought grapes to southern Italy, Sicily, and southern France. Wine growing moved further north with the Roman Empire, helping to create wine cultures in all of Europe, especially in what is now France, which was long the major producer of wine in Europe (and sometimes beyond European borders as well). By the Middle Ages, wine was France's premier export crop, and by the 1800s most regions of [France] boasted vineyards. By 1900 wine production had extended to North and South America, South Africa, and other areas."[25]

On moderate red wine intake, the *Archives of Internal Medicine*, in a study published in 1999 that evaluated prospectively the health risk of wine and beer drinking in 36,250 healthy middle-aged men in the area of Nancy, France, found that "for all-cause mortality, only daily wine intake (22-32 g of alcohol) [0.8-1.1 ounces] was associated with a lower risk due to a lower incidence of cardiovascular diseases, cancers, violent deaths, and other causes."[26]

In "Alcohol and Mortality from All Causes," *Biological Research* reported in 2004: "A large number of prospective studies have observed an inverse relationship between a moderate intake of alcohol and coronary heart disease morbidity and mortality… In our prospective studies in France on 35,000 middle-aged men, we observed that only wine at moderate intake was associated with a protective effect on all-cause mortality."[27]

In a 2009 study of long-term wine consumption and cardiovascular mortality and life expectancy, the *Journal of Epidemiology and Community Health* reported: "[L]ong-term wine consumption of, on average, less than half a glass per day, was strongly and inversely associated with coronary heart disease and all-cause mortality. These results could not be explained by differences in socioeconomic status. Life expectancy was about 5 years longer in men who consumed wine compared with those who did not use alcoholic beverages."[28]

In "Modulation of Endogenous Antioxidant System by Wine Polyphenols in Human Disease," the *International Journal of Clinical Chemistry* reported in 2011: "Numerous studies indicate that moderate red wine consumption is associated with a protective effect against all-cause mortality."[29]

And scientists at the University of Bordeaux in France reported in *Oxidative Medicine and Cellular Longevity* that "there are numerous studies indicating that a moderate consumption of red wine provides certain health benefits, such as the protection against neurodegenerative diseases," and that "this protective effect is most likely due to the presence of phenolic compounds in wine." They also observed that red wines contain more polyphenols than white wines.[30]

In 2006, research scientists at the Mount Sinai School of Medicine in New York City, led by Giulio Pasinetti, while noting that "recent studies suggest that moderate red wine consumption reduces the incidence of Alzheimer's disease clinical dementia," and using a mouse model of AD, "tested whether moderate consumption of the red wine Cabernet Sauvignon modulates AD-type neuropathology and cognitive deterioration." Reporting in *FASEB Journal*, they concluded: "This study supports epidemiological evidence indicating that moderate wine consumption, within the range recommended by the FDA dietary guidelines of one drink per day for women and two for men, may help reduce the relative risk for AD clinical dementia."[31]

In 2009, Pasinetti's lab at Mount Sinai, in "Heterogeneity in Red Wine Polyphenolic Contents Differentially Influences Alzheimer's Disease-Type Neuropathology and Cognitive Deterioration," reported in the *Journal of Alzheimer's Disease* that "we previously found that treatment with Cabernet Sauvignon reduced the generation of AD-type amyloid-beta (Abeta) peptides," and that the "present study suggests that muscadine treatment attenuates Abeta neuropathology and Abeta-related cognitive deterioration" in a mouse model of AD. Muscadine grapes are grown mainly in the southeastern region of the United States, where muscadine wine is common. The researchers concluded: "Collectively, our observations suggest that distinct polyphenolic compounds from red wines may be bioavailable at the organism level and beneficially modulate AD phenotypes through multiple Abeta-related mechanisms. Results from these studies suggest the possibility of developing a 'combination' of dietary polyphenolic compounds for AD prevention and/or therapy by modulating multiple Abeta-related mechanisms."[32]

In 2013, researchers in the Department of Pharmacology at the University Complutense de Madrid in Spain evaluated the potential effects of "the extracts from free-run and pressed Merlot red wine" in PC12 neuron-like cells under oxidative stress. They found that "the major compounds found in Merlot red wine extract were quercetin, catechin, epicatechin, tyrosol, gallic acid, and procyanidins" and that "pre-treatments with these polyphenolic compounds significantly increased cell viability" of the hydrogen peroxide and Fenton−reaction treated PC12 cells. They concluded: "These results support the beneficial effects of red wine extracts and some of its polyphenols under oxidative stress conditions. This research provides evidences of the preventive properties of wine extracts and its major polyphenols under oxidative stress conditions."[33]

In the midst of these studies, in "Novel Role of Red Wine-Derived Polyphenols in the Prevention of Alzheimer's Disease Dementia and Brain Pathology: Experimental Approaches and Clinical Implications," Pasinetti wrote in *Planta Medica* that "studies suggest that dietary polyphenolics may benefit Alzheimer's disease by modulating multiple disease-modifying modalities," and that they "provide impetus for the development of polyphenolic compounds for Alzheimer's disease prevention and/or therapy."[34]

With regard to grapes and grape products, in "Biomedical Effects of Grape Products," a review article in *Nutrition Reviews* reported in 2010 that "accumulating evidence suggests that consumption of grapes and grape products can positively influence risk factors associated with cardiovascular health, cancer, neurodegenerative disease, and age-related cognitive decline," that "these effects are often attributed to the antioxidant activity and function of flavonoid compounds found in grapes," and that "the well-established health effects of grapes on cardiovascular disease risk, mainly on endothelial function, LDL oxidation, progression of atherosclerosis, and reduction in oxidative stress, have been clearly identified." The reviewers further observed that "emerging research has also demonstrated that grapes have beneficial effects on other chronic-degenerative diseases such as cancer, Alzheimer's disease, age-related cognitive decline, and diabetes," and that "further beneficial effects of grapes on oral health, immune function, and antiviral activity have also been reported."[35]

Also in 2010, Pasinetti's group at Mount Sinai Medical Center reported: "Grape seeds are a rich source of polyphenolic compounds, with proanthocyanidins being the primary grape seed polyphenolic constituents. Grape seed polyphenols have been reported to possess a broad spectrum of pharmacologic and medicinal properties, such as mitigating breast cell carcinogenesis, modulating blood pressure in individuals with pre-hypertension, maintaining glucose homeostasis in diabetic conditions, protecting against ischemia-related injuries, promoting dermal wound healing, and protecting against diabetic nephropathy." The group also reported that "recent observations from in vitro experimental studies and preclinical studies demonstrated that grape seed polyphenols may interfere with specific neuropathogenic mechanisms underlying AD [Alzheimer's disease] and suggest a potential novel role of grape seed polyphenols for treating AD."[36] With respect to resveratrol, a grape-derived polyphenol, the Mount Sinai group reported in 2012 that "there is mounting evidence that dietary polyphenols, including resveratrol, may beneficially influence AD."[37]

In 2014, in *Frontiers in Aging Neuroscience*, researchers from the Departments of Neurology and Psychiatry at the Mount Sinai Medical Center, the Geriatric Research, Education and Clinical Center at the James J. Peters Veterans Affairs Medical Center in the Bronx, New York, and the Department of Foods and Nutrition at Purdue University reported: "There is growing consensus that successful treatment [of Alzheimer's disease] will rely on simultaneously targeting multiple pathological features of AD. Polyphenol compounds have many proven health benefits. In this study, we tested the hypothesis that combining three polyphenolic preparations (grape seed extract, resveratrol, and Concord grape juice extract), with different polyphenolic compositions and partially redundant bioactivities, may simultaneously and synergistically mitigate amyloid-β (Aβ) mediated neuropathology and cognitive impairments in a mouse model of AD." The researchers found that "combination treatment" of grape seed extract, resveratrol, and Concord grape juice extract "resulted in better protection against cognitive impairments compared to individual treatments." They concluded that "our studies provided experimental evidence that application of polyphenols targeting multiple disease-mechanisms may yield a greater likelihood of therapeutic efficacy."[38]

And in 2015, researchers at Mount Sinai Medical Center and the James J. Peters Veterans Affairs Medical Center reported that "there is mounting evidence that dietary polyphenols, including resveratrol, may beneficially influence AD," and that "based on this consideration, several studies reported in the last few years were designed to validate sensitive and reliable translational tools to mechanistically characterize brain-bioavailable polyphenols as disease-modifying agents to help prevent the onset of AD dementia and other neurodegenerative disorders."[37]

OLIVE OIL

According to the International Olive Council (IOC), an intergovernmental organization, "the existence of the olive tree dates back to the twelfth millennium B.C." Its original home "extended from the southern Caucasus to the Iranian plateau," with its cultivation likely originating on "the Mediterranean coasts of Syria and Palestine." The cultivated olive tree then spread "to Cyprus and on towards Anatolia or from the island of Crete towards Egypt." The IOC continued its detailed history of the olive tree:

> In the sixteenth century B.C. the Phoenicians started disseminating the olive tree throughout the Greek isles, later introducing it to the Greek mainland between the fourteenth and twelfth centuries B.C., where its cultivation increased and gained great importance in the fourth century B.C., when Solon (an Athenian) issued decrees regulating olive planting. From the sixth century B.C. onwards, the olive tree spread throughout the Mediterranean countries reaching Tripoli, Tunis, and the island of Sicily. From there, it moved to southern Italy.
>
> The Romans continued the expansion of the olive tree to the countries bordering the Mediterranean. It was introduced in Marseille around 600 B.C. and spread from there to the whole of Gaul [now roughly northern Italy, France, Luxembourg, and Belgium]. The olive tree made its appearance in Sardinia in Roman times, while in Corsica it is said to have been brought by the Genoese (from the Italian city of Genoa) after the fall of the Roman Empire.
>
> Olive growing was introduced to Spain during the maritime domination of the Phoenicians (1050 B.C.). After the Third Punic War [between Carthage and the Roman Republic], olive trees occupied a large stretch of the Baetica valley (in southern Spain) and spread towards the central and Mediterranean coastal areas of the Iberian Peninsula, including Portugal. The Arabs brought their varieties with them to the south of Spain and influenced the spread of cultivation.
>
> With the discovery of America, olive farming spread beyond its Mediterranean confines. The first olive trees were carried from Seville to the West Indies and later to the American continent. By 1560 olive groves were being cultivated in Mexico, then later in Peru, California, Chile, and Argentina.[39]

The IOC also observed that "virgin olive oils are the oils obtained from the fruit of the olive tree solely by mechanical or other physical means under conditions, particularly thermal conditions, that do not lead to alterations in the oil, and which have not undergone any treatment other than washing, decantation, centrifugation and filtration."[40]

In 2002, researchers in the department of pharmacological sciences at the University of Milan in Italy, in *Critical Reviews in Food Science and Nutrition*, reported: "Olive oil is the principal source of fat in the Mediterranean diet, which has been associated with a lower incidence of coronary heart disease and certain cancers. Extra-virgin olive oil contains a considerable amount of phenolic compounds, for example, hydroxytyrosol and oleuropein, that are responsible for its peculiar taste and for its high stability. Evidence is accumulating to demonstrate that olive oil phenolics are powerful antioxidants, both in vitro and in vivo; also, they exert other potent biological activities that could partially account for the observed healthful effects of the Mediterranean diet."[41]

Pursuant to an international conference on the health effects of virgin olive oil, the *European Journal of Clinical Investigation* reported in 2005:

> 1. Ageing represents a great concern in developed countries because [of] the number of people involved and the pathologies related with it, like atherosclerosis, morbus Parkinson, Alzheimer's

disease, vascular dementia, cognitive decline, diabetes and cancer. 2. Epidemiological studies suggest that a Mediterranean diet (which is rich in virgin olive oil) decreases the risk of cardiovascular disease. 3. The Mediterranean diet, rich in virgin olive oil, improves the major risk factors for cardiovascular disease, such as the lipoprotein profile, blood pressure, glucose metabolism and antithrombotic profile. Endothelial function, inflammation and oxidative stress are also positively modulated. Some of these effects are attributed to minor components of virgin olive oil. Therefore, the definition of the Mediterranean diet should include virgin olive oil. 4. Different observational studies conducted in humans have shown that the intake of monounsaturated fat may be protective against age-related cognitive decline and Alzheimer's disease. 5. Micro constituents from virgin olive oil are bioavailable in humans and have shown antioxidant properties and capacity to improve endothelial function. Furthermore, they are also able to modify the haemostasis [regulation of blood clotting], showing antithrombotic properties. 6. In countries where the populations fulfilled a typical Mediterranean diet, such as Spain, Greece and Italy, where virgin olive oil is the principal source of fat, cancer incidence rates are lower than in northern European countries. 7. The protective effect of virgin olive oil can be most important in the first decades of life, which suggests that the dietetic benefit of virgin olive oil intake should be initiated before puberty, and maintained through life. 8. The more recent studies consistently support that the Mediterranean diet, based in virgin olive oil, is compatible with a healthier ageing and increased longevity.[42]

In "Chemistry and Health of Olive Oil Phenolics," *Critical Reviews in Food Science and Nutrition* reported in 2009: "The Mediterranean diet is associated with a lower incidence of atherosclerosis, cardiovascular disease, and certain types of cancer. The apparent health benefits have been partially attributed to the dietary consumption of virgin olive oil by Mediterranean populations. Most recent interest has focused on the biologically active phenolic compounds naturally present in virgin olive oils. Studies (human, animal, in vivo and in vitro) have shown that olive oil phenolics have positive effects on certain physiological parameters, such as plasma lipoproteins, oxidative damage, inflammatory markers, platelet and cellular function, and antimicrobial activity."[43]

In "Olive Oil and Health: Summary of the II International Conference on Olive Oil and Health Consensus Report," *Nutrition, Metabolism, and Cardiovascular Diseases* reported in 2010:

Olive oil is the most representative food of the traditional Mediterranean Diet. Increasing evidence suggests that monounsaturated fatty acids as a nutrient, olive oil as a food, and the Mediterranean Diet as a food pattern are associated with a decreased risk of cardiovascular disease, obesity, metabolic syndrome, type 2 diabetes and hypertension. A Mediterranean Diet rich in olive oil… has been shown to improve cardiovascular risk factors, such as lipid profiles, blood pressure, postprandial hyperlipidemia, endothelial dysfunction, oxidative stress, and antithrombotic profiles. Some of these beneficial effects can be attributed to the olive oil minor components… Observational studies from Mediterranean cohorts have suggested that dietary monounsaturated fatty acids may be protective against age-related cognitive decline and Alzheimer's disease. Recent studies consistently support the concept that the olive oil-rich Mediterranean Diet is compatible with healthier aging and increased longevity… Experimental and human cellular studies have provided new evidence on the potential protective effect of olive oil on cancer. Furthermore, results of case-control and cohort studies suggest that monounsaturated fatty acids intake including olive oil is associated with a reduction in cancer risk (mainly breast, colorectal and prostate cancers).[44]

In "The Role of Olive Oil in Disease Prevention: A Focus on the Recent Epidemiological Evidence from Cohort Studies and Dietary Intervention Trials," *The British Journal of Nutrition* reported in 2015: "Consumption of olive oil within the Mediterranean diet has been long known to have many health benefits. However, only over the last decade has epidemiological research confirmed its protective role against developing several chronic diseases. The objective of this review was to give an overview of the state of art of epidemiological evidence concerning the relationship between olive oil and key public health outcomes including mortality, cardiovascular disease, diabetes, metabolic syndrome, obesity and cancer, with a particular focus on recent results from cohort studies and dietary intervention trials. Recent epidemiological research has shown that regular consumption of olive oil is associated with increased longevity."[45]

In 2015, scientists at the University of Florence in Italy reported as follows in the *Journal of Alzheimer's Disease*: "The present review summarizes the findings on the beneficial effects against neurodegeneration and other peripheral inflammatory and degenerative diseases of oleuropein aglycone (OLE), a natural phenol abundant in the extra virgin olive oil. The data presently available suggest that OLE could provide a protective and therapeutic effect against a number of pathologies, including AD as well as obesity, type 2 diabetes, non-alcoholic hepatitis, and other natural or experimentally-induced pathological conditions. Such a protection could result, at least in part, in a remarkable improvement of the pathological signs arising from stress conditions including oxidative stress, an excessive inflammatory response, and the presence of cytotoxic aggregated material. In particular, the recent data on the cellular and molecular correlates of OLE neuroprotection suggest it could also play a therapeutic role against AD."[46]

And Stefania Rigacci, a scientist at the University of Florence and a leading authority on olive oil and AD, reported in 2015 in *Advances in Experimental Medicine and Biology* that "studies conducted both in vivo and in vitro have started to reveal the great potential of the phenolic components of extra virgin olive oil (mainly oleuropein aglycone and oleocanthal) in counteracting amyloid aggregation and toxicity, with a particular emphasis on the pathways involved in the onset and progression of Alzheimer's disease."[47]

TURMERIC (*Curcuma longa*)

The National Center for Complementary and Alternative Medicine (NCCAM) of the US National Institutes of Health describes turmeric as "a shrub related to ginger [that] is grown throughout India, other parts of Asia, and Africa" that is characterized by "its warm, bitter taste and golden color" and "commonly used in fabric dyes and foods such as curry powders, mustards, and cheeses." NCCAM also reported: "In traditional Chinese medicine and Ayurvedic medicine, turmeric has been used to aid digestion and liver function, relieve arthritis pain, and regulate menstruation. Historically, turmeric has also been applied directly to the skin for eczema and wound healing. Today, traditional or folk uses of turmeric include heartburn, stomach ulcers, gallstones, inflammation, and cancer. Turmeric's finger-like underground stems (rhizomes) are dried and taken by mouth as a powder or in capsules, teas, or liquid extracts. Turmeric can also be made into a paste and used on the skin."[48]

The American Botanical Council describes turmeric this way: "Turmeric is a perennial rhizomatous shrub native to southern Asia, extensively cultivated in all parts of India, mainly in Madras, Bengal, and Bombay. It is also cultivated in southern mainland China, Taiwan, Japan, Burma, Indonesia, and throughout the African continent. The material of commerce in Europe is obtained mainly from India, Indonesia, and somewhat from China. India produces most of the world supply. Turmeric is an herb of major importance in the East and, until recently, one of relatively minor importance in the West… Its modern approved applications in European medicine stem from its traditional uses in Asia. Turmeric is used extensively in the Indian systems of medicine (Ayurveda, Unani, and Siddha)."[49]

Curcuminoids, which constitute 1%–2% of turmeric dry powder—curcumin, demethoxycurcumin, and bisdemethoxycurcumin—are derived from the rhizome of turmeric. The emerging roles of curcumin in neuroprotection and in dementia received heightened attention when epidemiology showed that there was a lower prevalence of AD in the population of India, which consumes a diet that is rich in curcumin as an ingredient in curry.[50] A rural area in northern India may have the lowest prevalence of AD in the world (but dietary consumption of turmeric in that area has not been recorded).[51] It is known, however, that urban populations in Asia have lower intake of turmeric (200 mg/day) than rural populations (600 mg/day).[52]

The turmeric-derived curcuminoids can be consumed safely by humans at reasonable doses for long-term consumption with extensive evidence of health-related benefits. For example, *Current Drug Targets* reported in 2011: "[D]ue to its efficacy and regulation of multiple targets, as well as its safety for human use, curcumin has received considerable interest as a potential therapeutic agent for the prevention and/or treatment of various malignant diseases, arthritis, allergies, Alzheimer's disease, and other inflammatory illnesses."[53]

In addition, *Biotechnology Advances* (2014) reports: "Although the history of the golden spice turmeric (*Curcuma longa*) goes back thousands of years, it is only within the past century that we learned about the chemistry of its active component, curcumin. More than 6,000 articles published within the past two decades have discussed the molecular basis for the antioxidant, anti-inflammatory, antibacterial, antiviral, antifungal, and anticancer activities assigned to this nutraceutical."[54]

In "The Multifaceted Role of Curcumin in Cancer Prevention and Treatment," *Molecules* reported in 2015: "Despite significant advances in treatment modalities over the last decade, neither the incidence of the disease nor the mortality due to cancer has altered in the last thirty years. Available anti-cancer drugs exhibit limited efficacy, associated with severe side effects, and are also expensive. Thus, identification of pharmacological agents that do not have these disadvantages is required. Curcumin, a polyphenolic compound derived from turmeric (*Curcumin longa*), is one such agent that has been extensively studied over the last three to four decades for its potential anti-inflammatory and/or anti-cancer effects."[55]

In 2016, in *Asian Pacific Journal of Cancer Prevention* researchers at the Institute of Molecular Biology and Biotechnology at Bahauddin Zakariya University in Multan, Pakistan, reported: "Curcumin, a polyphenol from *Curcuma longa* (the turmeric plant), belongs to the ginger family which has long been used in Ayurveda medicines to treat various diseases such as asthma, anorexia, coughing, hepatic diseases, diabetes, heart diseases, wound healing and Alzheimer's. Various studies have shown that curcumin has anti-infectious, anti-inflammatory, anti-oxidant, hepato-protective, thrombo-suppressive, cardio protective, anti-arthritic, chemopreventive, and anti-carcinogenic activities. It may suppress both initiation and progression stages of cancer. Anticancer activity of curcumin is due to negative regulation of inflammatory cytokines, transcription factors, protein kinases, reactive oxygen species (ROS) and oncogenes. This review focuses on the different targets of curcumin to treat cancer."[56]

In "Clinical Development of Curcumin in Neurodegenerative Disease" published in *Expert Review of Neurotherapeutics*, scientists at the University of California at Los Angeles and the Los Angeles Veteran's Administration Healthcare System reported in 2015 as follows: "Curcumin, a polyphenolic antioxidant derived from the turmeric root, has undergone extensive preclinical development, showing remarkable efficacy in wound repair, cancer and inflammatory disorders. This review addresses the rationale for its use in neurodegenerative disease, particularly Alzheimer's disease. Curcumin is a pleiotropic molecule which not only directly binds to and limits aggregation of the β-sheet conformations of amyloid characteristic of many neurodegenerative diseases, but also restores homeostasis of the inflammatory system, boosts the heat shock system to enhance clearance of toxic aggregates, scavenges free radicals, chelates iron and induces anti-oxidant response elements."[57] Furthermore, a major curcumin metabolite, tetrahydrocurcumin, has antiinflammatory and antioxidant properties and considerably reduces toxic soluble aggregates of amyloid.[58]

Ginkgo biloba EXTRACT

According to the American Botanical Council, the Ginkgo tree is the oldest living species of tree on earth, dating back to the Paleozoic period more than 225 million years ago.[59] The University of Maryland Medical Center observes that "a single [*Ginkgo biloba*] tree can live as long as 1,000 years and grow to a height of 120 feet," and adds: "Although Chinese herbal medicine has used both the ginkgo leaf and seed for thousands of years, modern research has focused on the standardized *Ginkgo biloba* extract made from the dried green leaves. This standardized extract is highly concentrated and seems to treat health problems (particularly circulatory problems) better than the non-standardized leaf alone."[60]

The Physicians' Desk Reference for Herbal Medicines lists the known medical uses of the leaf of the *Ginkgo biloba* tree: "The Ginkgo tree is indigenous to China, Japan, and Korea, and is also found in Europe and the United States. The medicinal parts are the fresh or dried leaves, which have shown anti-inflammatory, cognitive-promoting, antioxidant, and vascular effects. Ginkgo has proved effective against peripheral occlusive arterial disease. It has demonstrated antioxidant activity, inhibition of platelet aggregation, enhancement of coronary blood flow, vasodilation, and a decrease in human blood pressure... The majority of studies have demonstrated a benefit with Ginkgo supplementation on cognition. In dementia due to neuronal loss and impaired neurotransmission, there is a decrease in oxygen and glucose and a release of free radicals and lipid peroxidation. The active ingredients in Ginkgo, the flavonoids (ginkgo-flavone glycosides) and terpenoids (ginkgolides and bilobalide) probably affect the progression of dementia in several ways, such as reducing neutrophil infiltration and lipid peroxidation, increasing blood flow, antagonizing platelet-activation factor, and changing neuron metabolism."[61]

The most studied formula of *Ginkgo biloba* extract is identified in the scientific literature as *Ginkgo biloba* extract 761 (EGb 761) and is sold in the United States as "Ginkgold" by Nature's Way. About this product, researchers from the Department of Pharmacology at the University of Toledo in Ohio reported in 2015 in *Integrative Medicine Insights* as follows: "*Ginkgo biloba* extract is an alternative medicine available as a standardized formulation, EGb 761(®), which consists of ginkgolides, bilobalide, and flavonoids. The individual constituents have varying therapeutic mechanisms that contribute to the pharmacological activity of the extract as a whole. Recent studies show anxiolytic properties of ginkgolide A, migraine with aura treatment by ginkgolide B, a reduction in ischemia-induced glutamate excitotoxicity by bilobalide, and an alternative antihypertensive property of quercetin, among others. These findings have been observed in EGb 761 as well and have led to clinical investigation into its use as a therapeutic for conditions such as cognition, dementia, cardiovascular, and cerebrovascular diseases."[62]

In 2015, scientists at the University of Bari Aldo Moro in Italy reported in the *Journal of Alzheimer's Disease*: "Among nutraceuticals and nutritional bioactive compounds, the standardized *Ginkgo biloba* extract EGb 761 is the most extensively clinically tested herbal-based substance for cognitive impairment, dementia, and Alzheimer's disease (AD)... Meta-analytic findings suggested overall benefits of EGb 761 for stabilizing or slowing decline in cognition of subjects with cognitive impairment and dementia. The safety and tolerability of EGb 761 appeared to be excellent at different doses."[63]

Also in 2015, in "Efficacy and Adverse Effects of *Ginkgo biloba* for Cognitive Impairment and Dementia: A Systematic Review and Meta-Analysis," multiple scientists from the Department of Neurology at Qingdao Municipal Hospital in China, the College of Medicine and Pharmaceutics at Ocean University of China, and the Department of Neurology at the University of California at San Francisco reported in the *Journal of Alzheimer's Disease*: "EGb761 at 240 mg/day is able to stabilize or slow decline in cognition, function, behavior, and global change at 22-26 weeks in cognitive impairment and dementia, especially for patients with neuropsychiatric symptoms."[64]

It seems evident that drinking green tea or about 4–5 ounces of red wine per day, eating grape products or turmeric, or incorporating extra virgin olive oil into one's diet would help to enhance health and increase longevity. While each of these beverages and food items provides numerous health-related benefits as indicated in the reports above, the pleiotropic properties of each, at least in some cases, would redundantly and synergistically augment the multiple therapeutic benefits of the other if they were combined as part of a healthy diet to prevent the major chronic diseases, including possibly AD. Eating such a diet, itself augmented with other foods that have been shown to play a role in the prevention of age-related diseases—including docosahexaenoic acid found in cold-water fish,[65] blueberries,[66] and pomegranates[67] as frozen or fresh fruit, and cocoa (lavado)[68] and cinnamon (zeylanicum)[69] as dietary spices—would further contribute to an increase in health span and life span while reducing the likelihood of neurodegenerative diseases, among which AD is the most common.

The polyphenolic compounds featured in this book are the biologically active, health-promoting compounds contained in green tea, red wine, grape products, olive oil, and turmeric, in addition to the green, fanlike leaves of the *Ginkgo biloba* tree. As sketched above, the health-related effects of these compounds have been studied at major medical centers and university-based labs and published in leading bioscience and medical journals. While the health benefits of these foods have been confirmed largely in epidemiological and population studies, the polyphenols they contain and that are featured in this book have been meticulously studied for their mechanistic effects on the cellular and molecular pathogenic processes that are likely involved from the start of the AD-related disease process, and thus constitute an important additional nutritional model in applied research for preventing and treating AD.

REFERENCES

1. Ramassamy. Emerging role of polyphenolic compounds in the treatment of neurodegenerative diseases: a review of their intracellular targets. *European Journal of Pharmacology* September 1, 2006;**545**(1): 51–64.
2. Mandel, Weinreb, Youdim. Understanding the broad-spectrum neuroprotective action profile of green tea polyphenols in aging and neurodegenerative diseases. *Journal of Alzheimer's Disease* 2011;**25**(2): 187–208.
3. Malar, Devi. Dietary polyphenols for treatment of Alzheimer's disease—future research and development. *Current Pharmaceutical Biotechnology* 2014;**15**(4):330–42.
4. Siddiqui, Adhami, Saleem, Mukhtar. Beneficial effects of tea and its polyphenols against prostate cancer. *Molecular Nutrition & Food Research* February 2006;**50**(2):130–43.

5. Kiple, Ornelas, editors. *The Cambridge world history of food, Part I.* Cambridge, England: Cambridge University Press; 2008. p. 712. Online Edition.
6. Herbal Medicine: Tea (Black and Green). American Botanical Council. At http://cms.herbalgram.org/ healthyingredients/Tea28BlackandGreen29.html.
7. Tannahill. *Food in history.* New York: Three Rivers Press; 1988. p. 267–8.
8. America is slowly—but surely—becoming a nation of tea drinkers. *The Washington Post* September 3, 2014.
9. Mak. Potential role of green tea catechins in various disease therapies: progress and promise. *Clinical and Experimental Pharmacology & Physiology* March 2012;**39**(3):265–73.
10. Cabrera, Artacho, Giménez. Beneficial effects of green tea—a review. *Journal of the American College of Nutrition* April 2006;**25**(2):79–99.
11. Cooper, Morré, Morré. Medicinal benefits of green tea: Part I. Review of noncancer health benefits. *Journal of Alternative and Complementary Medicine* June 2005;**11**(3):521–8; Cooper, Morré, Morré. Medicinal benefits of green tea: Part II. Review of anticancer properties. *Journal of Alternative and Complementary Medicine* August 2005;**11**(4):639–52.
12. Kuriyama, Shimazu, Ohmori, et al. Green tea consumption and mortality due to cardiovascular disease, cancer, and all causes in Japan: the Ohsaki study. *Journal of the American Medical Association* September 13, 2006;**296**(10):1255–65.
13. Suzuki, Yorifuji, Takao, et al. Green tea consumption and mortality among Japanese elderly people: the prospective Shizuoka elderly cohort. *Annals of Epidemiology* October 2009;**19**(10):732–9.
14. Saito, Inoue, Sawada, et al. Association of green tea consumption with mortality due to all causes and major causes of death in a Japanese population: the Japan public health center-based prospective study (JPHC study). *Annals of Epidemiology* July 2015;**25**(7). 512.e3–518.e3.
15. Mandel, Youdim. Catechin polyphenols: neurodegeneration and neuroprotection in neurodegenerative diseases. *Free Radical Biology & Medicine* August 1, 2004;**37**(3):304–17.
16. Mandel, Avramovich-Tirosh, Reznichenko, et al. Multifunctional activities of green tea catechins in neuroprotection. Modulation of cell survival genes, iron-dependent oxidative stress and PKC signaling pathway. *Neuro-signals* 2005;**14**(1–2):46–60.
17. Mandel, Amit, Reznichenko, et al. Green tea catechins as brain-permeable, natural iron chelators-antioxidants for the treatment of neurodegenerative disorders. *Molecular Nutrition & Food Research* February 2006;**50**(2): 229–34.
18. Mandel, Amit, Bar-Am, Youdim, Iron dysregulation in Alzheimer's disease: multimodal brain permeable iron chelating drugs, possessing neuroprotective-neurorescue and amyloid precursor protein-processing regulatory activities as therapeutic agents. *Progress in Neurobiology* August 2007;**82**(6):348–60; Weinreb, Mandel, Amit, Youdim. Neurological mechanisms of green tea polyphenols in Alzheimer's and Parkinson's diseases. *The Journal of Nutritional Biochemistry* September 2004;**15**(9):506–16; Mandel, Amit, Kalfon, et al. Targeting multiple neurodegenerative disease etiologies with multimodal-acting green tea catechins. *The Journal of Nutrition* August 2008;**138**(8):1578S–83S; Mandel, Amit, Weinreb, et al. Simultaneous manipulation of multiple brain targets by green tea catechins: a potential neuroprotective strategy for Alzheimer and Parkinson diseases. *CNS Neuroscience & Therapeutics* Winter 2008;**14**(4): 352–65.
19. Weinreb, Amit, Mandel, Youdim. Neuroprotective molecular mechanisms of (-)-epigallocatechin-3-gallate: a reflective outcome of its antioxidant, iron chelating and neuritogenic properties. *Genes & Nutrition* December 2009;**4**(4):283–96.
20. Bastianetto, Krantic, Chabot, Quirion. Possible involvement of programmed cell death pathways in the neuroprotective action of polyphenols. *Current Alzheimer Research* August 2011;**8**(5):445–51.
21. Mandel, Youdim. In the rush for green gold: can green tea delay age-progressive brain neurodegeneration? *Recent Patents on CNS Drug Discovery* December 2012;**7**(3):205–17.
22. Trubek. The revolt against homogeneity. In: Freedman, Chaplin, Albala, editors. *Food in time and place: the American historical association companion to food history.* Berkeley: University of California Press; 2014. p. 307.
23. Toussaint-Samat. *A history of food.* 2nd ed. Hoboken, N.J.: Wiley-Blackwell; 2008. p. 223.
24. Tannahill. *Food in history.* p. 63.
25. Trubek. The revolt against homogeneity. In: *Food in time and place.* p. 307.

26. Renaud, Guéguen, Siest, Salamon. Wine, beer, and mortality in middle-aged men from Eastern France. *Archives of Internal Medicine* September 13, 1999;**159**(16):1865–70.

27. Renaud, Lanzmann-Petithory, Gueguen, Conard. Alcohol and mortality from all causes. *Biological Research* 2004;**37**(2):183–7.

28. Streppel, Ocké, Boshuizen, et al. Long-term wine consumption is related to cardiovascular mortality and life expectancy independently of moderate alcohol intake: the Zutphen study. *Journal of Epidemiology and Community Health* July 2009;**63**(7):534–40.

29. Rodrigo, Miranda, Vergara. Modulation of endogenous antioxidant system by wine polyphenols in human disease. *Clinica Chimica Acta: International Journal of Clinical Chemistry* February 20, 2011;**412**(5–6):410–24.

30. Basli, Soulet, Chaher, et al. Wine polyphenols: potential agents in neuroprotection. *Oxidative Medicine and Cellular Longevity* 2012;**2012**:805762; Granzotto, Zatta. Resveratrol and Alzheimer's disease: message in a bottle on red wine and cognition. *Frontiers in Aging Neuroscience* May 14, 2014;**6**:95.

31. Wang, Ho, Zhao, et al. Moderate consumption of Cabernet Sauvignon attenuates Abeta neuropathology in a mouse model of Alzheimer's disease. *FASEB Journal* November 2006;**20**(13):2313–20; Russo, Palumbo, Aliano, et al. Red wine micronutrients as protective agents in Alzheimer-like induced insult. *Life Sciences* April 11, 2003;**72**(21):2369–79.

32. Ho, Chen, Wang, et al. Heterogeneity in red wine polyphenolic contents differentially influences Alzheimer's disease-type neuropathology and cognitive deterioration. *Journal of Alzheimers Disease* 2009;**16**(1):59–72.

33. Martín, González-Burgos, Carretero, Gómez-Serranillos. Protective effects of Merlot red wine extract and its major polyphenols in PC12 cells under oxidative stress conditions. *Journal of Food Science* January 2013;**78**(1):H112–8.

34. Pasinetti. Novel role of red wine-derived polyphenols in the prevention of Alzheimer's disease dementia and brain pathology: experimental approaches and clinical implications. *Planta Medica* October 2012;**78**(15):1614–9.

35. Vislocky, Fernandez. Biomedical effects of grape products. *Nutrition Reviews* November 2010;**68**(11):656–70.

36. Pasinetti, Ho. Role of grape seed polyphenols in Alzheimer's disease neuropathology. *Nutrition and Dietary Supplements* August 1, 2010;**2010**(2):97–103.

37. Pasinetti, Wang, Ho, et al. Roles of resveratrol and other grape-derived polyphenols in Alzheimer's disease prevention and treatment. *Biochimica et Biophysica Acta* June 2015;**1852**(6):1202–8.

38. Wang, Cheng, Freire, et al. Targeting multiple pathogenic mechanisms with polyphenols for the treatment of Alzheimer's disease—experimental approach and therapeutic implications. *Frontiers in Aging Neuroscience* March 14, 2014;**6**:42.

39. *The olive tree.* International Olive Council. At http://www.internationaloliveoil.org/estaticos/view/76-the-olive-tree.

40. *Designations and definitions of olive oils.* International Olive Council. At http://www.internationaloliveoil.org/estaticos/view/83-designations-and-definitions-of-olive-oils.

41. Visioli, Galli. Biological properties of olive oil phytochemicals. *Critical Reviews in Food Science and Nutrition* 2002;**42**(3):209–21.

42. Perez-Jimenez, Alvarez de Cienfuegos, Badimon, et al. International conference on the healthy effect of virgin olive oil. *European Journal of Clinical Investigation* July 2005;**35**(7):421–4.

43. Cicerale, Conlan, Sinclair, Keast. Chemistry and health of olive oil phenolics. *Critical Reviews in Food Science and Nutrition* March 2009;**49**(3):218–36.

44. López-Miranda, Pérez-Jiménez, Ros, et al. Olive oil and health: summary of the II international conference on olive oil and health consensus report, Jaén and Córdoba (Spain) 2008. *Nutrition, Metabolism, and Cardiovascular Diseases* May 2010;**20**(4):284–94.

45. Buckland, Gonzalez. The role of olive oil in disease prevention: a focus on the recent epidemiological evidence from cohort studies and dietary intervention trials. *The British Journal of Nutrition* April 2015;**113**(Suppl. 2):S94–101.

46. Casamenti, Grossi, Rigacci, et al. Oleuropein aglycone: a possible drug against degenerative conditions. In vivo evidence of its effectiveness against Alzheimer's disease. *Journal of Alzheimer's Disease* 2015;**45**(3):679–88.

47. Rigacci. Olive oil phenols as promising multi-targeting agents against Alzheimer's disease. *Advances in Experimental Medicine and Biology* 2015;**863**:1–20.

48. *Herbs at a glance: turmeric*. National Center for Complementary and Alternative Medicine, U.S. National Institutes of Health. At https://nccih.nih.gov/health/turmeric/ataglance.htm.

49. *Herbal medicine: turmeric root*. American Botanical Council. At http://cms.herbalgram.org/expandedE/Turmericroot.html?ts=1459707838&signature=ce2688de95a3ff2acbb1376511f404b5.

50. Brondino, Re, Boldrini, et al. Curcumin as a therapeutic agent in dementia: a mini systematic review of human studies. *Scientific World Journal* January 22, 2014;**2014**:174282; Ganguli, Chandra, Kamboh, et al. Apolipoprotein E polymorphism and Alzheimer disease: The Indo-US cross-national dementia study. *Archives of Neurology* June 2000;**57**(6):824–30.

51. Tripathi, Gupta, et al. Risk factors of dementia in north India: a case-control study. *Aging & Mental Health* 2012;**16**(2):228–35.

52. Prasad, Aggarwal. Turmeric, the golden spice: from traditional medicine to modern medicine. In: Iris, Benzie, Wachtel-Galor, editors. *Herbal medicine: biomolecular and clinical aspects*. Boca Raton: CRC Press; 2011. p. 270.

53. Zhou, Beevers, Huang. The targets of curcumin. *Current Drug Targets* March 1, 2011;**12**(3):332–47.

54. Prasad, Gupta, Tyagi, Aggarwal. Curcumin, a component of golden spice: from bedside to bench and back. *Biotechnology Advances* November 1, 2014;**32**(6):1053–64.

55. Shanmugam, Rane, Kanchi, et al. The multifaceted role of curcumin in cancer prevention and treatment. *Molecules: A Journal of Synthetic Chemistry and Natural Product Chemistry* February 5, 2015;**20**(2):2728–69.

56. Qadir, Naqvi, Muhammad. Curcumin: a polyphenol with molecular targets for cancer control. *Asian Pacific Journal of Cancer Prevention* 2016;**17**(6):2735–9.

57. Hu, Maiti, Ma, et al. Clinical development of curcumin in neurodegenerative disease. *Expert Review of Neurotherapeutics* June 2015;**15**(6):629–37.

58. Begum, Jones, Lim, et al. Curcumin structure-function, bioavailability, and efficacy in models of neuro-inflammation and Alzheimer's disease. *The Journal of Pharmacology and Experimental Therapeutics* July 2008;**326**(1):196–208.

59. Ginkgo: Clinical Overview: The ABC Clinical Guide to Herbs. At http://abc.herbalgram.org/site/DocServer/Ginkgo.pdf?docID=166.

60. Ginkgo biloba: University of Maryland Medical Center. At http://umm.edu/health/medical/altmed/herb/ginkgo-biloba.

61. *PDR for herbal medicines*. 4th ed. Thomson-Reuters; 2007. p. 372–3.

62. Nash, Shah. Current perspectives on the beneficial role of *Ginkgo biloba* in neurological and cerebrovascular disorders. *Integrative Medicine Insights* November 9, 2015;**10**:1–9.

63. Solfrizzi, Panza. Plant-based nutraceutical interventions against cognitive impairment and dementia: meta-analytic evidence of efficacy of a standardized *Gingko biloba* extract. *Journal of Alzheimer's Disease* 2015;**43**(2):605–11.

64. Tan, Yu, Tan, et al. Efficacy and adverse effects of *Ginkgo biloba* for cognitive impairment and dementia: a systematic review and meta-analysis. *Journal of Alzheimer's Disease* 2015;**43**(2):589–603.

65. Eriksdotter, Vedin, Falahati, et al. Plasma fatty acid profiles in relation to cognition and gender in Alzheimer's disease patients during oral omega-3 fatty acid supplementation: the omegAD study. *Journal of Alzheimer's Disease* 2015;**48**(3):805–12; Teng, Taylor, Bilousova, et al. Dietary DHA supplementation in an APP/PS1 transgenic rat model of AD reduces behavioral and Aβ pathology and modulates Aβ oligomerization. *Neurobiology of Disease* October 2015;**82**:552–60; Philips, Childs, Calder, Rogers. Lower omega-3 fatty acid intake and status are associated with poorer cognitive function in older age: A comparison of individuals with and without cognitive impairment and Alzheimer's disease. *Nutritional Neuroscience* November 2012;**15**(6):271–7.

66. Joseph, Shukitt-Hale, Lau. Fruit polyphenols and their effects on neuronal signaling and behavior in senescence. *Annals of the New York Academy of Sciences* April 2007;**1100**:470–85; Lau, Shukitt-Hale, Joseph. Nutritional intervention in brain aging: reducing the effects of inflammation and oxidative stress. *Sub-Cellular Biochemistry* 2007;**42**:299–318; Lau, Shukitt-Hale, Joseph. The beneficial effects of fruit polyphenols on brain aging. *Neurobiology of Aging* December 2005;**26**(Suppl. 1):128–32; Joseph, Denisova, Arendash, et al. Blueberry supplementation enhances signaling and prevents behavioral deficits in an Alzheimer's disease model. *Nutritional Neuroscience* June 2003;**6**(3):153–62.

67. Subash, Braidy, Essa, et al. Long-term (15 mo) dietary supplementation with pomegranates from Oman attenuates cognitive and behavioral deficits in a transgenic mice model of Alzheimer's disease. *Nutrition* January 2015;**31**(1):223–29; Essa, Subash, Akbar, et al. Long-term dietary supplementation of pomegranates, figs and dates alleviate neuroinflammation in a transgenic mouse model of Alzheimer's disease. *PLoS One* March 25, 2015;**10**(3):e0120964; Ahmed, Subaiea, Eid, et al. Pomegranate extract modulates processing of amyloid-β precursor protein in an aged Alzheimer's disease animal model. *Current Alzheimer Research* 2014;**11**(9):834–43; Subash, Essa, Al-Asmi, et al. Pomegranate from Oman alleviates the brain oxidative damage in transgenic mouse model of Alzheimer's disease. *Journal of Traditional and Complementary Medicine* October 2014;**4**(4):232–8; Rojanathammanee, Puig, Combs. Pomegranate polyphenols and extract inhibit nuclear factor of activated T-cell activity and microglial activation in vitro and in a transgenic mouse model of Alzheimer disease. *The Journal of Nutrition* May 2013;**143**(5):597–605; Hartman, Shah, Fagan, et al. Pomegranate juice decreases amyloid load and improves behavior in a mouse model of Alzheimer's disease. *Neurobiology of Disease* December 2006;**24**(3):506–15.

68. Moreira, Diógenes, de Mendonça, et al. Chocolate consumption is associated with a lower risk of cognitive decline. *Journal of Alzheimer's Disease* May 6, 2016;**53**(1):85–93; Grassi, Ferri, Desideri. Brain protection and cognitive function: cocoa flavonoids as nutraceuticals. *Current Pharmaceutical Design* 2016;**22**(2):145–51; Dubner, Wang, Ho, et al. Recommendations for development of new standardized forms of cocoa breeds and cocoa extract processing for the prevention of Alzheimer's disease: role of cocoa in promotion of cognitive resilience and healthy brain aging. *Journal of Alzheimer's Disease* 2015;**48**(4):879–89; Wang, Varghese, Ono, et al. Cocoa extracts reduce oligomerization of amyloid-β: implications for cognitive improvement in Alzheimer's disease. *Journal of Alzheimer's Disease* 2014;**41**(2):643–50.

69. Madhavadas, Subramanian. Cognition enhancing effect of the aqueous extract of *Cinnamomum zeylanicum* on non-transgenic Alzheimer's disease rat model: biochemical, histological, and behavioural studies. *Nutritional Neuroscience* June 16, 2016:1–12; Malik, Munjal, Deshmukh. Attenuating effect of standardized lyophilized *Cinnamomum zeylanicum* bark extract against streptozotocin-induced experimental dementia of Alzheimer's type. *Journal of Basic and Clinical Physiology and Pharmacology* May 2015;**26**(3):275–85; Anderson, Qin, Canini, et al. Cinnamon counteracts the negative effects of a high fat/high fructose diet on behavior, brain insulin signaling and Alzheimer-associated changes. *PLoS One* December 13, 2013;**8**(12):e83243; George, Lew, Graves. Interaction of cinnamaldehyde and epicatechin with tau: implications of beneficial effects on modulating Alzheimer's disease pathogenesis. *Journal of Alzheimer's Disease* 2013;**36**(1):21–40.

PART II

Pleiotropism and Alzheimer's Disease

CHAPTER 3

Primary Prevention of Alzheimer's Disease

It is best to avoid the beginnings of evil.

Henry David Thoreau

Contents

The homepage of the National Institute on Aging (NIA) describes its mission as follows: "NIA, one of the 27 Institutes and Centers of NIH, leads a broad scientific effort to understand the nature of aging and to extend the healthy, active years of life. NIA is the primary Federal agency supporting and conducting Alzheimer's disease research."[1]

In its featured publication on Alzheimer's disease (AD) titled, "Alzheimer's Disease: Unraveling the Mystery," the NIA itemizes the five broad phases of the AD–related disease process as follows:

Stages of the AD–Related Disease Process (NIA)
1. Preclinical Alzheimer's Disease [Pathophysiological; Presymptomatic].
2. Mild Cognitive Impairment [Pathophysiological; Symptomatic].
3. Mild Alzheimer's Disease [Clinically Diagnosed AD].
4. Moderate Alzheimer's Disease.
5. Severe Alzheimer's Disease.[2]

The NIA also observed: "NIH-supported discoveries that Alzheimer's pathology is present in the brain many years before symptoms appear has opened the door for early intervention and raised hopes that primary prevention of the disease may be possible."[3] This statement appears to indicate that intervention at the preclinical stage to prevent

A Paradigm Shift to Prevent and Treat Alzheimer's Disease
ISBN 978-0-12-812259-4
http://dx.doi.org/10.1016/B978-0-12-812259-4.00003-5

AD-related symptoms constitutes "primary prevention," although onset of AD pathology has already occurred at the stage referenced here by the NIA.

The Mayo Clinic similarly lists the stages of AD, although it bundles the three phases of AD—mild AD, moderate AD, and severe AD—into a single "dementia phase." Like NIA, it also identifies preclinical AD as "a platform" for "primary prevention therapies."

Stages of the AD-Related Disease Process (Mayo Clinic)

1. An asymptomatic preclinical phase, intended as a platform from which to develop primary prevention therapies.
2. A symptomatic, predementia phase, in which mild cognitive impairment (MCI) consistent with AD is present.
3. A dementia phase, in which clinical symptoms are present.[4]

It thus appears that both the NIA and the Mayo Clinic view medical intervention at the pathophysiologically present but symptomatically absent preclinical stage of AD as "primary prevention."

Several sources define "primary prevention" similarly, including as follows:

- "avoidance of the onset of disease;"[5]
- "to prevent disease or injury before it ever occurs;"[6]
- "the first level of health care designed to prevent the occurrence of disease and promote health;"[7]
- "methods to avoid occurrence of disease."[8]

The same sources define "secondary prevention":

- "the avoidance or alleviation of disease by early detection and appropriate management;"[5]
- "aims to reduce the impact of a disease or injury that has already occurred;"[6]
- "the second level of health care, based on the earliest possible identification of disease so that it can be more readily treated or managed;"[7]
- "methods to detect and address an existing disease prior to the appearance of symptoms."[8]

Thus, primary prevention constitutes the first level of health care with the goal of preventing onset of the pathogenic processes of a given disease, while secondary prevention represents a second level of postonset early detection and treatment. And since preclinical AD is marked by the asymptomatic onset of the disease process, as we shall see, preventive measures undertaken at this stage should be regarded as secondary prevention. Yet, they are commonly characterized as "prevention" or "primary prevention" in the scientific literature on AD.

The confusion, in one respect, has to do with whether we define AD more along the lines of brain pathophysiology or clinical presentation. In 2011, a group of prominent AD researchers, chaired by Reisa A. Sperling of Harvard Medical School and sponsored by the Alzheimer's

Association and the NIA, revised the diagnostic guidelines for preclinical AD. About the question of pathophysiology versus clinical symptoms, the group suggested the adoption of a bipartite designation for AD: AD-pathophysiological (AD-P) and AD-clinical (AD-C). In its report, the group concluded that "it was important to define AD as encompassing the underlying pathophysiological disease process, as opposed to having 'AD' connote only the clinical [symptomatic] stages of the disease."[10] Pursuant to this authoritatively presented distinction, intervention at the preclinical stage—where the onset of AD-P but not AD-C has begun—should now be routinely cited as secondary prevention, not primary prevention.

To further underscore the distinction, we need to establish that preclinical AD is a pathophysiological stage. This is easily done, since there is broad consensus on the issue. For example, the NIA describes the disease process involved in preclinical AD as follows:

> Preclinical AD. *AD begins deep in the brain, in the entorhinal cortex, a brain region that is near the hippocampus and has direct connections to it. Healthy neurons in this region begin to work less efficiently, lose their ability to communicate, and ultimately die. This process gradually spreads to the hippocampus, the brain region that plays a major role in learning and is involved in converting short-term memories to long-term memories. Affected regions begin to atrophy. Ventricles, the fluid-filled spaces inside the brain, begin to enlarge as the process continues. Scientists believe that these changes begin 10 to 20 years before any clinically detectable signs or symptoms of forgetfulness appear.*[11]

The Mayo Clinic also summarized the progressive pathogenic features of preclinical AD:

> Stage 1 [of preclinical AD] is characterized by asymptomatic amyloidosis.
> Stage 2 [of preclinical AD] adds evidence of synaptic dysfunction and neurodegeneration.
> Stage 3 [of preclinical AD] marks the addition of subtle cognitive changes not overtly evident in day-to-day behavior.

The Mayo Clinic additionally reported that preclinical AD is marked by "an ordered pattern of progression in AD biomarkers in which amyloid deposition reaches a peak before symptom onset."[4]

Furthermore, the high-level group led by Sperling described preclinical AD-P as follows: "AD-P [AD-Pathophysiological] is thought to begin years before the emergence of AD-C [AD-Clinical]" with "a time lag of a decade or more between the beginning of the pathological cascade of AD and the onset of clinically evident impairment."[10] Thus, the frequent mislabeling of AD-related secondary prevention as "primary prevention" reflects a significant and consequential oversight within key segments of AD research.

Even accurate labeling of medical intervention at the preclinical stage as "secondary prevention" can neglect to acknowledge the potential utility of AD-related primary prevention. In "Testing the Right Target and the Right Drug at the Right Stage"—published in 2011 by *Science Translational Medicine*, a publication of the American Association for the Advancement of Science—Sperling, Clifford R. Jack Jr., and Paul S. Aisen (all prominent AD experts) explained as follows the relative absence of AD-related primary prevention as a focus of applied research: "Although primary prevention would be ideal, the prospect of large primary prevention trials for late-onset sporadic AD remains daunting and likely unfeasible given the length of treatment required to achieve a clinical endpoint. Furthermore, there is considerable concern over the cost and safety of treating thousands of individuals who may never develop AD pathology. At present, the earliest feasible stage for therapeutic trials in sporadic AD is likely to be at the stage of asymptomatic amyloid accumulation [preclinical AD], based on the (still unproven) hypothesis that brain amyloidosis is indeed indicative of early-stage AD. These studies may be considered 'secondary prevention' trials, aimed at preventing or slowing the progression of the clinical syndrome in those in whom the pathophysiological process of AD has already begun."[12] While the concerns here about AD-related primary prevention are perhaps valid within the realm of pharma-based AD therapeutics, they are less compelling within the domain of plant polyphenols that may be capable of safely and inexpensively preventing onset of AD-P.

For example, we have already observed that the application of the "Gandy test" to pharmaceuticals would constitute a major obstacle to a viable long-term primary-prevention program, as it apparently has, and as Sperling, Jack, and Aisen implicitly acknowledge. In contrast, epidemiological and experimental evidence of the long-term safety and overall health benefits of the plant polyphenols reviewed in this volume mark them as highly attractive candidates to prevent the onset of the AD-related disease process. In addition, the pleiotropic therapeutic capabilities of certain plant polyphenols mitigate somewhat the necessity to precisely map "the right target and the right drug at the right stage" of this multidimensional disease, given evidence that knowledgably formulated combinations of plant polyphenols can safely and effectively address most of the hallmark pathogenic processes of AD (see Chapters 3–6).

While the focus of pharmaceutically based monotargeting resembles a hedgehog-like search for one drug for one target, pleiotropic theory stands on the plant polyphenols "knowing many things," like a fox.[13] Although there is no consensus opinion on the relative therapeutic merits of the two approaches, there is an emerging body of scientific agreement that clearly favors the pleiotropic, plant–polyphenol based model, including as applied, for example, to AD-related primary prevention, the basic mechanics of which are described below.

Suppose we revisit the NIA's five-stage summary of the AD-related disease process and amended it to insert a prepathophysiology "zero stage" as follows:

Stages of the AD-Related Disease Process (Revised)
0. Zero AD [No Pathophysiology; No Symptoms].
1. Preclinical [Pathophysiological; Presymptomatic].
2. Mild Cognitive Impairment due to AD [Pathophysiological; Symptomatic].
3. Mild AD [Clinically Diagnosed AD].
4. Moderate AD.
5. Severe AD.

Like the number zero, which denotes "the absence of all magnitude or quantity,"[14] yet following its invention has played major roles in mathematics and computer science,[15] the addition of a prepathophysiological zero stage would permit an official category of AD-related primary prevention, while tightly securing interventions during preclinical AD to secondary prevention, as follows:

0. Zero AD→Primary Prevention of AD
1. Preclinical AD→Secondary Prevention of AD

The insertion of a zero stage in this way would introduce a population of disease-free people for which primary prevention measures could be initiated prior to the onset of the AD-related disease process. Also, by identifying "asymptomatic amyloidosis" as the initial pathogenic phase of preclinical AD,[4] the Mayo Clinic (among others) has identified "amyloidosis" as a focus of our primary-prevention efforts.

Because the pathogenic course of amyloidosis likely begins with the proteolytic enzymatic processing of amyloid precursor protein (APP), the preferred primary prevention agents would need to be capable of safely and effectively influencing the nonpathogenic processing of APP, probably over a long period of time. They would also need to be abundantly available and reasonably affordable to make them as accessible as possible for as many people as possible. Given the reported therapeutic capabilities of the plant polyphenols in our featured ancient foods, their abundance worldwide, and relatively low cost, the standardized polyphenolic compounds from green tea, red wine, turmeric, olive oil, and red wine and other grape products would meet these standards.

Furthermore, because it likely would be difficult, logistically and financially, and in advance of symptoms, to differentiate large populations of people with no AD-related pathology from people with presymptomatic, preclinical AD, the ideal primary-prevention agents also would need to function simultaneously as secondary-prevention agents. More specifically, the agents would need to be capable not only of inhibiting the amyloidogenic processing of APP as a primary-prevention requirement, but of inhibiting the progressive aggregation and synaptic toxicity of Abeta monomers and oligomers, in addition to increasing clearance of soluble and insoluble Abeta aggregates from the brain. To our knowledge, this degree of therapeutic pleiotropism does not exist within the pharmaceutical framework; on the other hand, it is common among the plant polyphenols featured in this volume.

In "Alzheimer's Disease: Unraveling the Mystery," the NIA includes a section titled, "The Hallmarks of AD," where a "hallmark" refers to a major pathophysiological characteristic of the disease. The first AD-related hallmark listed by the NIA is "Amyloid Plaques." About amyloid plaques, which are found in the brains of people with AD, the NIA observes: "They were first described by Dr. Alois Alzheimer in 1906. Plaques consist of largely insoluble deposits of an apparently toxic protein peptide, or fragment, called amyloid-beta."[16]

A basic objective in the primary prevention of AD would be to promote the non-pathogenic processing of APP by favorably influencing the activity of the proteolytic enzymes involved. This could occur by upregulating the activity of one such enzyme, alpha-secretase, and downregulating the activity of beta-secretase and gamma-secretase, two other such enzymes, to preclude the production of the amyloid-beta peptide. Barring a major revision of the current understanding of the first steps of AD pathogenesis, influencing the processing of APP in this way would likely prevent AD.[17]

The NIA provides brief, nicely rendered portraits of each of the differentiated elements involved in APP processing—APP, alpha-secretase (and sAPPα), beta-secretase, and gamma-secretase—which are presented below:

Amyloid Precursor Protein. *Amyloid precursor protein (APP), the starting point for amyloid plaques, is one of many proteins associated with the cell membrane, the barrier that encloses the cell. As it is being made inside the cell, APP becomes embedded in the membrane, like a toothpick stuck through the skin of an orange.*

In a number of cell compartments, including the outermost cell membrane, specific enzymes snip, or cleave, APP into discrete fragments [peptides]. In 1999 and 2000, scientists identified the enzymes responsible for cleaving APP. These enzymes are called alpha-secretase, beta-secretase, and gamma-secretase. In a major breakthrough, scientists then discovered that, depending on which enzyme is involved and the segment of APP where the cleaving occurs, APP processing can follow one of two pathways that have very different consequences for the cell.

Alpha-Secretase. *In the benign [non-amyloidogenic] pathway, alpha-secretase cleaves the APP molecule within the portion that has the potential to become beta-amyloid. This eliminates the production of the beta-amyloid peptide and the potential for plaque buildup. The cleavage releases from the neuron a fragment called sAPPα [soluble amyloid precursor protein alpha], which has beneficial properties, such as promoting neuronal growth and survival. The remaining APP fragment, still tethered in the neuron's membrane, is then cleaved by gamma-secretase at the end of the beta-amyloid segment. The smaller of the resulting fragments also is released into the space outside the neuron, while the larger fragment remains within the neuron and interacts with factors in the nucleus.*

Beta-Secretase. *In the harmful [amyloidogenic] pathway, beta-secretase [as opposed to alpha-secretase] first cleaves the APP molecule at one end of the beta-amyloid peptide, releasing sAPPβ from the cell.*

Gamma-Secretase. *Gamma-secretase then cuts the resulting APP fragment, still tethered in the neuron's membrane, at the other end of the beta-amyloid peptide.*[16]

In short, the initial cleaving of APP by alpha-secretase, as opposed to APP being first cut by beta-secretase, prevents the formation of the neurotoxic amyloid-beta peptides: "In the non-amyloidogenic pathway, alpha-secretase cleaves the amyloid precursor protein (APP) within the sequence of Abeta-peptides and precludes their formation."[18] Thus, from a primary-prevention standpoint, we would want to safely influence the upregulation of alpha-secretase activity and the downregulation of beta-secretase, and thereby reduce or largely prevent production of the pathogenic Abeta peptides.

While alpha-secretase is the enzyme that predominately cleaves APP at the neuronal membrane surface, there is competition from beta-secretase to be the first of the enzymes to cleave APP. Thus, as scientists who have conducted extensive research about this process have reported: "The observation that alpha-secretase cleaves APP at the cell surface, while beta-secretase can act earlier in the secretory pathway within the neuronal cell line, indicates that there must be strict control mechanisms in place to ensure that APP is normally cleaved primarily by alpha-secretase in the nonamyloidogenic pathway to produce the neuroprotective sAPPα."[19] Numerous other studies concur,[20] and thus support the upregulation of alpha-secretase and the downregulation of beta-secretase as a means of preempting production of the amyloid-beta peptide.

Key lines of evidence published in scientific and medical journals around the world indicate, as the abstracts (or excerpts of abstracts) below show, that certain nontoxic plant polyphenols possess an ability to upregulate the nonpathogenic alpha-secretase enzyme (and thereby generate the neuroprotective sAPPα), downregulate the pathogenic beta-secretase, and thus perform these key primary-prevention functions.

As we examine the evidence, we should note that each abstract provides the title of the journal in which the evidence or analysis was published, the country in which the journal is published, the year of publication, and the names and locations of the medical centers or university laboratories at which the researchers worked. This is to demonstrate at a glance not only the international scope and scale of the research, but the consistently high level of scientific evidence that has been generated on plant polyphenols and AD to date.

Note also, immediately below, that (1) alpha-secretase is functionally equivalent to ADAM10, (2) the generation of sAPPα (the cleavage product of alpha-secretase, also written as sAPPalpha) is neuroprotective, (3) beta-secretase is equivalent to BACE, (4) gamma-secretase is equivalent to presenilin, and (5) upward-pointing arrows signal increased or activated activity and downward-pointing arrows signal decreased or inactivated activity.

Ginkgo biloba EXTRACTS (↑) ALPHA-SECRETASE (ADAM10); sAPPα

Ginkgo biloba Extract 761 (↑) sAPPα
Neurobiology of Disease (United States) (2004)

Center of Excellence on Neurodegenerative Diseases and Department of Pharmacological Sciences, University of Milan, Italy.

[A]lpha-secretase, the enzyme regulating the non-amyloidogenic processing of APP and the release of alphaAPPs, the alpha-secretase metabolite, were studied in superfusates of hippocampal slices after EGb761 incubation, and in hippocampi and cortices of EGb761-treated rats... **EGb761 increases alphaAPPs release through a PKC-independent manner**. This effect is not accompanied by a modification of either APP forms or alpha-secretase expression.[21]

Ginkgo biloba Extract (Bilobalide) (↑) sAPPα
Neurochemistry International (England) (2011)
Department of Anatomy, Guangzhou Medical University, China.
Bilobalide (BB) is a sesquiterpenoid extracted from *Ginkgo biloba* leaves… Using differentiated SH-SY5Y cells, this study investigated whether BB modulation of intracellular signaling pathways, such as the protein kinase C (PKC) and PI3K pathways, contributes to amyloid precursor protein (APP) metabolism, a key event in the pathogenesis of Alzheimer's disease (AD). **We demonstrated in this study that BB enhanced the secretion of α-secretase-cleaved soluble amyloid precursor protein (sAPPα, a by-product of non-amyloidogenic processing of APP) and decreased the β amyloid protein (Aβ, a by-product of amyloidogenic processing of APP) via PI3K-dependent pathway…** Given the strong association between APP metabolism and AD pathogenesis, the ability of BB to regulate APP processing suggests its potential use in AD prevention.[22]

GREEN TEA EXTRACTS (↑) ALPHA-SECRETASE (ADAM10); sAPPα

Green Tea Extract (EGCG) (↑) sAPPα
FASEB Journal (United States) (2003)
Eve Topf and USA National Parkinson Foundation, Centers of Excellence for Neurodegenerative Diseases Research, Technion Faculty of Medicine, Haifa, Israel.
The central hypothesis guiding this study is that EGCG may play an important role in amyloid precursor protein (APP) secretion and protection against toxicity induced by beta-amyloid (Abeta). **The present study shows that EGCG enhances (approximately 6-fold) the release of the non-amyloidogenic soluble form of the amyloid precursor protein (sAPPalpha) into the conditioned media of human SH-SY5Y neuroblastoma and rat pheochromocytoma PC12 cells.**[23]

Green Tea Extract (EGCG) (↑) Alpha-Secretase
The Journal of Neuroscience (United States) (2005)
Silver Child Development Center, Department of Psychiatry and Behavioral Medicine, University of South Florida, Tampa, Florida.
Here, we report that (-)-epigallocatechin-3-gallate (EGCG), the main polyphenolic constituent of green tea, reduces Abeta generation in both murine neuron-like cells (N2a) transfected with the human "Swedish" mutant amyloid precursor protein (APP) and in primary neurons derived from Swedish mutant APP-overexpressing mice (Tg APPsw line 2576). In concert with these observations, **we find that EGCG markedly promotes cleavage of the alpha-C-terminal fragment of APP and elevates the N-terminal APP cleavage product, soluble**

APP-alpha. These cleavage events are associated with elevated alpha-secretase activity and enhanced hydrolysis of tumor necrosis factor alpha-converting enzyme, a primary candidate alpha-secretase. As a validation of these findings in vivo, we treated Tg APPsw transgenic mice overproducing Abeta with EGCG and found decreased Abeta levels and plaques associated with promotion of the nonamyloidogenic alpha-secretase proteolytic pathway. These data raise the possibility that EGCG dietary supplementation may provide effective prophylaxis for AD.[24]

GREEN TEA EXTRACT (EGCG) (↑) ALPHA-SECRETASE/ADAM10/sAPPα
The Journal of Biological Chemistry (United States) (2006)

Neuroimmunology Laboratory, Silver Child Development Center, Department of Psychiatry and Behavioral Medicine, The Byrd Alzheimer's Center and Research Institute, Tampa, Florida.

Recently, we have shown that green tea polyphenol (-)-epigallocatechin-3-gallate (EGCG) exerts a beneficial role on reducing brain Abeta levels, resulting in mitigation of cerebral amyloidosis in a mouse model of Alzheimer disease. **EGCG seems to accomplish this by modulating amyloid precursor protein (APP) processing, resulting in enhanced cleavage of the alpha-COOH-terminal fragment (alpha-CTF) of APP and corresponding elevation of the NH(2)-terminal APP product, soluble APP-alpha (sAPP-alpha). These beneficial effects were associated with increased alpha-secretase cleavage activity,** but no significant alteration in beta-or gamma-secretase activities… **Results show that EGCG treatment of N2a cells stably transfected with "Swedish" mutant human APP (SweAPP N2a cells) leads to markedly elevated active (approximately 60 kDa mature form) ADAM10 protein. Elevation of active ADAM10 correlates with increased alpha-CTF cleavage, and elevated sAPP-alpha.**[25]

GREEN TEA EXTRACT (EGCG) (↑) ALPHA-SECRETASE; (↓) BETA-SECRETASE; (↓) GAMMA-SECRETASE
The Journal of Nutrition (United States) (2009)

College of Pharmacy, Chungbuk National University 12, South Korea.

Abeta(1-42) induced [in mice] a decrease in brain alpha-secretase and increases in both brain beta- and gamma-secretase activities, which were reduced by EGCG… **Compared with untreated mutant PS2 AD mice, treatment with EGCG enhanced memory function and brain alpha-secretase activity [and] reduced brain beta- and gamma-secretase activities as well as Abeta levels…** These studies suggest that EGCG may be a beneficial agent in the prevention of development or progression of AD.[26]

GREEN TEA EXTRACT (EGCG) (↑) NONAMYLOIDOGENIC APP PROCESSING
Neurobiology of Aging (United States) (2009)

Institute of Biochemistry and Molecular Biology, College of Medicine, National Taiwan University, Taipei, Taiwan.

Here, we used a human neuronal cell line MC65 conditional expression of an amyloid precursor protein fragment (APP-C99) to investigate the protection mechanism of epigallocatechin gallate (EGCG), the main constituent of green tea. **We demonstrated that treatment with EGCG reduced the A beta levels by enhancing endogenous APP nonamyloidogenic proteolytic processing.**[27]

GREEN TEA EXTRACT (EGCG) (↑) ADAM10
FEBS Letters (The Netherlands) (2010)

Rashid Laboratory for Developmental Neurobiology, Silver Child Development Center, University of South Florida, Tampa.

Estrogen depletion following menopause has been correlated with an increased risk of developing Alzheimer's disease… **Our results in this study suggest that EGCG-mediated enhancement of non-amyloidogenic processing of APP is mediated by the maturation of ADAM10** via an estrogen receptor-α (ERα)/phosphoinositide 3-kinase/Ak-transforming dependent mechanism, independent of furin-mediated ADAM10 activation. These data support prior assertions that central selective estrogen receptor modulation could be a therapeutic target for AD and support the use of EGCG as a well-tolerated alternative to estrogen therapy in the prophylaxis and treatment of this disease.[28]

Some studies indicate that, to improve oral bioavailability in humans, EGCG may need to be available in a nanolipid formula or taken with DHA from fish oil.

GREEN TEA EXTRACT (EGCG) NANOLIPID (↑) ALPHA-SECRETASE
International Journal of Pharmaceutics (The Netherlands) (2010)

Center of Excellence for Aging and Brain Repair, Department of Neurosurgery and Brain Repair, University of South Florida, Tampa.

We found that forming nanolipidic EGCG particles improves the neuronal (SweAPP N2a cells) alpha-secretase enhancing ability in vitro by up to 91% (P<001) and its oral bioavailability in vivo by more than two-fold over free EGCG.[29]

GREEN TEA EXTRACT (EGCG) AND DHA (↑) sAPPα
Neuroscience Letters (Ireland) (2010)

Department of Psychiatry and Behavioral Medicine, Neuroimmunology Laboratory, University of South Florida, College of Medicine, Tampa.

In vitro co-treatment of N2a cells with fish oil and EGCG enhanced sAPP-alpha production compared to either compound alone (P < 0.001).[30]

Research also has indicated that DHA itself upregulates alpha-secretase and down-regulates beta-secretase and gamma-secretase.

> **DHA** (↑) **Alpha-Secretase**; (↓) **Beta-Secretase**; (↓) **Gamma-Secretase**
> **The Journal of Biological Chemistry (United States) (2011)**
> *Department of Neurodegeneration and Neurobiology, Deutsches Institut für Demenz Prävention, Saarland University, Homburg, Germany.*
> **DHA reduces amyloidogenic processing by decreasing β- and γ-secretase activity**, whereas the expression and protein levels of BACE1 and presenilin1 remain unchanged. **In addition, DHA increases protein stability of α-secretase resulting in increased nonamyloidogenic processing.**[31]

Other nontoxic plant polyphenols that have been shown to upregulate alpha-secretase and promote nonamyloidogenic processing of APP include olive oil extract (oleuropein) and red wine polyphenols.

OLIVE OIL EXTRACTS (↑) sAPPα

> **Olive Oil Extract (Oleuropein)** (↑) **Alpha-Secretase/sAPPα**
> **Cellular and Molecular Neurobiology (United States) (2013)**
> *Institute of Biology, National Centre for Scientific Research Demokritos, Athens, Greece.*
> **In the present study, it was demonstrated that oleuropein treatment of HEK293 cells stably transfected with the isoform 695 of human AβPP (APP695) leads to markedly elevated levels of sAPPα and to significant reduction of Aβ oligomers.** These effects were associated with increased activity of matrix metalloproteinase 9 (MMP-9), whereas no significant alterations in the expression of secretases TACE, ADAM-10 or BACE-1 were observed.[32]

RED WINE (CABERNET SAUVIGNON) (↑) ALPHA-SECRETASE

> **Red Wine (Cabernet Sauvignon)** (↑) **Alpha-Secretase**
> **FASEB Journal (United States) (2006)**
> *Department of Psychiatry, Mount Sinai School of Medicine, New York.*
> Our studies suggest Cabernet Sauvignon exerts a beneficial effect by promoting non-amyloidogenic processing of amyloid precursor protein, which ultimately prevents the generation of Abeta peptides… We also found that **moderate consumption of Cabernet Sauvignon promoted nonamyloidogenic, α-secretase-mediated APP processing, thereby preventing or delaying the generation of amyloidogenic Aβ peptide-associated cognitive deterioration.**[33]

While the studies above report that *Ginkgo biloba* extract, green tea extract (EGCG), olive oil extract (oleuropein), and red wine polyphenols upregulate alpha-secretase and sAPPα, the following studies found that some of these plant polyphenols and the curcuminoids downregulate beta-secretase activity.

CURCUMINOIDS (↓) BETA-SECRETASE (BACE-1)

CURCUMIN AND GREEN TEA EXTRACT (EGCG) (↓) BETA-SECRETASE (BACE-1)
Neuroreport (England) (2008)
Department of Neuroscience for Drug Discovery, Graduate School of Pharmaceutical Sciences, Kyoto University, Japan.

Beta-site APP cleaving enzyme-1 (BACE-1), is a rate-limiting enzyme for beta amyloid production. Beta amyloid induces the production of radical oxygen species and neuronal injury. Oxidative stress plays a key role in various neurological diseases such as ischemia and Alzheimer's disease. Recent studies suggest that oxidative stress induces BACE-1 protein upregulation in neuronal cells. **Here, we demonstrate that naturally occurring compounds (-)-epigallocatechin-3-gallate and curcumin suppress beta amyloid-induced BACE-1 upregulation.**[34]

CURCUMIN (↓) BETA-SECRETASE (BACE-1)
CNS Neuroscience and Therapeutics (England) (2010)
Department of Neurology and Neurobiology of Aging, Kanazawa University Graduate School of Medical Science, Japan.

In in vitro studies, curcumin has been reported to inhibit amyloid-β-protein (Aβ) aggregation, and Aβ-induced inflammation, as well as the activities of β-secretase and acetylcholinesterase. In in vivo studies, oral administration of curcumin has resulted in the inhibition of Aβ deposition, Aβ oligomerization, and tau phosphorylation in the brains of AD animal models, and improvements in behavioral impairment in animal models.[35]

CURCUMIN (↓) AMYLOIDOGENIC APP PROCESSING; BISDEMETHOXYCURCUMIN (↓) BACE-1
Neuroscience Letters (Ireland) (2010)
Department of Neurology, The First Affiliated Hospital, Sun Yat-sen University, Guangzhou, China.

The present study was conducted to examine the effects of curcumin mix and its different curcuminoids including curcumin (Cur), demethoxycurcumin (DMC) and bisdemethoxycurcumin (BDMC) on Aβ(42), APP and BACE1. We found that Cur was the most active curcuminoid fraction in suppressing Aβ(42) production and the order of inhibitory potency of other curcuminoids was DMC > curcumin mix > BDMC. Cur, but not other curcuminoids, could reduce APP protein expression and none of curcuminoids affected APP mRNA level. BDMC could reduce BACE1 mRNA and protein levels, while DMC only affected BACE1 mRNA expression. **Our data indicate that the anti-amyloidogenic effect of Cur may be mediated through the modulation of APP, while the anti-amyloidogenic effect of BDMC may be mediated through the modulation of BACE1.**[36]

CURCUMIN AND BISDEMETHOXYCURCUMIN (↓) BETA-SECRETASE (BACE-1)
BMC Complementary and Alternative Medicine (England) (2014)
WCU Biomodulation Major, Department of Agricultural Biotechnology, Seoul National University, South Korea.
Bisdemethoxycurcumin has the strongest inhibitory activity toward BACE-1 with 17 μM IC50, which was 20 and 13 times lower than those of curcumin and demethoxycurcumin respectively… **Remarkably, supplementing diet with either 1 mM bisdemethoxycurcumin or 1 mM curcumin rescued APP/BACE1-expressing flies and kept them from developing both morphological and behavioral defects.** Our results suggest that structural characteristics, such as degrees of saturation, types of carbon skeleton and functional group, and hydrophobicity appear to play a role in determining inhibitory potency of curcuminoids on BACE-1.[37]

CURCUMINOIDS (↓) BETA-SECRETASE (BACE-1)
Journal of Natural Medicines (Japan) (2015)
Faculty of Pharmacy, Kinki University, Osaka, Japan.
[W]e screened β-secretase and acetylcholinesterase inhibitors from curry spices. Amongst them, curry leaf, black pepper, and **turmeric extracts [curcuminoids] were effective to inhibit β-secretase.**[38]

Ginkgo biloba EXTRACTS (↓) BETA-SECRETASE

Ginkgo biloba EXTRACT (BILOBALIDE) (↓) BETA-SECRETASE
Neurochemical Research (United States) (2012)
Department of Anatomy, Guangzhou Medical University, China.
Using HT22 cells and SAMP8 mice (a senescence-accelerated strain of mice), **this study showed that bilobalide treatment reduced generation of two β-secretase cleavage products of APP, the amyloid β-peptide (Aβ) and soluble APPβ (sAPPβ), via PI3K-dependent pathway…** Bilobalide showed no significant effects on β-site APP cleaving enzyme 1 (BACE-1) or γ-secretase but inhibited the β-secretase activity of another protease cathepsin B, suggesting that bilobalide-induced Aβ reduction was probably mediated through modulation of cathepsin B rather than BACE-1.[39]

Ginkgo biloba EXTRACT 761 (↓) BETA-SECRETASE
Brain, Behavior, and Immunity (United States) (2015)
Department of Neurology, University of the Saarland, Homburg/Saar, Germany; German Institute for Dementia Prevention (DIDP), University of the Saarland, Homburg/Saar, Germany; The Institute of Neuroscience, Soochow University, Suzhou, China; Department of Pharmacy, Putuo People's Hospital, Shanghai, China; Dr. Willmar Schwabe GmbH & Co. KG, Karlsruhe, Germany; Institute for Clinical and Experimental Surgery, University of the Saarland, Homburg/Saar, Germany.
We gave APP-transgenic mice EGb 761 as a dietary supplement for 2 or 5 months... **[L]ong-term treatment with EGb 761 may reduce cerebral Aβ pathology by inhibiting β-secretase activity and Aβ aggregation.**[40]

GREEN TEA EXTRACTS (↓) BETA-SECRETASE (BACE-1)

GREEN TEA CATECHINS (↓) BETA-SECRETASE (BACE-1)
Bioorganic and Medicinal Chemistry Letters (England) (2003)
Division of Applied Biology and Chemistry, College of Agriculture and Life Sciences, Kyungpook National University, Daegu, South Korea.
In the course of searching for BACE1 (beta-secretase) inhibitors from natural products, the ethyl acetate soluble fraction of green tea, which was suspected to be rich in catechin content, showed potent inhibitory activity... **Catechin gallate, gallocatechin, and epigallocatechin significantly inhibited BACE1 activity.**[41]

Curcuminoids also downregulate gamma secretase, which plays a key supporting role to beta-secretase in the pathogenic processing of APP.[42]

CURCUMINOIDS (↓) GAMMA-SECRETASE (PRESENILIN-1)

CURCUMIN (↓) GAMMA-SECRETASE (PRESENILIN-1)
Pharmacological Reports (Poland) (2011)
Department of Pathology, Chongqing Medical University, China.
The present study aimed to investigate the effects of curcumin on the generation of Aβ in cultured neuroblastoma cells and on the in vitro expression of PS1 [presenilin-1] and GSK-3β... **Curcumin treatment was found to markedly reduce the production of Aβ(40/42). Treatment with curcumin also decreased both PS1 and GSK-3β mRNA and protein levels in a dose- and time-dependent manner.** Furthermore, curcumin increased the inhibitory phosphorylation of GSK-3β protein at Ser9. Therefore, we propose that curcumin decreases **Aβ production by inhibiting GSK-3β-mediated PS1 activation.**[43]

Curcumin (↓) Gamma Secretase (Presenilin-1)
Experimental and Therapeutic Medicine (Greece) (2011)
Department of Environmental Health Science, Nara Women's University, Nara, Japan.
Our results showed that curcumin, a component of turmeric, induced the down-regulation of presenilin 1 protein in Jurkat and K562 cell lines.[44]

Overall, these studies indicate that certain plant polyphenols—curcuminoids, *Ginkgo biloba* extract, green tea extract (EGCG), olive oil extract (oleuropein), and red wine polyphenols—possess an ability to upregulate alpha-secretase and sAPPα, and down-regulate beta-secretase and gamma-secretase. Several studies and expert commentary have indicated that these effects on APP processing have the potential to prevent AD.

For example, in "The Search for Alpha-Secretase and its Potential as a Therapeutic Approach to Alzheimer's Disease," *Current Medicinal Chemistry* reported in 2002: "In the nonamyloidogenic processing pathway the Alzheimer's amyloid precursor protein (APP) is proteolytically cleaved by alpha-secretase. As this cleavage occurs at the Lys16-Leu17 bond within the amyloid beta domain, it prevents deposition of intact amyloidogenic peptide. In addition, the large ectodomain sAPP(alpha) released by the action of alpha-secretase has several neuroprotective properties... As the alpha-secretase cleavage of APP both precludes the deposition of amyloid beta peptide and releases the neuroprotective sAPP(alpha), **pharmacological up-regulation of alpha-secretase may provide alternative therapeutic approaches for Alzheimer's disease.**"[45]

In "Alpha-Secretase as a Therapeutic Target," *Current Alzheimer Research* reported in 2007: "In the non-amyloidogenic pathway the alpha-secretase cleaves the amyloid precursor protein (APP) within the sequence of Abeta-peptides and precludes their formation. In addition, alpha-secretase cleavage releases an N-terminal extracellular domain [sAPPα] with neurotrophic and neuroprotective properties. The disintegrin metalloproteinase ADAM10 has been shown to act as alpha-secretase in vivo to prevent amyloid plaque formation and hippocampal defects in an Alzheimer disease mouse model. **An increase in alpha-secretase activity therefore is an attractive strategy for treatment of AD and may be achieved by modulating selective signaling pathways.**"[18]

And in "Activation of α-Secretase Cleavage," the *Journal of Neurochemistry* reported in 2012: "Alpha-secretase-mediated cleavage of the amyloid precursor protein (APP) releases the neuroprotective APP fragment sαAPP [sAPPα] and prevents amyloid β peptide (Aβ) generation... Assuming that Aβ is responsible for the development of Alzheimer's disease (AD), **activation of α-secretase should be preventive.**"[46]

Regarding the primary-prevention relevance of the downregulation of beta-secretase (BACE1), in "Inhibition of BACE1 for Therapeutic Use in Alzheimer's Disease," the *International Journal of Clinical and Experimental Pathology* reported in 2010: "Since BACE1 was reported as the beta-secretase in Alzheimer's disease (AD) over ten years ago, encouraging progress has been made toward understanding the cellular functions of BACE1. Genetic studies have further confirmed that BACE1 is essential for processing amyloid precursor protein (APP) at the beta-secretase site. **Only after this cleavage can the membrane-bound APP C-terminal fragment be subsequently cleaved by gamma-secretase to release so-called AD-causing Abeta peptides.**"[47]

In "BACE1 in Alzheimer's Disease," the *International Journal of Clinical Chemistry* reported in 2012: "Targeting BACE1 (β-site APP cleaving enzyme 1 or β-secretase) is the focus of Alzheimer's disease research because this aspartyl protease is involved in the abnormal production of β amyloid plaques (Aβ), the hallmark of its pathophysiology. Evidence suggests that there is a strong connection between AD and BACE1. As such, **strategies to inhibit Aβ formation in the brain should prove beneficial for AD treatment.**"[48]

In "BACE1 as a Therapeutic Target in Alzheimer's Disease: Rationale and Current Status," *Drugs and Aging* reported in 2013: "Aβ [amyloid beta], which is the major constituent of amyloid plaques, is a peptidic fragment derived from proteolytic processing of the amyloid precursor protein (APP) by sequential cleavages that involve β-site APP-cleaving enzyme 1 (BACE1) and γ-secretase [gamma-secretase]. **Targeting BACE1 is a rational approach as its cleavage of APP is the rate-limiting step in Aβ production and this enzyme is elevated in the brain of patients with AD.**"[49]

In "BACE1 (β-Secretase) Inhibitors for the Treatment of Alzheimer's Disease," *Chemical Society Reviews* reported in 2014: "**BACE1 (β-secretase, memapsin 2, Asp2) has emerged as a promising target for the treatment of Alzheimer's disease.** BACE1 is an aspartic protease which functions in the first step of the pathway leading to the production and deposition of amyloid-β peptide (Aβ)."[50]

And in "Targeting the β Secretase BACE1 for Alzheimer's Disease Therapy," *Lancet Neurology* reported in 2014: "The β secretase, widely known as β-site amyloid precursor protein cleaving enzyme 1 (BACE1), initiates the production of the toxic amyloid β (Aβ) that plays a crucial early part in Alzheimer's disease pathogenesis… **The potential of therapeutic BACE1 inhibition might prove to be a watershed in the treatment of Alzheimer's disease.**"[51]

Studies on gamma-secretase also have reported the therapeutic rationale for its downregulation. In "Gamma-Secretase as a Therapeutic Target in Alzheimer's Disease," *Current Drug Targets* reported in 2010: "**In Alzheimer's disease interest in gamma-secretase comes, in part, from the fact that this complex is responsible for the last cleavage step of the amyloid precursor protein (APP) that generates the amyloid-beta peptide (Abeta).** Abeta represents the primary component of the amyloid plaque, one of the main pathological hallmarks of AD."[52]

In "Interacting with γ-Secretase for Treating Alzheimer's Disease: From Inhibition to Modulation," *Current Medicinal Chemistry* reported in 2011: "**Compounds that inhibit or modulate γ-secretase, the pivotal enzyme that generates β-amyloid (Aβ), are potential therapeutics for AD.**"[53]

Studies and reviews also have revealed the treatment-related downside of *nonselective* gamma-secretase inhibitors, which undesirably interact with substrates of gamma-secretase other than APP. For example, in "γ-Secretase Inhibitors and Modulators," *Biochimica et Biophysica Acta* reported in 2013: "[D]ue to inherent mechanism based toxicity of non-selective inhibition of γ-secretase, clinical development of GSIs [gamma-secretase inhibitors] will require empirical testing with careful evaluation of benefit versus risk. In addition to GSIs, compounds referred to as γ-secretase modulators (GSMs) remain in development as AD therapeutics. GSMs do not inhibit γ-secretase, but modulate γ-secretase processivity and thereby shift the profile of the secreted amyloid β peptides (Aβ) peptides produced. Although GSMs are thought to have an inherently safe mechanism of action, their effects on substrates other than the amyloid β protein precursor (APP) have not been extensively investigated."[54]

In contrast to safety concerns about pharmaceutical products that modify the activity of the proteolytic enzymes involved in the processing of APP, few such concerns exist about the dietary polyphenols that have been shown to upregulate alpha-secretase activity and downregulate beta-secretase and gamma-secretase. For example, turmeric and green tea, from which the curcuminoids and EGCG, respectively, are extracted, are ancient foods with well-documented histories of long-term safety and numerous health benefits. The abilities of plant polyphenols to safely and inexpensively modulate the nonamyloidogenic processing of APP, and thus possibly prevent onset of the AD-related disease process, seems more promising than the ability of pharmaceutically derived products in this regard. And the polyphenols featured in this volume are far more likely to pass the "Gandy test" for long-term use as primary prevention agents. In many respects, they already have passed that test over hundreds of years of human consumption while demonstrating not only long-term safety but significant health benefits, as indicated in the previous chapter.

APP processing has been an AD-related pharmaceutical drug target in recent years, with little success to date. For example, in August 2010, the *New York Times* reported: "In yet another setback in efforts to treat Alzheimer's disease, Eli Lilly & Company announced that it had halted development of an experimental treatment after the compound actually made patients worse in two late-stage clinical trials. The company said the drug, semagacestat, a gamma-secretase inhibitor, did not slow progression of the disease and was associated with a worsening of cognition and the ability to perform the tasks of daily living… Moreover, those getting the drug had a higher risk of getting skin cancer." The *Times* also reported that "Lilly's drug was designed to reduce the body's production of the [amyloid] plaques by inhibiting the activity of an enzyme, called gamma-secretase, that is believed to play an important role in formation of the plaque."[55]

The drug involved in the trials, semagacestat, is a non-selective gamma-secretase inhibitor. In contrast to turmeric's curcuminoids, green tea's catechins, and the polyphenols from red wine and olive oil, humans had not consumed semagacestat prior to the initiation of recent experiments designed to test its safety and efficacy. In any event, the failed semagacestat trials were viewed as a watershed in AD-related applied research, judging by the high-end press coverage and commentary that followed.[56]

In arguably the most authoritative analysis of the failed semagacestat trials, Dennis Selkoe, Vincent and Stella Coates Professor of Neurologic Diseases at Harvard Medical School and Co-Director of the Center for Neurologic Diseases in the Department of Neurology at Brigham and Women's Hospital in Boston, who with Michael Wolfe and their research colleagues in 1999 discovered that presenilin was the gamma-secretase enzyme involved in the pathogenic processing of APP,[57] observed in a 2011 interview with *Science Watch*:

> Eli Lilly had one of the biggest disappointments in Alzheimer's treatment when it stopped a phase III trial of a new drug last September [2010]. Their drug was a gamma-secretase inhibitor, which is just what I've been saying I'd like to see. But it was not a good drug. It inhibited presenilin/gamma secretase from cutting many, many different proteins—other proteins—and one in particular is called Notch, which is a very famous molecule in biology. It's very important in fruit flies, worms, humans. It controls aspects of cell fate; how one cell becomes one thing and another cell becomes something else entirely. And you don't want to mess with that, even in adult humans, where it's important in the gastrointestinal tract and bone marrow. For example, to make the proper cell in the gut that makes acid, we need Notch… Well, Lilly announced in September that the patients didn't get better and they stopped the trial short and said some people in the trial had clear signs of Notch toxicity, for example, developing a type of skin cancer. This is not a minor matter. And they had other problems—gastrointestinal problems, etc. Lilly also said, by the way, when they cognitively tested patients during the trial, they actually seemed to do worse on the drug rather than better.[58]

Relatedly, writing in *Nature Medicine*, Selkoe observed further:

> Semagacestat, a γ-secretase inhibitor, had a therapeutic index of <3 (that is, the half-maximal inhibitory concentration for Aβ reduction was only two to three times lower than that for inhibiting Notch signaling). Consequently, patients in the phase 3 trial were dosed only half as frequently (once a day) as originally intended. Nonetheless, signs of apparent Notch toxicity accrued, such as gastrointestinal symptoms and skin cancer. The information released after trial cessation suggested that some treated patients actually declined in mental function. But it is very unlikely this was due to some adverse cognitive effect of lowering brain Aβ levels too much… Rather, the acknowledged occurrence of adverse events, including toxicity from chronically blocking the processing of Notch and other important γ-substrates, could cause a setback of cognition in subjects with Alzheimer's disease… Such unacceptable Notch effects could be avoided by using γ-secretase inhibitors with much higher therapeutic indexes or γ-secretase modulators that modify the processing of APP to Aβ without blocking γ-secretase cleavage in general.[59]

Upon reading about the worsening condition in cognition and the serious adverse effects from the human trials involving semagacestat, and while comparing those results to the long-term safety and health benefits of the plant polyphenols in curcuminoids, green tea, red wine, and olive oil, it seems likely that these natural products would function better overall as modulators of APP processing and AD-related primary prevention.

Not only would the plant polyphenols in question likely pass the Gandy test for long-term safety in a primary-prevention context, they also would likely pass the "Selkoe test" for modulators of APP processing, which requires "much higher therapeutic indexes" without harmfully influencing non-APP substrates. Given their long-observed record of overall health benefits and safety, it could be assumed that the plant polyphenols in question are naturally built with high therapeutic indexes and the innate ability to selectively avoid adversely affecting substrates that are not expressly targeted in an AD context.

Finally, in a section in his *Nature Medicine* piece titled, "The Importance of Treating Early," Selkoe invoked "the cardinal medical principle of intervening early if one hopes to arrest a chronic disease."[59] While "arresting" the AD-related disease process during preclinical AD would be critically important, preventing the onset of the disease process in the first place would be "ideal,"[12] as Reisa Sperling, Clifford Jack, and Paul Aisen have asserted while also discouraging the idea as unfeasible. However, it seems that the nontoxic, safe primary prevention of AD is decidedly possible with certain plant polyphenols.

REFERENCES

1. The Leader in Aging Research, The National Institute on Aging, U.S. Department of Health and Human Services. At: https://www.nia.nih.gov/.
2. Alzheimer's Disease: Unraveling the Mystery, National Institute on Aging, September 2011; 27–32.
3. Advancing Discovery: The National Plan to Address Alzheimer's Disease, Alzheimer's Disease Education and Referral Center, National Institute on Aging. At: https://www.nia.nih.gov/alzheimers/publication/2013-2014-alzheimers-disease-progress-report/advancing-discovery-national.
4. Preclinical Alzheimer's Disease, Clinical Updates for Medical Professionals, Mayo Clinic. At: http://www.mayoclinic.org/medical-professionals/clinical-updates/neurosciences/preclinical-alzheimers-disease.
5. *Concise medical dictionary (online).* Oxford University Press; 2010.
6. Institute for Work and Health. At: http://www.iwh.on.ca/wrmb/primary-secondary-and-tertiary-prevention.
7. Prevention, The Free Dictionary by Farlex. At: http://medical-dictionary.thefreedictionary.com/primary+prevention.
8. Preventive Healthcare, Wikipedia. At: https://en.wikipedia.org/wiki/Preventive_healthcare#Primary_prevention.
9. Deleted in review.
10. Sperling, Aisen, Beckett, et al. Toward defining the preclinical stages of Alzheimer's disease: recommendations from the National Institute on Aging-Alzheimer's Association Workgroups on diagnostic guidelines for Alzheimer's disease. *Alzheimer's and Dementia* May 2011;**7**(3):280–92.
11. Alzheimer's Disease: Unraveling the Mystery, National Institute on Aging, September 2011; 27.
12. Sperling, Jack, Aisen. Testing the right target and right drug at the right stage. *Science Translational Medicine* November 30, 2011;**3**(111):111cm33.
13. Berlin I. *The hedgehog and the fox: an essay on Tolstoy's view of history.* London: Weidenfeld & Nicolson; 1953.
14. Merriam Webster Dictionary. At: http://www.merriam-webster.com/dictionary/zero.
15. Who invented zero. *Live Science.* At: http://www.livescience.com/27853-who-invented-zero.html.
16. Alzheimer's Disease: Unraveling the Mystery.
17. Dennis Selkoe on the amyloid hypothesis of Alzheimer's disease: special topic of Alzheimer's disease interview. *Science Watch* March 2011. At: http://archive.sciencewatch.com/ana/st/alz2/11marSTAlz2Selk/. During the interview, Selkoe said: "You can't get Alzheimer's without presenilin [gamma-secretase] cutting APP".
18. Fahrenholz. Alpha-secretase as a therapeutic target. *Current Alzheimer Research* September 2007;**4**(4):412–7.
19. Parvathy. Cleavage of Alzheimer's amyloid precursor protein by alpha-secretase occurs at the surface of neuronal cells. *Biochemistry* July 1999;**38**(30):9728–34.

20. Hooper, Turner. The search for alpha-secretase and its potential as a therapeutic approach to Alzheimer's disease. *Current Medicinal Chemistry* June 2002;**9**(11):1107–19; Vassar. The beta-secretase, BACE: a prime drug target for Alzheimer's disease. *Journal of Molecular Neuroscience* October 2001;**17**(2):157–70; Roberds, Anderson, Basi, et al. BACE knockout mice are healthy despite lacking the primary beta-secretase activity in brain: implications for Alzheimer's disease therapeutics. *Human Molecular Genetics* June 1, 2001;**10**(12):1317–24; Arbel, Yacoby, Solomon. Inhibition of amyloid precursor protein processing by beta-secretase through site-directed antibodies. *Proceedings of the National Academy of Sciences* May 24, 2005;**102**(21):7718–23; Hussain, Hawkins, Harrison, et al. Oral administration of a potent and selective non-peptidic BACE-1 inhibitor decreases beta-cleavage of amyloid precursor protein and amyloid-beta production in vivo. *Journal of Neurochemistry* February 2007;**100**(3):802–9; Boddapati, Levites, and Sierks. Inhibiting β-secretase activity in Alzheimer's disease cell models with single-chain antibodies specifically targeting APP. *Journal of Molecular Biology* January 14, 2011;**405**(2):436–47.

21. Colciaghi, Borroni, Zimmermann, et al. Amyloid precursor protein metabolism is regulated toward alpha-secretase pathway by *Ginkgo biloba* extracts. *Neurobiology of Disease* July 2004;**16**(2):454–60.

22. Shi, Wu, Xu, Zou. Bilobalide regulates soluble amyloid precursor protein release via phosphatidyl inositol 3 kinase-dependent pathway. *Neurochemistry International* August 2011;**59**(1):59–64.

23. Levites, Amit, Mandel, Youdim. Neuroprotection and neurorescue against Abeta toxicity and PKC-dependent release of nonamyloidogenic soluble precursor protein by green tea polyphenol (-)-epigallocatechin-3-Gallate. *FASEB Journal* May 2003;**17**(8):952–4.

24. Rezai-Zadeh, Shytle, Sun, et al. Green tea epigallocatechin-3-gallate (EGCG) modulates amyloid precursor protein cleavage and reduces cerebral amyloidosis in Alzheimer transgenic mice. *The Journal of Neuroscience* September 21, 2005;**25**(38):8807–14.

25. Obregon, Rezai-Zadeh, Bai, et al. ADAM10 activation is required for green tea (-)-epigallocatechin-3-gallate-induced alpha-secretase cleavage of amyloid precursor protein. *The Journal of Biological Chemistry* June 16, 2006;**281**(24):16419–27.

26. Lee, Lee, Ban, et al. Green tea (-)-epigallocatechin-3-gallate inhibits beta-amyloid-induced cognitive dysfunction through modification of secretase activity via inhibition of ERK and NF-kappaB pathways in mice. *The Journal of Nutrition* October 2009;**139**(10):1987–93.

27. Lin, Chen, Chiu, et al. Epigallocatechin gallate (EGCG) suppresses beta-amyloid-induced neurotoxicity through inhibiting c-Abl/FE65 nuclear translocation and GSK3 beta activation. *Neurobiology of Aging* January 2009;**30**(1):81–92.

28. Fernandez, Rezai-Zadeh, Obregon, Tan. EGCG functions through estrogen receptor-mediated activation of ADAM10 in the promotion of non-amyloidogenic processing of APP. *FEBS Letters* October 8, 2010;**584**(19):4259–67.

29. Smith, Giunta, Bickford, et al. Nanolipidic particles improve the bioavailability and alpha-secretase inducing ability of epigallocatechin-3-gallate (EGCG) for the treatment of Alzheimer's disease. *International Journal of Pharmaceutics* April 15, 2010;**389**(1–2):207–12.

30. Giunta, Hou, Zhu, et al. Fish oil enhances anti-amyloidogenic properties of green tea EGCG in Tg2576 mice. *Neuroscience Letters* March 8, 2010;**471**(3):134–8.

31. Grimm, Kuchenbecker, Grösgen. Docosahexaenoic acid reduces amyloid beta production via multiple pleiotropic mechanisms. *The Journal of Biological Chemistry* April 22, 2011;**286**(16):14028–39.

32. Kostomoiri, Fragkouli, Sagnou, et al. Oleuropein, an anti-oxidant polyphenol constituent of olive, promotes α-secretase cleavage of the amyloid precursor protein (AβPP). *Cellular and Molecular Neurobiology* January 2013;**33**(1):147–54.

33. Wang, Ho, Zhao, et al. Moderate consumption of Cabernet Sauvignon attenuates Abeta neuropathology in a mouse model of Alzheimer's disease. *FASEB Journal* November 2006;**20**(13):2313–20.

34. Shimmyo, Kihara, Akaike, et al. Epigallocatechin-3-gallate and curcumin suppress amyloid beta-induced beta-site APP cleaving enzyme-1 upregulation. *Neuroreport* August 27, 2008;**19**(13):1329–33.

35. Hamaguchi, Ono, Yamada. Review: curcumin and Alzheimer's disease. *CNS Neuroscience & Therapeutics* October 2010;**16**(5):285–97.

36. Liu, Li, Qiu, et al. The inhibitory effects of different curcuminoids on β-amyloid protein, β-amyloid precursor protein and β-site amyloid precursor protein cleaving enzyme 1 in swAPP HEK293 cells. *Neuroscience Letters* November 19, 2010;**485**(2):83–8.

37. Wang, Kim, Lee, et al. Effects of curcuminoids identified in rhizomes of *Curcuma longa* on BACE-1 inhibitory and behavioral activity and lifespan of Alzheimer's disease *Drosophila* models. *BMC Complementary and Alternative Medicine* March 5, 2014;**14**:88.
38. Murata, Matsumura, Yoshioka, et al. Screening of β-secretase and acetylcholinesterase inhibitors from plant resources. *Journal of Natural Medicines* January 2015;**69**(1):123–9.
39. Shi, Zheng, Wu, et al. The phosphatidyl inositol 3 kinase-glycogen synthase kinase 3β pathway mediates bilobalide-induced reduction in amyloid β-peptide. *Neurochemical Research* February 2012;**37**(2):298–306.
40. Liu, Hao, Qin. Long-term treatment with *Ginkgo biloba* extract EGb 761 improves symptoms and pathology in a transgenic mouse model of Alzheimer's disease. *Brain, Behavior, and Immunity* May 2015;**46**:121–31.
41. Jeon, Bae, Seong, Song. Green tea catechins as a BACE1 (Beta-Secretase) inhibitor. *Bioorganic & Medicinal Chemistry Letters* November 17, 2003;**13**(22):3905–8.
42. Descamps, Spilman, Zhang, et al. AβPP-Selective BACE inhibitors (ASBI): novel class of therapeutic agents for Alzheimer's disease. *Journal of Alzheimer's Disease* 2013;**37**(2):343–55.
43. Xiong, Hongmei, Lu, Yu. Curcumin mediates presenilin-1 activity to reduce β-amyloid production in a model of Alzheimer's disease. *Pharmacological Reports* 2011;**63**(5):1101–8.
44. Yoshida, Okumura, Nishimura, et al. Turmeric and curcumin suppress presenilin 1 protein expression in Jurkat cells. *Experimental and Therapeutic Medicine* July 2011;**2**(4):629–32.
45. Hooper, Turner. The search for alpha-secretase and its potential as a therapeutic approach to Alzheimer's disease. *Current Medicinal Chemistry* June 2002;**9**(11):1107–19.
46. Postina. Activation of α-secretase cleavage. *Journal of Neurochemistry* January 2012;**120**(Suppl. 1):46–54.
47. Luo, Yan. Inhibition of BACE1 for therapeutic use in Alzheimer's disease. *International Journal of Clinical and Experimental Pathology* July 8, 2010;**3**(6):618–28.
48. Sathya, Premkumar, Karthick, et al. BACE1 in Alzheimer's disease. *Clinica Chimica Acta: International Journal of Clinical Chemistry* December 24, 2012;**414**:171–8.
49. Evin, Hince. BACE1 as a therapeutic target in Alzheimer's disease: rationale and current status. *Drugs & Aging* October 2013;**30**(10):755–64.
50. Ghosh, Osswald. BACE1 (β-Secretase) inhibitors for the treatment of Alzheimer's disease. *Chemical Society Reviews* October 7, 2014;**43**(19):6765–813.
51. Yan, Vassar. Targeting the β secretase BACE1 for Alzheimer's disease therapy. *Lancet Neurology* March 2014;**13**(3):319–29.
52. Guardia-Laguarta, Pera, Lleó. Gamma-secretase as a therapeutic target in Alzheimer's disease. *Current Drug Targets* April 2010;**11**(4):506–17.
53. Panza, Frisardi, Solfrizzi, et al. Interacting with γ-secretase for treating Alzheimer's disease: from inhibition to modulation. *Current Medicinal Chemistry* 2011;**18**(35):5430–47.
54. Golde, Koo, Felsenstein, et al. γ-secretase inhibitors and modulators. *Biochimica et Biophysica Acta* December 2013;**1828**(12):2898–907.
55. Lilly stops Alzheimer's drug trial. *New York Times* August 18, 2010.
56. Lilly stops Alzheimer's drug trial. *New York Times* August 18, 2010; Doubt on tactic in Alzheimer's battle. *New York Times* August 18, 2010; Sperling, Jack, and Aisen. Testing the right target and right drug at the right stage. *Science Translational Medicine* November 30, 2011;**3**(111):111cm33; Selkoe. Resolving controversies on the path to Alzheimer's therapeutics. *Nature Medicine* September 7, 2011;**17**(9):1060–5; Dennis Selkoe on the amyloid hypothesis of Alzheimer's disease: special topic of Alzheimer's disease interview. *Science Watch* March 2011. At: http://archive.sciencewatch.com/ana/st/alz2/11marSTAlz2Selk2/.
57. Wolfe, Xia, Moore, et al. Peptidomimetic probes and molecular modeling suggest that Alzheimer's gamma-secretase is an intramembrane-cleaving aspartyl protease. *Biochemistry* April 13, 1999;**38**(15): 4720–7; Wolfe, Los Angeles, Miller, et al. Are presenilins intramembrane-cleaving proteases? Implications for the molecular mechanism of Alzheimer's disease. *Biochemistry* August 31, 1999;**38**(35):11223–30; Wolfe, Xia, Ostaszewski, et al. Two transmembrane aspartates in presenilin-1 required for presenilin endoproteolysis and gamma-secretase activity. *Nature* April 8, 1999;**398**(6727):513–7.
58. Dennis Selkoe on the amyloid hypothesis of Alzheimer's disease: special topic of Alzheimer's disease interview. *Science Watch* March 2011.
59. Selkoe. Resolving controversies on the path to Alzheimer's therapeutics. *Nature Medicine* September 7, 2011;**17**(9):1060–5.

CHAPTER 4

Secondary Prevention of Alzheimer's Disease

Contents

We noted earlier that Mount Sinai neurologist, Sam Gandy, reported that Alzheimer's disease (AD)–related "amyloid could appear [in the brain] as early as age 30." But it could also appear at age 40 or 50. Or perhaps never in any pathogenic sense. Short of guidelines stipulating that everyone should begin to take measures to prevent AD on their 21st birthday, how might we go about deciding when to take such action? The answer speaks once again to the advantages of the pleiotropic capabilities of plant polyphenols.

As we have indicated, an effective AD-related primary prevention formula would upregulate alpha-secretase while downregulating beta-secretase and gamma-secretase. We also have observed that since both zero AD and preclinical AD populations would be asymptomatic and thus difficult to differentiate on an ongoing basis and on a large scale, an ideal AD prevention formula would need to perform both primary-prevention and secondary-prevention functions. Thus, whereas the previous chapter demonstrated the primary-prevention capabilities of certain plant polyphenols, this chapter will report

A Paradigm Shift to Prevent and Treat Alzheimer's Disease
ISBN 978-0-12-812259-4
http://dx.doi.org/10.1016/B978-0-12-812259-4.00004-7

detailed evidence that a similar grouping of polyphenols is capable of performing the key secondary-prevention functions of inhibiting Abeta aggregation, decreasing Abeta toxicity, and increasing Abeta clearance, while protecting synaptic function.

Another major advantage of the plant polyphenols featured in this volume, as compared to the pharma-based monotargeting products, is that pleiotropism, within the context provided here, is able to accommodate various nuanced views about AD pathogenesis. For example, in "Toward Defining the Preclinical Stages of Alzheimer's Disease" by the NIA-appointed workgroup led by Reisa Sperling, the distinguished panel of AD researchers wrote: "The proposed model of AD [by Sperling et al] views Aβ peptide accumulation as a key early event in the pathophysiological process of AD." This, indeed, is the likely starting point to the Abeta-based pathophysiological model of AD, otherwise known as the "amyloid hypothesis," that in part is applied to our investigation in this volume. "However," Sperling's group continued, "we acknowledge that the etiology of AD remains uncertain, and some investigators have proposed that synaptic, mitochondrial, metabolic, inflammatory, neuronal, cytoskeletal, and other age-related alterations may play an even earlier, or more central, role than Aβ peptides in the pathogenesis of AD. There also remains significant debate in the field as to whether abnormal processing versus clearance of Aβ42 is the etiologic event in sporadic, late-onset AD."[1]

From the standpoint of pleiotropic plant polyphenols, degrees of differences over AD-related etiology pose fewer prevention- and treatment-related riddles than they present to the pharma-based monotarget framework. For example, a debate as to whether abnormal APP processing or poor Abeta clearance is the true "etiologic event" in AD presents fewer prevention- and treatment-related challenges when our relatively small cadre of plant polyphenols can normalize APP processing *and* improve Abeta clearance. Likewise, certain pairs of plant polyphenols can also protect and improve synaptic and mitochondrial function while preventing and reversing cellular damage due to oxidative stress and aberrant neuroinflammation. And, as the following chapters illustrate, the windfall of pleiotropically driven benefits provided by the plant polyphenols is by no means limited to these already considerable capabilities. In short, the prevention- and treatment-related capacities of the plant polyphenols featured in this volume are largely undiminished by the otherwise daunting therapeutic implications of scientific debate about the etiology and pathogenic course of AD.

In the NIA's "Alzheimer's Disease: Unraveling the Mystery," and in a sidebar titled, "From APP to Beta-Amyloid Plaques," the NIA analysts described the biochemical events that follow the amyloidogenic processing of APP by beta-secretase and gamma-secretase:

> Beta-Amyloid Peptides. *Following the cleavages [of APP] at each end [by beta-secretase and gamma-secretase], the beta-amyloid peptide is released into the space outside the neuron and begins to stick to other beta-amyloid peptides.*

Beta-Amyloid-Oligomers. These small, soluble aggregates of two, three, four, or even up to a dozen beta-amyloid peptides are called oligomers. Specific sizes of oligomers may be responsible for reacting with receptors on neighboring cells and synapses, affecting [the] ability [of the cells and synapses] to function. It is likely that some oligomers are cleared from the brain. Those that cannot be cleared clump together with more beta-amyloid peptides.

Beta-Amyloid Fibrils. As the process continues, oligomers grow larger, becoming entities called protofibrils and fibrils.

Beta-Amyloid Plaques. Eventually, other proteins and cellular material are added, and these increasingly insoluble entities combine to become the well-known [amyloid] plaques that are characteristic of AD. For many years, scientists thought that plaques might cause all of the damage to neurons that is seen in AD. However, that concept has evolved greatly in the past few years. Many scientists now think that [Abeta] oligomers may be a major culprit.[2]

As in the primary prevention–related zero stage, where we identified certain plant polyphenols that could upregulate alpha-secretase activity and downregulate beta-secretase activity, during the secondary prevention–related preclinical stage we would want the same combination of plant polyphenols to (1) inhibit aggregation of the Abeta peptides, oligomers, and soluble protofibrils, (2) decrease the neurotoxicity of these Abeta formations, (3) increase clearance of Abeta peptides and oligomers from the brain, and (4) protect synaptic function while reversing synaptic dysfunction. Doing so across the board would potentially arrest the amyloid-related disease process and reverse the damage induced by the soluble Abeta oligomers at the synaptic gap between neurons, which by several authoritative accounts is responsible for the memory dysfunctions seen in mild cognitive impairment as an early symptomatic signal of the AD-related disease process.

This section presents a brief review of the therapeutic relevance of inhibiting Abeta aggregation, decreasing Abeta toxicity, and increasing Abeta clearance. We pick up the literature in 2009, when researchers at Kanazawa University in Japan reported in *The American Journal of Pathology* that "inhibition of amyloid-beta (Abeta) aggregation is an attractive therapeutic strategy for Alzheimer's disease."[3] Also in 2009 researchers at Northwestern University, in *Toxicology and Applied Pharmacology*, reported: "It now appears likely that soluble oligomers of amyloid-beta1–42 peptide, rather than insoluble fibrils, act as the primary neurotoxin in Alzheimer's disease. Consequently, compounds capable of altering the assembly state of these oligomers may have potential for AD therapeutics."[4]

In 2014, researchers at the University of Twente in The Netherlands reported in *Cellular and Molecular Life Sciences*: "The transition of an unstructured monomeric [Abeta] peptide into self-assembled and more structured aggregates is the crucial conversion from what appears to be a harmless polypeptide into a malignant form that causes synaptotoxicity and neuronal cell death."[5] Also in 2014, researchers at the

Department of Neurology at New York University reported in *Gerontology*: "The dominant theory for the causation of AD is the amyloid cascade hypothesis, which suggests that the aggregation of Aβ as oligomers and amyloid plaques is central to the pathogenesis of AD."[6]

By 2016, the idea that soluble Abeta oligomers were a major contributor to AD-related neuropathology appeared to be widely accepted, and that inhibiting Abeta aggregation, reducing Abeta toxicity (in particular the neurotoxicity of soluble Abeta oligomers), and enhancing Abeta clearance (of Abeta monomers, oligomers, fibrils, and plaques) were increasingly viewed as key therapeutic targets. For example, writing in *Acta Neuropathologica* in 2015, Kirsten Viola and William Klein of Northwestern University reported: "Protein aggregation is common to dozens of diseases including prionoses, diabetes, Parkinson's and Alzheimer's. Over the past 15 years, there has been a paradigm shift in understanding the structural basis for these proteinopathies. Precedent for this shift has come from investigation of soluble Aβ oligomers (AβOs), toxins now widely regarded as instigating neuron damage leading to Alzheimer's dementia… Because pathogenic AβOs appear early in the disease, they offer appealing targets for therapeutics and diagnostics."[7]

The 2011 high-level workgroup appointed by the NIA and the Alzheimer's Association on diagnostic guidelines for the preclinical stage of AD, led by Sperling, supported the idea of treating AD-related Abeta pathology at the preclinical stage. In this regard, the workgroup wrote: "Both laboratory work and recent disappointing clinical trial results raise the possibility that therapeutic interventions applied earlier in the course of AD would be more likely to achieve disease modification. Studies with transgenic mouse models suggest that Aβ-modifying therapies may have limited effect after neuronal degeneration has begun. Several recent clinical trials involving the stages of mild to moderate dementia have failed to demonstrate clinical benefit, even in the setting of biomarker or autopsy evidence of decreased Aβ burden. Although the field is already moving to earlier clinical trials at the stage of MCI, it is possible that similar to cardiac disease and cancer treatment, AD would be optimally treated before significant cognitive impairment, in the 'presymptomatic' or 'preclinical' stages of AD."[1]

In a 2011 interview for *Science Watch*, Dennis Selkoe stated that Abeta-related pathology played a "necessary" and possibly "sufficient" role in AD pathogenesis: "My opinion is that there is an imbalance between the production and the removal of a small hydrophobic protein called amyloid beta that triggers the process we call Alzheimer's. We believe that imbalance arises from a lot of different, more fundamental causes. What I'm saying is that amyloid beta is both necessary and, at least in some cases, sufficient to cause Alzheimer's disease, but there are many other factors. If we had to choose one, and I think the clearest, path to treatment, it would be targeting amyloid beta rather than any of these other factors, including tau, which looks like it comes downstream in the Alzheimer's cascade. So to summarize, my opinion is that it's an imbalance of amyloid beta protein in the brain that triggers or precipitates Alzheimer's."[8]

Aside from Selkoe's therapeutic focus on "one" target, which we view as unnecessarily narrow, the "imbalance" of amyloid beta in the brain to which he refers involves an excess of Abeta peptide production due to faulty APP processing combined with impaired Abeta clearance from the brain. The subsequent oligomer-induced damage to synaptic function may percolate asymptomatically during the preclinical phase, and emerge over time by way of memory-related first symptoms. Thus, the opportunity exists during the asymptomatic preclinical phase to arrest and perhaps reverse the Abeta pathogenic process by normalizing APP processing, enhancing Abeta clearance from the brain, and repairing synaptic function before the disease process progresses to the symptomatic MCI and AD dementia.

The aim of this chapter is to demonstrate, as the reports below indicate, that the same group of plant polyphenols that can safely normalize APP processing as a primary prevention function also can safely inhibit Abeta aggregation, reduce Abeta toxicity, increase Abeta clearance, prevent and repair synaptic dysfunction, and thus possibly impede and reverse progression to MCI.

CURCUMINOIDS (↓) ABETA AGGREGATION

CURCUMIN AND ROSMARINIC ACID (↓) ABETA AGGREGATION
Journal of Neuroscience Research (United States) (2004)
Department of Neurology and Neurobiology of Aging, Kanazawa University Graduate School of Medical Science, Kanazawa, Japan.
Inhibition of the accumulation of amyloid beta-peptide (Abeta) and the formation of beta-amyloid fibrils (fAbeta) from Abeta, as well as the destabilization of preformed fAbeta in the central nervous system, would be attractive therapeutic targets for the treatment of Alzheimer's disease (AD)... **Cur and RA dose-dependently inhibited fAbeta formation from Abeta(1–40) and Abeta(1–42), as well as their extension. In addition, they dose-dependently destabilized preformed fAbetas.**[9]

CURCUMIN (↓) ABETA AGGREGATION
The Journal of Biological Chemistry (United States) (2005)
Department of Medicine, University of California, Los Angeles.
Under aggregating conditions in vitro, curcumin inhibited aggregation (IC(50) = 0.8 microM) as well as disaggregated fibrillar Abeta40 (IC(50) = 1 microM), indicating favorable stoichiometry for inhibition. Curcumin was a better Abeta40 aggregation inhibitor than ibuprofen and naproxen, and prevented Abeta42 oligomer formation and toxicity between 0.1 and 1.0 microM... When fed to aged Tg2576 mice with advanced amyloid accumulation, curcumin labeled plaques and reduced amyloid levels and plaque burden. **Hence, curcumin directly binds small beta-amyloid species to block aggregation and fibril formation in vitro and in vivo.**[10]

Curcumin, Green Tea Extract (EGCG), and Grape Seed Extract (↓) Abeta Aggregation

Experimental Neurology (United States) (2010)

Centre for Research in Neurodegenerative Diseases, University of Toronto, Canada.

Additional small-molecule inhibitors, including polyphenolic compounds such as **curcumin, (-)-epigallocatechin gallate (EGCG), and grape seed extract have been shown to attenuate Abeta aggregation through distinct mechanisms, and have shown effectiveness at reducing amyloid levels when administered to transgenic mouse models of AD.**[11]

Curcumin (↓) Abeta Aggregation

ACS Chemical Neuroscience (United States) (2011)

Department of Chemistry, City University of New York at College of Staten Island, New York.

Curcumin inhibits amyloid-β and tau peptide aggregation at micromolar concentrations; the sugar-curcumin conjugate inhibits Aβ and tau peptide aggregation at concentrations as low as 8 nM and 0.1 nM, respectively.[12]

Curcumin (↓) Abeta Aggregation

Nanomedicine (United States) (2011)

Division of Biomedical and Life Sciences, School of Health and Medicine, University of Lancaster, England.

All nanoliposomes with curcumin, or the curcumin derivative, were able to inhibit the formation of fibrillar and/or oligomeric Aβ in vitro. Of the three forms of curcumin liposomes tested, the click-curcumin type was by far the most effective.[13]

Curcumin (↓) Abeta Aggregation

Neurochemical Research (United States) (2012)

College of Life Sciences and Technology, Beijing University of Chemical Technology, China.

The abilities of curcumin to scavenge free radicals and to inhibit the formation of β-sheeted aggregation are both beneficial to depress Aβ-induced oxidative damage.[14]

Curcumin (↓) Abeta Aggregation

Chemical Biology and Drug Design (England) (2015)

School of Pharmacy, University of Waterloo, Canada.

Curcumin, a chemical constituent present in the spice turmeric, is known to prevent the aggregation of amyloid peptide implicated in the pathophysiology of Alzheimer's disease.[15]

Ginkgo biloba EXTRACTS (↓) ABETA AGGREGATION

Ginkgo biloba EXTRACT 761 (↓) ABETA AGGREGATION.
Brain Research (The Netherlands) (2001)
Division of Hormone Research, Departments of Cell Biology, Pharmacology, and Neuroscience, Georgetown University Medical Center, Washington, D.C.
[I]n vitro reconstitution studies demonstrated that **EGb 761 inhibits, in a dose-dependent manner, the formation of beta-amyloid-derived diffusible neurotoxic soluble ligands (ADDLs) [soluble Abeta oligomers], suggested to be involved in the pathogenesis of Alzheimer's disease.**[16]

Ginkgo biloba EXTRACT 761 (↓) ABETA AGGREGATION
Proceedings of the National Academy of Sciences (United States) (2002)
Departments of Biological Sciences and Chemistry and Biochemistry, University of Southern Mississippi, Hattiesburg.
Using a neuroblastoma cell line stably expressing an AD-associated double mutation, **we report that EGb 761 inhibits formation of amyloid-beta (Abeta) fibrils, which are the diagnostic, and possibly causative, feature of AD.**[17]

Ginkgo biloba Extract 761 (↓) Abeta Aggregation
FASEB Journal (United States) (2007)
Department of Pharmaceutical Sciences, School of Pharmacy, University of Maryland.
Administration of EGb 761 reduces Abeta oligomers and restores CREB phosphorylation in the hippocampus of these mice.[18]

GRAPE SEED EXTRACT (↓) ABETA AGGREGATION

GRAPE SEED EXTRACT (↓) ABETA AGGREGATION
The Journal of Neuroscience (United States) (2008)
Department of Psychiatry, Mount Sinai School of Medicine, New York.
In this study, we found that a naturally derived grape seed polyphenolic extract can significantly inhibit amyloid beta-protein aggregation into high-molecular-weight oligomers in vitro. When orally administered to Tg2576 mice, this polyphenolic preparation significantly attenuates AD-type cognitive deterioration coincidentally with reduced high-molecular-weight soluble oligomeric Abeta in the brain.[19]

GRAPE SEED EXTRACT (↓) ABETA AGGREGATION
The Journal of Biological Chemistry (United States) (2008)
Department of Neurology, David Geffen School of Medicine, University of California, Los Angeles.

Initial studies revealed that **MN [a commercially available grape seed polyphe-nolic extract, MegaNatural-AZ] blocked Abeta fibril formation**. Subsequent evaluation of the assembly stage specificity of the effect showed that **MN was able to inhibit protofibril formation, pre-protofibrillar oligomerization**, and initial coil-->alpha-helix/beta-sheet secondary structure transitions.[20]

GRAPE SEED EXTRACT (GALLIC ACID) (↓) ABETA AGGREGATION
Bioorganic and Medicinal Chemistry Letters (England) (2013)
School of Chemistry and Physics, The University of Adelaide, Australia.
[G]allic acid [an active component of grape seed extract] effectively inhibited fibril formation by the amyloid-beta peptide, the putative causative agent in Alzheimer's disease.[21]

GRAPE SEED EXTRACT (↓) ABETA AGGREGATION
Journal of Neurochemistry (England) (2015)
Department of Neurology, David Geffen School of Medicine, University of California, Los Angeles; Medical Scientist Training Program, Neuroscience Interdepartmental Ph.D. Program, University of California, Los Angeles; Departments of Psychiatry and Neuroscience, Icahn School of Medicine at Mount Sinai, New York; Department of Neurology, Icahn School of Medicine at Mount Sinai, New York; Geriatric Research, Education and Clinical Center (GRECC), James J. Peters Veterans Affairs Medical Center, Bronx, New York; Molecular Biology Institute (MBI), and Brain Research Institute (BRI), David Geffen School of Medicine, University of California, Los Angeles.
Epidemiological evidence that red wine consumption negatively correlates with risk of Alzheimer's disease has led to experimental studies demonstrating that grape seed extracts inhibit the aggregation and oligomerization of Aβ in vitro and ameliorate neuropathology and behavioral deficits in a mouse model of Alzheimer's disease.[22]

GREEN TEA EXTRACTS (↓) ABETA AGGREGATION

GREEN TEA EXTRACT (EGCG) (↓) ABETA AGGREGATION
The European Journal of Neuroscience (France) (2006)
Douglas Hospital Research Centre, Department of Psychiatry, McGill University, Montreal, Canada.
[T]he protective effect of EGCG is likely to be associated, at least in part, with its inhibitory action on Abeta fibrils/oligomers formation.[23]

GREEN TEA EXTRACT (EGCG) (↓) ABETA AGGREGATION
Phytomedicine (Germany) (2010)

Institute of Pharmacy and Molecular Biotechnology, Department of Biology, Heidelberg University, Germany.
EGCG reduced beta amyloid (Abeta) deposits and inhibited Abeta oligomerization in transgenic *C. elegans* (CL2006).[24]

GREEN TEA EXTRACT (EGCG) (↓) ABETA AGGREGATION
The Journal of Physical Chemistry. B (United States) (2011)
Department of Biochemical Engineering and Key Laboratory of Systems Bioengineering of the Ministry of Education, School of Chemical Engineering and Technology, Tianjin University, China.
Considerable experimental evidence indicates that (-)-epigallocatechin-3-gallate (EGCG) inhibits the fibrillogenesis of Aβ(42) and alleviates its associated cytotoxicity.[25]

GREEN TEA EXTRACT (EGCG) (↓) ABETA AGGREGATION
The Journal of Physical Chemistry. B (United States) (2013)
School of Biological Sciences, Nanyang Technological University, Singapore.
Growing evidence supports that amyloid β (Aβ) oligomers are the major causative agents leading to neural cell death in Alzheimer's disease. **The polyphenol (-)-epigallocatechin gallate (EGCG) was recently reported to inhibit Aβ fibrillization and redirect Aβ aggregation into unstructured, off-pathway oligomers...** In the presence of EGCG, the Aβ structures are characterized by increased inter-center-of-mass distances, reduced interchain and intrachain contacts, reduced β-sheet content, and increased coil and α-helix contents. **Analysis of the free energy surfaces reveals that the Aβ dimer with EGCG adopts new conformations, affecting therefore its propensity to adopt fibril-prone states.**[26]

GREEN TEA EXTRACT (EGCG) (↓) ABETA AGGREGATION
ACS Applied Materials and Interfaces (United States) (2014)
Department of Chemistry, Jinan University, Guangzhou, China.
Evidence is increasingly showing that epigallocatechin-3-gallate (EGCG) can partly protect cells from Aβ-mediated neurotoxicity by inhibiting Aβ aggregation.[27]

OLIVE OIL EXTRACTS (↓) ABETA AGGREGATION

OLIVE OIL EXTRACT (OLEUROPEIN) (↓) ABETA AGGREGATION
Current Alzheimer Research (United Arab Emirates) (2011)
Department of Biochemical Sciences, University of Florence, Italy.
Here we report that oleuropein aglycon also hinders amyloid aggregation of Aβ(1-42) and its cytotoxicity, suggesting a general effect of such

polyphenol… **We here report that the polyphenol [oleuropein aglycon] eliminates the appearance of early toxic oligomers favouring the formation of stable harmless protofibrils, structurally different from the typical Aβ(1–42) fibrils.**[28]

RED WINE (↓) ABETA AGGREGATION

Red Wine (Muscadine) (↓) Abeta Aggregation
Journal of Alzheimer's Disease (The Netherlands) (2009)
Department of Psychiatry, Mount Sinai School of Medicine, New York.
[E]vidence from our present study suggests that muscadine [red wine] treatment attenuates Abeta neuropathology and Abeta-related cognitive deterioration in Tg2576 mice by interfering with the oligomerization of Abeta molecules to soluble high-molecular-weight Abeta oligomer species that are responsible for initiating a cascade of cellular events resulting in cognitive decline.[29]

CURCUMIN (↓) ABETA TOXICITY

CURCUMINOIDS (↓) ABETA TOXICITY
Food and Chemical Toxicology (England) (2008)
Environmental Toxico-Genomic and Proteomic Center, College of Medicine, Korea University, Seoul, South Korea.
Taken together, these data indicate that curcumin protected PC12 cells against Abeta-induced neurotoxicity through the inhibition of oxidative damage, intracellular calcium influx, and tau hyperphosphorylation.[30]

CURCUMIN (↓) ABETA TOXICITY
Neuroscience Letters (Ireland) (2009)
Laboratory of Biotechnology and State Key Laboratory of Chinese Ethnic Minority Traditional Medicine, College of Life and Environmental Science, Minzu University of China, Beijing.
The results suggest that curcumin protects cultured rat primary prefrontal cortical neurons against Abeta-induced cytotoxicity, and both Bcl2 and caspase-3 are involved in the curcumin-induced protective effects.[31]

TETRAHYDROCURCUMIN (↓) ABETA TOXICITY
Neuroreport (England) (2011)
National Brain Research Centre of India, Manesar, Haryana, India.
[W]e show that tetrahydrocurcumin, a metabolite of curcumin, shows a protective effect against oligomeric amyloid-β-induced toxicity.[32]

CURCUMIN (↓) ABETA TOXICITY
International Journal of Clinical and Experimental Medicine (United States) (2012)
State Key Laboratory of Biomembrane and Membrane Biotechnology, College of Life Sciences, Peking University, Beijing, China.
We find that at low dosages, curcumin effectively inhibits intracellular Aβ toxicity.[33]

CURCUMIN (↓) ABETA TOXICITY
European Review for Medical and Pharmacological Sciences (Italy) (2012)
Tianjin Key Laboratory of TCM Pharmacology, Tianjin Key Laboratory of TCM Chemistry and Analysis, Chinese Materia Medica College, Tianjin University of Traditional Chinese Medicine, China.
The results suggest that curcumin has a protective effect against Abeta-induced toxicity in cultured rat cortical neurons.[34]

CURCUMIN (↓) ABETA TOXICITY
International Journal of Molecular Sciences (Switzerland) (2012)
School of Physical and Mathematical Sciences, Nanyang Technological University, Singapore.
We further illustrate that the molecule curcumin is a potential Aβ toxicity inhibitor as a β-sheet breaker by having a high propensity to interact with certain Aβ residues without binding to them.[35]

CURCUMIN (↓) ABETA TOXICITY
Langmuir (United States) (2013)
Department of Chemical and Nuclear Engineering and the Center for Biomedical Engineering, University of New Mexico, Albuquerque, New Mexico.
Curcumin dose-dependently ameliorates Aβ-induced neurotoxicity... Our work reveals a novel molecular mechanism by which curcumin reduces Aβ-related pathology and toxicity and suggests a therapeutic strategy for preventing or treating AD by targeting the inhibition of Aβ-induced membrane disruption.[36]

CURCUMIN (↓) ABETA TOXICITY
Neurobiology of Learning and Memory (United States) (2013)
Programa de Pós-Graduação em Bioquímica, Departamento de Bioquímica, Universidade Federal do Rio Grande do Sul, Porto Alegre, Brazil.
Our findings demonstrate that administration of curcumin was effective in preventing behavioral impairments, neuroinflammation, tau hyperphosphorylation as well as cell signaling disturbances triggered by Aβ in vivo.[37]

CURCUMIN (↓) ABETA TOXICITY
Biochemical and Biophysical Research Communications (United States) (2014)
Department of Human Anatomy and Histo-Embryology, Xi'an Jiaotong University Health Science Center, China.
[P]retreatment of curcumin prevented the cultured cortical neurons from Aβ25–35-induced cell toxicity.[38]

CURCUMIN (↓) ABETA TOXICITY
ACS Chemical Neuroscience (United States) (2015)
Department of Chemical and Biological Engineering and the Center for Biomedical Engineering, and Department of Cell Biology and Physiology, University of New Mexico Health Sciences Center, University of New Mexico, Albuquerque.
Taken together, our study shows that curcumin exerts its neuroprotective effect against Aβ induced toxicity through at least two concerted pathways, modifying the Aβ aggregation pathway toward the formation of nontoxic aggregates and ameliorating Aβ-induced toxicity possibly through a non-specific pathway.[39]

Ginkgo biloba EXTRACTS (↓) ABETA TOXICITY

Ginkgo biloba EXTRACT 761 (↓) ABETA TOXICITY
The European Journal of Neuroscience (France) (2000)
Douglas Hospital Research Centre, Department of Psychiatry, McGill University, Montreal, Québec, Canada.
A co-treatment with EGb 761 [*Ginkgo biloba* extract 761] dose-dependently (10–100 microg/mL) protected hippocampal neurons against toxicity induced by Abeta fragments, with a maximal and complete protection at the highest concentration tested.[40]

Ginkgo biloba EXTRACT 761 (↓) ABETA TOXICITY
Cellular and Molecular Biology (France) (2002)
Douglas Hospital Research Centre, Department of Psychiatry and Pharmacology and Therapeutics, McGill University, Montreal, Québec, Canada.
[T]hese results and those obtained by other groups highlight the neuro-protective abilities of EGb 761 against dysfunction and death of neurons caused by Abeta deposits.[41]

Ginkgo biloba EXTRACT 761 (↓) ABETA TOXICITY
Neurobiology of Aging (United States) (2002)

Department of Psychiatry and Pharmacology and Therapeutics, Douglas Hospital Research Centre, McGill University, Québec, Canada.

We found that EGb 761, possibly through the antioxidant properties of its flavonoids, was able to protect hippocampal cells against toxic effects induced by Abeta peptides.[42]

Ginkgo biloba EXTRACT 761 (↓) ABETA TOXICITY
Cellular and Molecular Biology (France) (2002)
Douglas Hospital Research Centre, Department of Psychiatry and Pharmacology and Therapeutics, McGill University, Verdun, Canada.

A co-treatment with EGb 761 protected cells against toxicity induced by Abeta fragments in a concentration dependent manner.[43]

Ginkgo biloba EXTRACT 761 (↓) ABETA TOXICITY
Pharmacopsychiatry (Germany) (2003)
Department of Pharmacology, Biocenter, J. W. Goethe University of Frankfurt, Germany.

Here, we will report that EGb 761 was able to protect mitochondria from the attack of hydrogen peroxide, antimycin, and Abeta. Furthermore, EGb 761 reduced ROS levels and ROS-induced apoptosis in lymphocytes from aged mice treated orally with EGb 761 for 2 weeks. **Our data further emphasize neuroprotective properties of EGb 761, such as protection against Abeta-toxicity,** and anti-apoptotic properties, which are probably due to its preventive effects on mitochondria.[44]

Ginkgo biloba EXTRACTS (GINKGOLIDES) (↓) ABETA TOXICITY
Journal of Neuroinflammation (England) (2004)
Department of Veterinary Pathology, Glasgow University Veterinary School, Glasgow, United Kingdom.

Nanomolar concentrations of the ginkgolides protect neurons against the otherwise toxic effects of amyloid-beta1–42 or sPrP106.[45]

Ginkgo biloba EXTRACT 761 (↓) ABETA TOXICITY
Free Radical Biology and Medicine (United States) (2006)
INRS, Institut Armand Frappier, Université du Québec, Pointe Claire, Univ. Laval, Quebec, Canada.

Our study demonstrates that the protection of neuronal cells by EGb 761 against Abeta could involve different mechanisms as the regulation of several key intracellular pathways and the inhibition of fAbeta [Abeta fibril] formation and implicate more than its free radical scavenging property.[46]

Ginkgo biloba Extract 761 (↓) Abeta Toxicity
The Journal of Neuroscience (United States) (2006)
Department of Pharmaceutical Sciences, School of Pharmacy, University of Maryland, Baltimore.
These findings suggest that EGb 761 suppresses Abeta-related pathological behaviors, (2) the protection against Abeta toxicity by EGb 761 is mediated primarily by modulating Abeta oligomeric species, and (3) ginkgolide A has therapeutic potential for prevention and treatment of Alzheimer's disease.[47]

Ginkgo biloba Extract (↓) Abeta Toxicity
Zhongguo Zhong Yao Za Zhi (China) (2011)
Research Center, Xiyuan Hospital, China Academy of Chinese Medical Sciences, Beijing, China.
We conclude that Abeta40 could significantly induce primary cultured neurons to apoptosis in vitro, and that *Ginkgo biloba* extract showed beneficial neuroprotective effects against neuronal apoptosis, which might be due to improving the structures of neuron and its subcellular organelles, enhancing cellular proliferative activity, and inhibiting caspase-3 overexpression in neurons.[48]

GREEN TEA EXTRACTS (↓) ABETA TOXICITY

Green Tea Extract (EGCG) (↓) Abeta Toxicity
FASEB Journal (United States) (2003)
Eve Topf and USA National Parkinson Foundation, Centers of Excellence for Neurodegenerative Diseases Research, Technion Faculty of Medicine, Haifa, Israel.
EGCG is not only able to protect, but it can rescue PC12 cells against the beta-amyloid (Abeta) toxicity in a dose-dependent manner... EGCG has protective effects against Abeta-induced neurotoxicity and regulates secretory processing of non-amyloidogenic APP via PKC pathway.[49]

Green Tea Extracts (Green and Black Tea, EGCG) (↓) Abeta Toxicity
The European Journal of Neuroscience (France) (2006)
Douglas Hospital Research Centre, Department of Psychiatry, McGill University, Montreal, Québec, Canada.
Both green and black tea extracts (5-25 microg/mL) displayed neuroprotective action against Abeta toxicity. These effects were shared by gallic acid (1-20 microm), epicatechin gallate (ECG; 1-20 microM) and epigallocatechin gallate (EGCG; 1-10 microM), the former being the most potent flavan-3-ol... **Moreover, the protective effect of EGCG is likely to be associated, at least in part, with its inhibitory action on Abeta fibrils/oligomers formation.** These data also support the hypothesis that not only green but also black teas may reduce age-related neurodegenerative diseases, such as AD.[23]

GREEN TEA EXTRACT (EGCG) (↓) ABETA TOXICITY
Journal of Alzheimer's Disease (The Netherlands) (2011)

Department of Cell Biology, Microbiology, and Molecular Biology, University of South Florida, Tampa.

The results of this study lend further credence to the notion that EGCG and other flavonoids, such as luteolin, are "multipotent therapeutic agents" that not only reduce toxic levels of brain Aβ, but also hold the potential to protect neuronal mitochondrial function in AD.[50]

OLIVE OIL EXTRACTS (↓) ABETA TOXICITY

OLIVE OIL EXTRACT (OLEUROPEIN AGLYCONE) (↓) ABETA TOXICITY
Neuroscience Letters (Ireland) (2014)

Department of Neuroscience, Psychology, Drug Research and Child Health, Division of Pharmacology and Toxicology, University of Florence, Italy; Department of Experimental and Clinical Biomedical Sciences, University of Florence, Italy.

Altogether, these data provide additional support to the anti-aggregation, neuroprotective and anti-inflammatory activities of this natural phenol [oleuropein aglycone], confirming its beneficial properties against neurodegeneration.[51]

OLIVE OIL EXTRACT (OLEUROPEIN AGLYCONE) (↓) ABETA TOXICITY
Journal of Alzheimer's Disease (The Netherlands) (2015)

Department of Neuroscience, Psychology, Drug Research and Child Health, Division of Pharmacology and Toxicology, University of Florence, Italy; Department of Experimental and Clinical Biomedical Sciences "Mario Serio," University of Florence, Italy.

The present review summarizes the findings on the beneficial effects against neurodegeneration and other peripheral inflammatory and degenerative diseases of oleuropein aglycone, a natural phenol abundant in the extra virgin olive oil… Such a protection could result, at least in part, in a remarkable improvement of the pathological signs arising from stress conditions including oxidative stress, an excessive inflammatory response, and the presence of cytotoxic aggregated material.[52]

OLIVE OIL EXTRACTS (OLEUROPEIN AGLYCONE AND OLEOCANTHAL) (↓) ABETA TOXICITY
Advances in Experimental Medicine and Biology (United States) (2015)

Department of Experimental and Clinical Biomedical Sciences "Mario Serio," University of Florence, Italy.

[S]tudies conducted both in vivo and in vitro have started to reveal the great potential of the phenolic component of extra virgin olive oil (mainly

oleuropein aglycone and oleocanthal) in counteracting amyloid aggregation and toxicity.[53]

RESVERATROL (↓) ABETA TOXICITY

RESVERATROL (↓) ABETA TOXICITY
Free Radical Biology and Medicine (United States) (2003)
Research Institute of Pharmaceutical Sciences, College of Pharmacy, Seoul National University, South Korea.
Resveratrol attenuated beta-amyloid-induced cytotoxicity, apoptotic features, and intracellular reactive oxygen intermediates.[54]

RESVERATROL (↓) ABETA TOXICITY
British Journal of Pharmacology (England) (2004)
Department of Psychiatry, Douglas Hospital Research Centre, McGill University, Montreal, Québec, Canada.
Pre-, co- and post-treatment with resveratrol significantly attenuated Abeta-induced cell death in a concentration-dependent manner.[55]

RESVERATROL (↓) ABETA TOXICITY
The Journal of Biological Chemistry (United States) (2005)
Litwin-Zucker Research Center for the Study of Alzheimer's Disease and Memory Disorders, North Shore-Long Island Jewish Institute for Medical Research, Manhasset, New York
Resveratrol does not inhibit Abeta production, because it has no effect on the Abeta-producing enzymes beta- and gamma-secretases, but **promotes instead intracellular degradation of Abeta via a mechanism that involves the proteasome.**[56]

RESVERATROL (↓) ABETA TOXICITY
Biological and Pharmaceutical Bulletin (Japan) (2007)
College of Pharmacy, Seoul National University, South Korea.
These results suggest that [the resveratrol compounds] vitisin A and heyneanol A prevent Abeta-induced neurotoxicity through attenuating oxidative stress induced by Abeta, and may be useful as potential preventive or therapeutic agents for AD.[57]

RESVERATROL (↓) ABETA TOXICITY
Molecular Aspects of Medicine (England) (2008)
Department of Biochemistry, Yong Loo Lin School of Medicine, National University of Singapore.

Induction of SIRT1 expression also attenuates neuronal degeneration and death in animal models of Alzheimer's disease and Huntington's disease. **SIRT1 induction, either by sirtuin activators such as resveratrol or metabolic conditioning associated with caloric restriction, could be neuroprotective in several ways.**[58]

RESVERATROL (↓) ABETA TOXICITY
Neurotoxicology (The Netherlands) (2009)
Tsinghua University School of Medicine, Beijing, China.
In conjunction with the concept that Abeta oligomers are linked to Abeta toxicity, we speculate that aside from potential antioxidant activities, **resveratrol may directly bind to Abeta42, interfere in Abeta42 aggregation, change the Abeta42 oligomer conformation and attenuate Abeta42 oligomeric cytotoxicity.**[59]

RESVERATROL (↓) ABETA TOXICITY
Journal of Neurochemistry (England) (2009)
Department of Neuroscience, Mario Negri Institute for Pharmacological Research, Milan, Italy.
We conclude that SIRT1 activation by resveratrol can prevent in our neuroblastoma model the deleterious effects triggered by oxidative stress or alpha-synuclein(A30P) aggregation, while **resveratrol displayed a SIRT1-independent protective action against Abeta42.**[60]

RESVERATROL (↓) ABETA TOXICITY
PLoS One (United States) (2011)
Centro Nazionale delle Ricerche, Istituto Tecnologie Biomediche, Metalloproteins Unit, Department of Biology, University of Padova, Padova, Italy.
In this paper, we demonstrate that resveratrol is cytoprotective in human neuroblastoma cells exposed to Aβ and or to Aβ-metal complex. Our findings suggest that resveratrol acts not through anti-aggregative pathways but mainly via its scavenging properties.[61]

RESVERATROL (↓) ABETA TOXICITY
PLoS One (United States) (2011)
Department of Biochemical Science and Biotechnology, National Chia-Yi University, Chia-Yi, Taiwan.
The present study is aimed to elucidate the cellular effect of resveratrol, a natural phytoestrogen with neuroprotective activities, on Aβ-induced hippocampal neuron loss and memory impairment… These findings strongly implicate that iNOS [inducible nitric oxide synthase] is involved in the Aβ-induced lipid peroxidation and HO-1 [heme oxygenase-1] downregulation, and **resveratrol protects animals from Aβ-induced neurotoxicity by suppressing iNOS production.**[62]

RESVERATROL (↓) ABETA TOXICITY
PLoS One (United States) (2013)

Key Laboratory of the Ministry of Education for Experimental Teratology and Department of Anatomy, School of Medicine, Shandong University, China.

We observed that resveratrol increased cell viability. Flow cytometry indicated resveratrol-induced reduction of cell apoptosis. **Resveratrol also stabilized the intercellular Ca(2+) homeostasis and attenuated Aβ(25–35) neurotoxicity.**[63]

CURCUMINOIDS (↑) ABETA CLEARANCE

CURCUMIN (↑) ABETA CLEARANCE.
The Journal of Neuroscience (United States) (2001)

Departments of Medicine and Neurology, University of California, Los Angeles.

With low-dose but not high-dose curcumin treatment, the astrocytic marker GFAP [glial fibrillary acidic protein] was reduced, and insoluble beta-amyloid, soluble Abeta, and plaque burden were significantly decreased by 43–50% [in an Alzheimer transgenic APPsw mouse model (Tg2576)]... Microgliosis was also suppressed in neuronal layers but not adjacent to plaques.[64]

CURCUMINOIDS (↑) ABETA CLEARANCE
Journal of Alzheimer's Disease (The Netherlands) (2006)

Department of Medicine, Greater Los Angeles Veterans Administration Medical Center; School of Medicine, University of California, Los Angeles.

AD patients have defects in phagocytosis of amyloid-beta (1–42) (Abeta) in vitro by the innate immune cells, monocyte/macrophages and in clearance of Abeta plaques. **The natural product curcuminoids enhanced brain clearance of Abeta in animal models.**[65]

BISDEMETHOXYCURCUMIN (↑) ABETA CLEARANCE
Proceedings of the National Academy of Sciences (United States) (2007)

Department of Medicine, Greater Los Angeles Veterans Affairs Medical Center; School of Medicine, University of California at Los Angeles.

We have tested a hypothesis that the natural product curcuminoids, which has epidemiologic and experimental rationale for use in AD, may improve the innate immune system and increase amyloid-beta (Abeta) clearance from the brain of patients with sporadic Alzheimer's disease (AD)... **In mononuclear cells of some AD patients, the curcuminoid compound bisdemethoxycurcumin may enhance defective phagocytosis of Abeta,** the transcription of MGAT3 and TLRs, and the translation of TLR2-4.[66]

CURCUMIN (↑) ABETA CLEARANCE
Journal of Neurochemistry (England) (2007)
Department of Neurology, Alzheimer's Disease Research Laboratory, Massachusetts General Hospital, Charlestown, Massachusetts.

[S]ystemic treatment of mice with curcumin for 7 days clears and reduces existing [Abeta] plaques, as monitored with longitudinal imaging, suggesting a potent disaggregation effect. Curcumin also led to a limited, but significant reversal of structural changes in dystrophic dendrites, including abnormal curvature and dystrophy size. Together, these data suggest that curcumin reverses existing amyloid pathology and associated neurotoxicity in a mouse model of AD.[67]

CURCUMINOIDS AND VITAMIN D3 (↑) ABETA CLEARANCE
Journal of Alzheimer's Disease (The Netherlands) (2009)
School of Medicine, University of California at Los Angeles.

Patients with Alzheimer's disease (AD) suffer from brain amyloidosis related to defective clearance of amyloid-beta (Abeta) by the innate immune system… **[Vitamin] D3 interacts with curcuminoids to stimulate amyloid-beta clearance by macrophages of Alzheimer's disease patients**.[68]

CURCUMIN (↑) ABETA CLEARANCE
The Journal of Biological Chemistry (United States) (2010)
Genetics and Aging Research Unit, MassGeneral Institute for Neurodegenerative Disease, Department of Neurology, Massachusetts General Hospital and Harvard Medical School, Charlestown, Massachusetts.

Here, we investigated the effects of curcumin on Abeta levels and APP processing in various cell lines and mouse primary cortical neurons. **We show for the first time that curcumin potently lowers Abeta levels by attenuating the maturation of APP in the secretory pathway**. These data provide a mechanism of action for the ability of curcumin to attenuate amyloid-beta pathology.[69]

CURCUMIN (↑) ABETA CLEARANCE
Brain Research (The Netherlands) (2010)
Department of Anatomy and Neurobiology, Zhongshan School of Medicine, Sun Yat-Sen University, Guangzhou, Guangdong, China.

Curcumin can bind senile plaques and promote disaggregation of existing amyloid deposits and prevent aggregation of new amyloid deposits. Curcumin can also reverse distorted and curvy neurites around senile plaques and repair the neuritic abnormalities.[70]

CURCUMIN (↑) ABETA CLEARANCE.
Journal of Alzheimer's Disease (The Netherlands) (2011)
Molecular Neuroscience Research Center, Shiga University of Medical Science, Japan.
These results strongly suggested that curcumin binds to Aβ oligomers and to Aβ fibrils. The association of curcumin with Aβ oligomers may contribute to the therapeutic effect on AD.[71]

CURCUMINOIDS (↑) ABETA CLEARANCE
Neurodegenerative Diseases (Switzerland) (2012)
Human BioMolecular Research Institute, San Diego, California.
Clearance of Aβ or SOD-1 [superoxide dismutase-1 gene] by the innate immune system may be important for controlling or preventing disease onset. **Curcuminoids restore Aβ phagocytosis by peripheral blood mononuclear cells from AD patients and Aβ clearance with upregulation of key genes including MGAT3, vitamin D receptor, and Toll-like receptors.**[72]

BISDEMETHOXYCURCUMIN AND VITAMIN D3 (↑) ABETA CLEARANCE
Journal of Alzheimer's Disease (The Netherlands) (2012)
Department of Biochemistry, University of California at Riverside.
Brain clearance of amyloid-β (Aβ42) by innate immune cells is necessary for maintenance of normal brain function… **Bisdemethoxycurcumin is a vitamin D receptor ligand and additive with 1,25D3 in promoting Aβ42 phagocytosis by Type I, but not by Type II macrophages.**[73]

CURCUMIN (↑) ABETA CLEARANCE
PLoS One (United States) (2012)
Bio Nano Electronics Research Center, Graduate School of Interdisciplinary New Science, Toyo University, Japan.
Anti-amyloid activity and anti-oxidant activity of curcumin is highly beneficial for the treatment of Alzheimer's disease… **Our results suggest that curcumin-encapsulated PLGA nanoparticles are able to destroy amyloid aggregates, exhibit anti-oxidative property, and are non-cytotoxic.**[74]

CURCUMIN (↑) ABETA CLEARANCE
Journal of Neuroscience Research (United States) (2014)
Key Laboratory of Chinese Internal Medicine, Ministry of Education, Beijing University of Chinese Medicine, China; Key Laboratory of Pharmacology of Dongzhimen Hospital, State Administration of Traditional Chinese Medicine, Beijing, China.
This evidence suggests that curcumin, as a potential AD therapeutic method, can reduce β-amyloid pathological aggregation, possibly through

mechanisms that prevent its production by inhibiting presenilin-2 and/or by accelerating its clearance by increasing degrading enzymes such as insulin-degrading enzyme and neprilysin.[75]

CURCUMIN (↑) ABETA CLEARANCE
Expert Review of Neurotherapeutics (England) (2015)
Greater Los Angeles Healthcare System, Veteran's Administration, Geriatric Research Education and Clinical Center, Los Angeles, California.

Curcumin is a pleiotropic molecule, which not only directly binds to and limits aggregation of the β-sheet conformations of amyloid characteristic of many neurodegenerative diseases but also restores homeostasis of the inflammatory system, boosts the heat shock system to enhance clearance of toxic aggregates, scavenges free radicals, chelates iron and induces anti-oxidant response elements.[76]

Ginkgo biloba EXTRACTS (↑) ABETA CLEARANCE

Ginkgo biloba EXTRACT 761 (↑) ABETA CLEARANCE
Acta Neuropathologica (Germany) (2008)
Department of Neurology, Zhongda Hospital, Southeast University, Nanjing, China.

Decreased clearance of Abeta from [the] brain is the main cause of Abeta accumulation in sporadic AD… **Our findings suggest that EGb 761 favor[s] clearance of Abeta via regulating the expression of RAGE and LRP-1 during brain ischemia.**[77]

GRAPE SEED EXTRACT (↑) ABETA CLEARANCE

GRAPE SEED EXTRACT AND CURCUMINOIDS (↑) ABETA CLEARANCE.
Neurotoxicity Research (United States) (2009)
Department of Human Physiology and Centre for Neuroscience, Flinders University, Adelaide, Australia.

The Abeta levels in the brain and serum of the mice fed with grape seed extract were reduced by 33% and 44%, respectively, compared with the Alzheimer's mice fed with the control diet. Amyloid plaques and microgliosis in the brain of Alzheimer's mice fed with grape seed extract were also reduced by 49% and 70%, respectively. Curcumin also significantly reduced brain Abeta burden and microglia activation.[78]

GRAPE SEED EXTRACT (EPICATECHIN) (↑) ABETA CLEARANCE
Frontiers in Neurology (Switzerland) (2014)
Key Laboratory of Stem Cells and Regenerative Medicine, Institute of Molecular and Clinical Medicine, Kunming Medical University, China; Department of Neurology, Centre for Clinical

Neuroscience, Daping Hospital, Third Military Medical University, Chongqin, China; School of Pharmacy and Medical Sciences, Sansom Institute, University of South Australia, Adelaide.

(-)Epicatechin [a polyphenolic compound in grape seed extract] significantly reduced [in APP/PS1 transgenic mice] total Aβ in brain and serum by 39% and 40%, respectively, compared with control diet.[79]

GREEN TEA EXTRACTS (↑) ABETA CLEARANCE

GREEN TEA EXTRACT (EGCG) (↓) ABETA BURDEN
Brain Research (The Netherlands) (2008)
Rashid Laboratory for Developmental Neurobiology, Silver Child Development Center, Department of Psychiatry and Behavioral Medicine, and Department of Neurosurgery, University of South Florida, Tampa.

Here, we find that EGCG administered orally in drinking water (50 mg/ kg) similarly reduces Abeta deposition in these mice. Following a six-month treatment of an 8-month old cohort, immunohistochemical analysis of coronal sections reveals that **plaque burdens were reduced in the cingulate cortex, hippocampus, and entorhinal cortex by 54%, 43%, and 51%, respectively.**[80]

OLIVE OIL EXTRACTS (↑) ABETA CLEARANCE

OLIVE OIL EXTRACT (OLEUROPEIN AGLYCONE) (↑) ABETA CLEARANCE
PLoS One (United States) (2013)
Department of Molecular Biochemistry and Pharmacology, Istituto di Ricerche Farmacologiche Mario Negri, Milan, Italy.

There is increasing evidence to suggest that natural polyphenols may prevent the formation of toxic amyloid aggregates; this applies also to oleuropein aglycone (OLE), the most abundant polyphenol in extra virgin olive oil, previously shown to hinder amylin and Aβ aggregation... OLE-fed [*C. elegans*] displayed reduced Aβ plaque deposition, less abundant toxic Aβ oligomers, remarkably decreased paralysis and increased lifespan with respect to untreated animals.[81]

OLIVE OIL EXTRACT (OLEOCANTHAL) (↑) ABETA CLEARANCE
ACS Chemical Neuroscience (United States) (2013)
Department of Basic Pharmaceutical Science, College of Pharmacy, University of Louisiana at Monroe.

Here, we provide in vitro and in vivo evidence for the potential of oleocanthal to enhance Aβ clearance from the brain via up-regulation of P-glycoprotein (P-gp) and LDL lipoprotein receptor related protein-1 (LRP1), major Aβ transport proteins, at the blood-brain barrier (BBB). Results from in vitro and in vivo studies demonstrated similar and consistent pattern of oleocanthal in controlling Aβ levels… **In conclusion, these findings provide experimental support that potential reduced risk of AD associated with extra-virgin olive oil could be mediated by enhancement of Aβ clearance from the brain.**[82]

Olive Oil Extract (Oleocanthal) (↑) Abeta Clearance
ACS Chemical Neuroscience (United States) (2015)

Department of Basic Pharmaceutical Sciences, School of Pharmacy, University of Louisiana at Monroe; Pennington Biomedical Research Center, Louisiana State University at Baton Rouge.
Mice treatment for 4 weeks with oleocanthal significantly decreased amyloid load in the hippocampal parenchyma and micro-vessels. This reduction was associated with enhanced cerebral clearance of Aβ across the blood-brain barrier. Further mechanistic studies demonstrated oleocanthal to increase the expression of important amyloid clearance proteins at the blood-brain barrier, including P-glycoprotein and LRP1, and to activate the ApoE-dependent amyloid clearance pathway in the mice brains.[83]

Olive Oil (↑) Abeta Clearance
The Journal of Nutritional Biochemistry (United States) (2015)

Department of Basic Pharmaceutical Sciences, School of Pharmacy, University of Louisiana at Monroe; Sanders-Brown Center on Aging, University of Kentucky, Lexington, Kentucky; Pennington Biomedical Research Center, Louisiana State University, Baton Rouge, Louisiana.
[A]lthough feeding mice with extra-virgin olive oil-enriched diet for 3 months, beginning at an age after Aβ accumulation starts, showed improved clearance across the blood-brain barrier and significant reduction in Aβ levels, it did not affect tau levels or improve cognitive functions of TgSwDI mouse.[84]

RESVERATROL (↑) ABETA CLEARANCE

Resveratrol (↑) Abeta Clearance
Neurochemistry International (England) (2009)

Department of Neurology and Neurosciences, Weill Medical College of Cornell University, Burke Medical Research Institute, White Plains, New York.
Mice were fed clinically feasible dosages of resveratrol for forty-five days. Neither **resveratrol nor its conjugated metabolites were detectable in brain. Nevertheless, resveratrol diminished [Abeta] plaque formation in a region specific manner. The largest reductions in the percent area occupied by plaques were observed in medial cortex (-48%), striatum (-89%) and hypothalamus (-90%).** The changes occurred without detectable activation of SIRT-1 or alterations in APP processing.[85]

RESVERATROL (↑) ABETA CLEARANCE

Current Pharmaceutical Design (The Netherlands) (2012)

School of Traditional Chinese Pharmacy, Shanghai University of Traditional Chinese Medicine, China.

Resveratrol promotes the non-amyloidogenic cleavage of the amyloid precursor protein, enhances clearance of amyloid beta-peptides, and reduces neuronal damage.[86]

RESVERATROL (↑) ABETA CLEARANCE

Neurochemistry International (England) (2012)

School of Life Science, University of Science and Technology of China, Hefei, Anhui, China.

It has been shown that resveratrol may attenuate amyloid β peptide-induced toxicity, promote Aβ clearance and reduce senile plaques.[87]

Upon reviewing the research, it seems apparent that the plant polyphenols featured in this volume possess the ability to favorably influence the hallmark pathogenic targets in stages of the disease that are relevant to the primary prevention and secondary prevention of AD. Given evidence of their long-term safety and collateral health benefits, skillfully formulated combinations of plant polyphenols might constitute an effective means of AD-related prevention.

We define "effective" as satisfying the previously identified criteria for the primary prevention and secondary prevention of AD. Thus, individual polyphenolic compounds and certain combinations of such compounds have shown an ability to (1) effectively target the key pathogenic processes of AD-related primary prevention, (2) therapeutically target the key pathogenic processes of AD-related secondary prevention, (3) meet the Gandy test for long-term safety, (4) satisfy the Selkoe Test for selectively targeting APP with biologically active agents with a high therapeutic index without adversely influencing non-APP, non-AD substrates, (5) help prevent non-AD chronic disease, including cancer, (6) cost relatively little for long-term daily use, and (7) apply to AD-asymptomatic adults worldwide, including zero AD and preclinical AD populations.

REFERENCES

1. Sperling, Aisen, Beckett, et al. Toward defining the preclinical stages of Alzheimer's disease: recommendations from the National Institute on Aging-Alzheimer's Association Workgroups on diagnostic guidelines for Alzheimer's disease. *Alzheimer's and Dementia* May 2011;**7**(3):280–92.
2. Alzheimer's Disease: Unraveling the Mystery. p. 22.
3. Hamaguchi, Ono, Murase, Yamada. Phenolic compounds prevent Alzheimer's pathology through different effects on the amyloid-beta aggregation pathway. *American Journal of Pathology* December 2009;**175**(6):2557–65.
4. Pitt, Roth, Lacor, et al. Alzheimer's-associated Abeta oligomers show altered structure, immunoreactivity and synaptotoxicity with low doses of oleocanthal. *Toxicology and Applied Pharmacology* October 15, 2009;**240**(2):189–97.

5. Hubin, van Nuland, Broersen, Pauwels. Transient dynamics of Aβ contribute to toxicity in Alzheimer's disease. *Cellular and Molecular Life Sciences* September 2014;**71**(18):3507–21.

6. Boutajangout, Wisniewski. Tau-based therapeutic approaches for Alzheimer's disease—a mini-review. *Gerontology* 2014;**60**(5):381–5.

7. Viola, Klein. Amyloid β oligomers in Alzheimer's disease pathogenesis, treatment, and diagnosis. *Acta Neuropathologica* February 2015;**129**(2):183–206.

8. Dennis Selkoe on the amyloid hypothesis of Alzheimer's disease: special topic of Alzheimer's disease interview. *Sciencewatch* March 2011. at http://archive.sciencewatch.com/ana/st/alz2/11marSTAlz2Selk/.

9. Ono, Hasegawa, Naiki, Yamada. Curcumin has potent anti-amyloidogenic effects for Alzheimer's beta-amyloid fibrils in vitro. *Journal of Neuroscience Research* March 15, 2004;**75**(6):742–50.

10. Yang, Lim, Begum, et al. Curcumin inhibits formation of amyloid beta oligomers and fibrils, binds plaques, and reduces amyloid in vivo. *The Journal of Biological Chemistry* February 18, 2005;**280**(7): 5892–901.

11. Dasilva, Shaw, McLaurin. Amyloid-beta fibrillogenesis: structural insight and therapeutic intervention. *Experimental Neurology* June 2010;**223**(2):311–21.

12. Dolai, Shi, Corbo, et al. Clicked sugar-curcumin conjugate: modulator of amyloid-β and tau peptide aggregation at ultralow concentrations. *ACS Chemical Neuroscience* December 21, 2011;**2**(12):694–9.

13. Taylor, Moore, Mourtas, et al. Effect of curcumin-associated and lipid ligand-functionalized nanoliposomes on aggregation of the Alzheimer's Aβ peptide. *Nanomedicine: Nanotechnology, Biology, and Medicine* October 2011;**7**(5):541–50.

14. Huang, Chang, Dai, Jiang. Protective effects of curcumin on amyloid-β-induced neuronal oxidative damage. *Neurochemical Research* July 2012;**37**(7):1584–97.

15. Rao, Mohamed, Teckwani, Tin. Curcumin binding to beta amyloid: a computational study. *Chemical Biology & Drug Design* March 16, 2015. [Epub ahead of print].

16. Yao, Drieu, Papadopoulos. The *Ginkgo biloba* extract EGb 761 rescues the PC12 neuronal cells from beta-amyloid-induced cell death by inhibiting the formation of beta-amyloid-derived diffusible neurotoxic ligands. *Brain Research* January 19, 2001;**889**(1–2):181–90.

17. Luo, Smith, Paramasivam, et al. Inhibition of amyloid-beta aggregation and Caspase-3 activation by the *Ginkgo biloba* extract EGb 761. *Proceedings of the National Academy of Sciences* September 17, 2002;**99**(19): 12197–202.

18. Tchantchou, Xu, Wu, et al. EGb 761 enhances adult hippocampal neurogenesis and phosphorylation of CREB in transgenic mouse model of Alzheimer's disease. *FASEB Journal* August 2007;**21**(10): 2400–8.

19. Wang, Ho, Zhao, et al. Grape-derived polyphenolics prevent Abeta oligomerization and attenuate cognitive deterioration in a mouse model of Alzheimer's disease. *The Journal of Neuroscience* June 18, 2008;**28**(25):6388–92.

20. Ono, Condron, Ho, et al. Effects of grape seed-derived polyphenols on amyloid beta-protein self-assembly and cytotoxicity. *The Journal of Biological Chemistry* November 21, 2008;**283**(47):32176–87.

21. Liu, Pukala, Musgrave, et al. Gallic acid is the major component of grape seed extract that inhibits amyloid fibril formation. *Bioorganic & Medicinal Chemistry Letters* December 1, 2013;**23**(23):6336–40.

22. Hayden, Yamin, Beroukhim, et al. Inhibiting amyloid β-protein assembly: size-activity relationships among grape seed-derived polyphenols. *Journal of Neurochemistry* October 2015;**135**(2):416–30.

23. Bastianetto, Yao, Papadopoulos, Quirion. Neuroprotective effects of green and black teas and their catechin gallate esters against beta-amyloid-induced toxicity. *The European Journal of Neuroscience* January 2006;**23**(1):55–64.

24. Abbas, Wink. Epigallocatechin gallate inhibits beta amyloid oligomerization in *Caenorhabditis elegans* and affects the daf-2/insulin-like signaling pathway. *Phytomedicine: International Journal of Phytotherapy and Phytopharmacology* September 2010;**17**(11):902–9.

25. Liu, Dong, He, et al. Molecular insight into conformational transition of amyloid β-peptide 42 inhibited by (-)-epigallocatechin-3-gallate probed by molecular simulations. *The Journal of Physical Chemistry B* October 20, 2011;**115**(41):11879–87.

26. Zhang, Zhang, Derreumaux, Mu. Molecular mechanism of the inhibition of EGCG on the Alzheimer Aβ(1-42) dimer. *The Journal of Physical Chemistry B* April 18, 2013;**117**(15):3993–4002.

27. Zhang, Zhou, Yu, et al. Epigallocatechin-3-gallate (EGCG)-stabilized selenium nanoparticles coated with Tet-1 peptide to reduce amyloid-β aggregation and cytotoxicity. *ACS Applied Materials & Interfaces* June 11, 2014;**6**(11):8475–87.

28. Rigacci, Guidotti, Bucciantini, et al. Aβ(1-42) aggregates into non-toxic amyloid assemblies in the presence of the natural polyphenol oleuropein aglycon. *Current Alzheimer Research* December 2011;**8**(8): 841–52.

29. Ho, Chen, Wang, et al. Heterogeneity in red wine polyphenolic contents differentially influences Alzheimer's disease-type neuropathology and cognitive deterioration. *Journal of Alzheimer's Disease* 2009;**16**(1):59–72.

30. Park, Kim, Cho, et al. Curcumin protected PC12 cells against beta-amyloid-induced toxicity through the inhibition of oxidative damage and tau hyperphosphorylation. *Food and Chemical Toxicology* August 2008;**46**(8):2881–7.

31. Qin, Cheng, Cui, et al. Potential protection of curcumin against amyloid beta-induced toxicity on cultured rat prefrontal cortical neurons. *Neuroscience Letters* October 2, 2009;**463**(2):158–61.

32. Mishra, Mishra, Seth, Sharma. Tetrahydrocurcumin confers protection against amyloid β-induced toxicity. *Neuroreport* January 5, 2011;**22**(1):23–7.

33. Ye, Zhang. Curcumin protects against intracellular amyloid toxicity in rat primary neurons. *International Journal of Clinical and Experimental Medicine* 2012;**5**(1):44–9.

34. Wang, Zhang, Du. The protective effect of curcumin on Aβ induced aberrant cell cycle reentry on primary cultured rat cortical neurons. *European Review for Medical and Pharmacological Sciences* April 2012;**16**(4):445–54.

35. Zhao, Long, Mu, LY. The toxicity of amyloid β oligomers. *International Journal of Molecular Sciences* 2012;**13**(6):7303–27.

36. Thapa, Vernon, De la Peña, et al. Membrane-mediated neuroprotection by curcumin from amyloid-β-peptide-induced toxicity. *Langmuir: The ACS Journal of Surfaces and Colloids* September 17, 2013;**29**(37):11713–23.

37. Hoppe, Coradini, Frozza, et al. Free and nanoencapsulated curcumin suppress β-amyloid-induced cognitive impairments in rats: involvement of BDNF and Akt/GSK-3β signaling pathway. *Neurobiology of Learning and Memory* November 2013;**106**:134–44.

38. Sun, Jia, Wang, et al. Activation of SIRT1 by curcumin blocks the neurotoxicity of amyloid-β25-35 in rat cortical neurons. *Biochemical and Biophysical Research Communications* May 23, 2014;**448**(1):89–94.

39. Thapa, Jett, Chi. Curcumin attenuates amyloid-β aggregate toxicity and modulates amyloid-β aggregation pathway. *ACS Chemical Neuroscience* November 17, 2015. [Epub ahead of print].

40. Bastianetto, Ramassamy, Doré, et al. The *Ginkgo biloba* extract (EGb 761) protects hippocampal neurons against cell death induced by beta-amyloid. *The European Journal of Neuroscience* June 2000;**12**(6): 1882–90.

41. Bastianetto, Quirion. EGb 761 is a neuroprotective agent against beta-amyloid toxicity. *Cellular and Molecular Biology* September 2002;**48**(6):693–7.

42. Bastianetto, Quirion. Natural extracts as possible protective agents of brain aging. *Neurobiology of Aging* September–October, 2002;**23**(5):891–7.

43. Bastianetto, Quirion. EGb 761 is a neuroprotective agent against beta-amyloid toxicity. *Cellular and Molecular Biology* September 2002;**48**(6):693–7.

44. Eckert, Keil, Kressmann, et al. Effects of EGb 761 *Ginkgo biloba* extract on mitochondrial function and oxidative stress. *Pharmacopsychiatry* June 2003;**36**(Suppl. 1):S15–23.

45. Bate, Salmona, Williams. Ginkgolide B inhibits the neurotoxicity of prions or amyloid-beta 1-42. *Journal of Neuroinflammation* May 11, 2004;**1**(1):4.

46. Longpré, Garneau, Christen, Ramassamy. Protection by EGb 761 against beta-amyloid-induced neurotoxicity: involvement of NF-kappaB, SIRT1, and MAPKs pathways and inhibition of amyloid fibril formation. *Free Radical Biology and Medicine* December 15, 2006;**41**(12):1781–94.

47. Wu, Wu, Butko, et al. Amyloid-beta-induced pathological behaviors are suppressed by *Ginkgo biloba* extract EGb 761 and ginkgolides in transgenic *Caenorhabditis elegans*. *The Journal of Neuroscience* December 13, 2006;**26**(50):13102–13.

48. Cong, Sheng, Li, et al. Protective effects of ginseng-ginko extracts combination on rat primary cultured neurons induced by Abeta(1-40). *Zhongguo Zhong Yao Za Zhi* April 2011;**36**(7):908–11.

49. Levites, Amit, Mandel, Youdim. Neuroprotection and neurorescue against Abeta toxicity and PKC-dependent release of nonamyloidogenic soluble precursor protein by green tea polyphenol (-)-Epigallocatechin-3-Gallate. *FASEB Journal* May 2003;**17**(8):952–4.

50. Dragicevic, Smith, Lin, et al. Green tea epigallocatechin-3-gallate (EGCG) and other flavonoids reduce Alzheimer's amyloid-induced mitochondrial dysfunction. *Journal of Alzheimer's Disease* 2011;**26**(3):507–21.

51. Luccarini, Dami, Grossi, et al. Oleuropein aglycone counteracts Aβ42 toxicity in the rat brain. *Neuroscience Letters* January 13, 2014;**558**:67–72.

52. Casamenti, Grossi, Rigacci, et al. Oleuropein aglycone: a possible drug against degenerative conditions. In vivo evidence of its effectiveness against Alzheimer's disease. *Journal of Alzheimer's Disease* 2015;**45**(3):679–88.

53. Rigacci. Olive oil phenols as promising multi-targeting agents against Alzheimer's disease. *Advances in Experimental Medicine and Biology* 2015;**863**:1–20.

54. Jang, Surh. Protective effect of resveratrol on beta-amyloid-induced oxidative PC12 cell death. *Free Radical Biology and Medicine* April 15, 2003;**34**(8):1100–10.

55. Han, Zheng, Bastianetto, et al. Neuroprotective effects of resveratrol against beta-amyloid-induced neurotoxicity in rat hippocampal neurons: involvement of protein kinase C. *British Journal of Pharmacology* March 2004;**141**(6):997–1005.

56. Marambaud, Zhao, Davies. Resveratrol promotes clearance of Alzheimer's diseases amyloid-beta peptides. *The Journal of Biological Chemistry* November 11, 2005;**280**(45):37377–82.

57. Jang, Piao, Kim, et al. Resveratrol oligomer from *Vitis amurensis* attenuate beta-amyloid-induced oxidative stress in PC12 cells. *Biological & Pharmaceutical Bulletin* June 2007;**30**(6):1130–4.

58. Tang, Chua. SIRT1 and neuronal diseases. *Molecular Aspects of Medicine* June 2008;**29**(3):187–200.

59. Feng, Wang, Yang, et al. Resveratrol inhibits beta-amyloid oligomeric cytotoxicity but does not prevent oligomer formation. *Neurotoxicology* November 2009;**30**(6):986–95.

60. Albani, Polito, Batelli, et al. The SIRT1 activator resveratrol protects SK-N-BE cells from oxidative stress and against toxicity caused by alpha-synuclein or amyloid-beta (1-42) peptide. *Journal of Neurochemistry* September 2009;**110**(5):1445–56.

61. Granzotto, Zatta. Resveratrol acts not through anti-aggregative pathways but mainly via its scavenging properties against Aβ and Aβ-metal complexes toxicity. *PLoS One* 2011;**6**(6):e21565.

62. Huang, Lu, Wo, et al. Resveratrol protects rats from Aβ-induced neurotoxicity by the reduction of iNOS expression and lipid peroxidation. *PLoS One* 2011;**6**(12):e29102.

63. Feng, Liang, Zhu, et al. Resveratrol inhibits β-amyloid-induced neuronal apoptosis through regulation of SIRT1-ROCK1 signaling pathway. *PLoS One* 2013;**8**(3):e59888.

64. Lim, Chu, Yang, et al. The curry spice curcumin reduces oxidative damage and amyloid pathology in an Alzheimer's transgenic mouse. *The Journal of Neuroscience* November 1, 2001;**21**(21):8370–7.

65. Zhang, Fiala, Cashman, et al. Curcuminoids enhance amyloid-beta uptake by macrophages of Alzheimer's disease patients. *Journal of Alzheimer's Disease* September 2006;**10**(1):1–7.

66. Fiala, Liu, Espinosa-Jeffrey, et al. Innate immunity and transcription of MGAT-III and toll-like receptors in Alzheimer's disease patients are improved by bisdemethoxycurcumin. *Proceedings of the National Academy of Sciences* July 31, 2007;**104**(31):12849–54.

67. Garcia-Alloza, Borrelli, Rozkalne, et al. Curcumin labels amyloid pathology in vivo, disrupts existing plaques, and partially restores distorted neurites in an Alzheimer mouse model. *Journal of Neurochemistry* August 2007;**102**(4):1095–104.

68. Masoumi, Goldenson, Ghirmai, et al. 1alpha,25-Dihydroxyvitamin D3 interacts with curcuminoids to stimulate amyloid-beta clearance by macrophages of Alzheimer's disease patients. *Journal of Alzheimer's Disease* 2009;**17**(3):703–17.

69. Zhang, Browne, Child, Tanzi. Curcumin decreases amyloid-beta peptide levels by attenuating the maturation of amyloid-beta precursor protein. *The Journal of Biological Chemistry* September 10, 2010;**285**(37):28472–80.

70. Xiao, Lin, Liu, et al. Potential therapeutic effects of curcumin: relationship to microtubule-associated proteins 2 in Aβ1-42 insult. *Brain Research* November 18, 2010;**1361**:115–23.

71. Yanagisawa, Taguchi, Yamamoto, et al. Curcuminoid binds to amyloid-β1-42 oligomer and fibril. *Journal of Alzheimer's Disease* 2011;**24**(Suppl. 2):33–42.

72. Cashman, Gagliardi, Lanier, et al. Curcumins promote monocytic gene expression related to β-amyloid and superoxide dismutase clearance. *Neuro-degenerative Diseases* 2012;**10**(1–4):274–6.

73. Mizwicki, Menegaz, Zhang, et al. Genomic and nongenomic signaling induced by 1α,25(OH)2-vitamin D3 promotes the recovery of amyloid-β phagocytosis by Alzheimer's disease macrophages. *Journal of Alzheimer's Disease* 2012;**29**(1):51–62.

74. Mathew, Fukuda, Nagaoka, et al. Curcumin loaded-PLGA nanoparticles conjugated with Tet-1 peptide for potential use in Alzheimer's disease. *PLoS One* 2012;**7**(3):e32616.

75. Wang, Su, Li, et al. Mechanisms and effects of curcumin on spatial learning and memory improvement in appswe/PS1dE9 mice. *Journal of Neuroscience Research* February 2014;**92**(2):218–31.

76. Hu, Maiti, Ma, et al. Clinical development of curcumin in neurodegenerative disease. *Expert Review of Neurotherapeutics* June 2015;**15**(6):629–37.

77. Yan, Zheng, Zhao. Effects of *Ginkgo biloba* extract EGb 761 on expression of RAGE and LRP-1 in cerebral microvascular endothelial cells under chronic hypoxia and hypoglycemia. *Acta Neuropathologica* November 2008;**116**(5):529–35.

78. Wang, Thomas, Zhong, et al. Consumption of grape seed extract prevents amyloid-beta deposition and attenuates inflammation in brain of an Alzheimer's disease mouse. *Neurotoxicity Research* January 2009;**15**(1):3–14.

79. Zeng, Wang, Zhou. Effects of (-)Epicatechin on the pathology of APP/PS1 transgenic mice. *Frontiers in Neurology* May 9, 2014;**5**:69.

80. Rezai-Zadeh, Arendash, Hou, et al. Green tea epigallocatechin-3-gallate (EGCG) reduces beta-amyloid mediated cognitive impairment and modulates tau pathology in Alzheimer transgenic mice. *Brain Research* June 12, 2008;**1214**:177–87.

81. Diomede, Rigacci, Romeo, et al. Oleuropein aglycone protects transgenic *C. elegans* strains expressing Aβ42 by reducing plaque load and motor deficit. *PLoS One* 2013;**8**(3):e58893.

82. Abuznait, Qosa, Busnena, et al. Olive-oil-derived oleocanthal enhances β-amyloid clearance as a potential neuroprotective mechanism against Alzheimer's disease: in vitro and in vivo studies. *ACS Chemical Neuroscience* June 19, 2013;**4**(6):973–82.

83. Qosa, Batarseh, Mohyeldin, et al. Oleocanthal enhances amyloid-β clearance from the brains of TgSwDI mice and in vitro across a human blood–brain barrier model. *ACS Chemical Neuroscience* September 16, 2015. [Epub ahead of print].

84. Qosa, Mohamed, Batarseh, et al. Extra-virgin olive oil attenuates amyloid-β and tau pathologies in the brains of TgSwDI mice. *The Journal of Nutritional Biochemistry* August 13, 2015. [Epub ahead of print].

85. Karuppagounder, Pinto, Xu, et al. Dietary supplementation with resveratrol reduces plaque pathology in a transgenic model of Alzheimer's disease. *Neurochemistry International* February 2009;**54**(2):111–8.

86. Li, Gong, Dong, Shi. Resveratrol, a neuroprotective supplement for Alzheimer's disease. *Current Pharmaceutical Design* 2012;**18**(1):27–33.

87. Ge, Qiao, Qi, et al. The binding of resveratrol to monomer and fibril amyloid beta. *Neurochemistry International* December 2012;**61**(7):1192–201.

CHAPTER 5

Treatment Mechanisms in Mild to Moderate Alzheimer's Disease

Contents

In this chapter we transition from a prevention-oriented focus to treatment of clinical (symptomatic) Alzheimer's disease (AD) which, for our purposes, begins with mild cognitive impairment (MCI) due to AD and proceeds to mild AD and moderate AD. While Chapters 3 and 4 focused on primary prevention and secondary prevention, this chapter will address how the plant polyphenols therapeutically interact with the hallmark

A Paradigm Shift to Prevent and Treat Alzheimer's Disease
ISBN 978-0-12-812259-4
http://dx.doi.org/10.1016/B978-0-12-812259-4.00005-9

neuropathological features of clinical AD, including synaptic dysfunction induced by neurotoxic soluble Abeta oligomers, tau pathology, aberrant neuroinflammation, and AD-related oxidative stress.

As with preclinical AD, the National Institute on Aging and the Alzheimer's Association charged a workgroup with the task of developing diagnostic criteria for "the symptomatic predementia phase of Alzheimer's disease." The result was the 2011 report, "The Diagnosis of Mild Cognitive Impairment Due to Alzheimer's Disease."[1] According to the report, the clinical criteria for diagnosing MCI are (1) "concern regarding a change in cognition," (2) "impairment in one or more cognitive domains," (3) "preservation of independence in functional abilities," and (4) "not demented." These are summarized below.

Concern regarding a change in condition. This MCI-related condition requires "evidence of concern about a change in cognition, in comparison with the person's previous level."

Impairment in one or more cognitive domains. This criterion requires the following conditions: "[E]vidence of lower performance in one or more cognitive domains that is greater than would be expected for the patient's age and educational background… This change can occur in a variety of cognitive domains, including memory, executive function, attention, language, and visuospatial skills. An impairment in episodic memory (i.e., the ability to learn and retain new information) is seen most commonly in MCI patients who subsequently progress to a diagnosis of AD dementia."

Preservation of independence in functional abilities. This was described as follows: "Persons with MCI commonly have mild problems performing complex functional tasks which they used to perform previously, such as paying bills, preparing a meal, or shopping. They may take more time, be less efficient, and make more errors at performing such activities than in the past. Nevertheless, they generally maintain their independence of function in daily life, with minimal aids or assistance."

Not demented. "These cognitive changes should be sufficiently mild that there is no evidence of a significant impairment in social or occupational functioning."[1]

According to the Mayo Clinic: "MCI is an intermediate stage between the expected cognitive decline of normal aging and the more serious decline of dementia. It can involve problems with memory, language, thinking and judgment that are greater than normal age-related changes. If you have mild cognitive impairment, you may be aware that your memory or mental function has 'slipped.' Your family and close friends also may notice a change. But generally these changes aren't severe enough to interfere with your day-to-day life and usual activities."[2]

The Mayo Clinic details MCI-related symptoms further as follows: "You forget things more often. You forget important events such as appointments or social engagements. You lose your train of thought or the thread of conversations, books or movies. You feel increasingly overwhelmed by making decisions, planning steps to accomplish a task or interpreting instructions. You start to have trouble finding your way around familiar environments. You become more impulsive or show increasingly poor judgment. Your family and friends notice any of these changes." Other symptoms may include "depression, irritability and aggression, anxiety, apathy."[3]

Key players in the pathogenesis of MCI appear to be soluble Abeta oligomers and neuronal synapses, which are "the tiny gaps between nerve cells across which neurotransmitters pass."[4] There is a good deal of evidence identifying the damaging effects that neurotoxic Abeta oligomers have on normal synaptic function. For example, in 1998 neuroscientists based at Northwestern University reported that "neuron damage leading to AD is instigated by small toxic oligomers of the Aβ peptide."[5] In *Proceedings of the National Academy of Sciences* that year, the same scientists reported: "We hypothesize that impaired synaptic plasticity and associated memory dysfunction during early stage Alzheimer's disease and severe cellular degeneration and dementia during end stage could be caused by the biphasic impact of Abeta-derived diffusible ligands [soluble Abeta oligomers] acting upon particular neural signal transduction pathways."[6] Large oligomers, dodecamers or larger, called Abeta*56 by Kathleen Zahs and Karen Ashe at the University of Minnesota, are also thought to be a major toxic version of Abeta.[7]

In 2002 *Science* published a review article titled, "Alzheimer's Disease is a Synaptic Failure," by Dennis Selkoe, who wrote that symptomatic AD "usually begins with a remarkably pure impairment of cognitive function." He continued:

> Patients with this devastating disorder of the limbic and association cortices lose their ability to encode new memories, first of trivial and then of important details of life. The insidious dissolution of the ability to learn new information evolves in an individual whose motor and sensory functions are very well preserved and who is otherwise neurologically intact. Over time, both declarative and nondeclarative memory become profoundly impaired, and the capabilities for reasoning, abstraction, and language slip away. But the subtlety and variability of the earliest amnestic symptoms, occurring in the absence of any other clinical signs of brain injury, suggest that something is discretely, perhaps intermittently, interrupting the function of synapses that help encode new declarative memories. A wealth of evidence now suggests that this "something" is the amyloid β protein, [including] a 42-residue hydrophobic peptide with an ominous tendency to assemble into long-lived oligomers and polymers.

Still writing in *Science*, in a section titled "Synapses as the Initial Target in Alzheimer's Disease," Selkoe also observed that these early symptoms "appear to correlate with dysfunction of cholinergic and glutamatergic synapses," and that "indicators suggest that AD represents, at least initially, an attack on synapses."[8]

In 2007, *Biochemical Society Transactions* reported: "There is growing evidence that mild cognitive impairment in early AD may be due to synaptic dysfunction caused by the accumulation of non-fibrillar, oligomeric Abeta, long before widespread synaptic loss and neurodegeneration occurs. Soluble Abeta oligomers can rapidly disrupt synaptic memory mechanisms at extremely low concentrations via stress-activated kinases and oxidative/nitrosative stress mediators."[9]

In 2008, in *Behavioural Brain Research*, Selkoe reported: "[D]uring the last two decades, it became apparent that the key challenge for understanding and ultimately treating AD was to focus not on what was killing neurons over the course of the disease but rather on what was interfering subtly and intermittently with episodic declarative memory well before widespread neurodegeneration had occurred. In other words, one wishes to understand the factors underlying early synaptic dysfunction in the hippocampus and then attempt to neutralize these as soon as feasible, perhaps even before a definitive diagnosis of AD can be made."[10]

Likewise, in 2008, *Brain* reported "long before the onset of clinical Alzheimer's disease non-fibrillar, soluble assembly states of amyloid-beta peptides are believed to cause cognitive problems by disrupting synaptic function in the absence of significant neurodegeneration."[11]

In 2009, *Reviews in the Neurosciences* reported that there is "ample evidence that Abeta oligomers do not affect neuronal viability in general, but interfere specifically with synaptic function" even in the absence of toxicity. It also reported: "Long-term neurophysiological impairment ultimately causes degeneration of synapses, which becomes most apparent on the morphological level by retraction of dendritic spines. Loss of meaningful synaptic connections in the brain of patients with AD will shatter their capacity to encode and retrieve memories."[12]

In 2013, William Klein, a neuroscientist at Northwestern University and among the group of scientists that promulgated the theory involving Abeta oligomers and early synaptic dysfunction, reported in the *Journal of Alzheimer's Disease* that the oligomer hypothesis of AD "was based on evidence that oligomers could exist free of amyloid fibrils, that fibril-free oligomer solutions rapidly inhibited long term potentiation, and that oligomers ultimately caused a highly selective nerve cell death." In short: "Fibrils no longer were the only toxins made by amyloid-β, and likely not the most important ones." Furthermore: "Oligomers provided a new basis for instigating AD. Since introduction of the hypothesis, more than 1500 articles on oligomers have been published… These and related findings from many groups have helped establish oligomers as central to the mechanism of AD pathogenesis. Comprising a ligand-based attack on specific synapses, the action of toxic oligomers gives a molecular basis to account for key features of AD neuropathology and to explain why early disease targets memory."[13]

Also in 2013, researchers at Columbia University reported in *Neuroscience*: "Alzheimer's disease is a highly prevalent neurodegenerative disorder characterized by a progressive loss of cognition and the presence of two hallmark lesions, senile plaques and neurofibrillary tangles, which result from the accumulation and deposition of the β-amyloid peptide (Aβ) and the aggregation of hyperphosphorylated tau protein, respectively. Initially, it was thought that Aβ fibrils, which make up senile plaques, were the root cause of the massive neurodegeneration usually found in AD brains. Over time, the longstanding emphasis on fibrillar Aβ deposits and neuronal death slowly gave way to a new paradigm involving soluble oligomeric forms of Aβ, which play a prominent role in triggering the cognitive deficits by specifically targeting synapses and disrupting synaptic signaling pathways."[14] In 2014, *Molecular Neurodegeneration* reported that "mounting evidence suggests that soluble oligomers of amyloid-β represent the pertinent synaptotoxic form of Aβ in sporadic [late-onset] Alzheimer's disease."[15]

Again, in 2014, *Current Pharmaceutical Design* reported: "In the last decade numerous studies have proposed small soluble aggregates of Aβ, known as oligomers, as the species responsible for synaptic dysfunction, memory loss and neurodegeneration typical of AD. In vitro and in vivo experiments have identified Aβ oligomers as the elements that can alter synaptic function by a reversible mechanism, which gradually becomes permanent when exposure is continuous."[16]

In 2015, *Frontiers in Chemistry* reported: "Amyloid-beta (Aβ) peptide oligomers are believed to be the causative agents of Alzheimer's disease. Though post-mortem examination shows that insoluble fibrils are deposited in the brains of AD patients in the form of intracellular (tangles) and extracellular (plaque) deposits, it has been observed that cognitive impairment is linked to synaptic dysfunction in the stages of the illness well before the appearance of these mature deposits. Increasing evidence suggests that the most toxic forms of Aβ are soluble low-oligomer ligands whose amounts better correlate with the extent of cognitive loss in patients than the amounts of fibrillar insoluble forms."[17]

Also in 2015, Kirsten Viola and William Klein reported in *Acta Neuropathologica*: "The AD-like cellular pathologies induced by Aβ oligomers suggest their impact provides a unifying mechanism for AD pathogenesis, explaining why early stage disease is specific for memory and accounting for major facets of AD neuropathology."[18]

The likely best way to address the disruptive effects of Abeta oligomers on synaptic function as a key transitional stage from preclinical AD to MCI would be to inhibit Abeta production in the first place among disease-free zero-stage populations by influencing the upregulation of alpha-secretase activity and the downregulation of beta-secretase activity as a primary-prevention measure (see Chapter 3). Another level of protection would be to inhibit Abeta aggregation from Abeta peptides to Abeta oligomers, while enhancing Abeta clearance from the brain (and into the blood stream) as a function of AD-related secondary prevention (see Chapter 4). A third level would be to directly protect synaptic function at the point of attack by Abeta oligomers, as indicated in the studies below, by deploying our now-familiar troupe of plant polyphenols with the objectives of protecting synaptic function and reversing synaptic damage, thereby potentially reversing MCI-related memory impairment while preventing progression to clinical AD.

CURCUMINOIDS (↑) SYNAPTIC FUNCTION

CURCUMIN (↑) SYNAPTIC FUNCTION
Neurobiology of Aging (United States) (2001)
Veterans Administration Greater Los Angeles Healthcare System, Geriatric Research Education and Clinical Center; Departments of Medicine and Neurology, University of California at Los Angeles.

Dietary curcumin (2000 ppm), but not ibuprofen, suppressed oxidative damage (isoprostane levels) and synaptophysin loss [in 22-month Sprague-Dawley rats]... In a second group of middle-aged female SD rats, **500 ppm dietary curcumin prevented Abeta-infusion induced spatial memory deficits in the Morris Water Maze and post-synaptic density (PSD)-95 loss and reduced Abeta deposits.**[19]

CURCUMIN (↑) SYNAPTIC FUNCTION
Annals of the New York Academy of Sciences (United States) (2004)
Greater Los Angeles Healthcare System, Veterans Administration Medical Center, North Hills, California.

At sustainable doses designed to mimic protective consumption in the epidemiology, ibuprofen reduces amyloid accumulation but suppresses a surprisingly limited subset of inflammatory markers in APPsw transgenic mice. **Curcumin lowered oxidative damage, cognitive deficits, synaptic marker loss, and amyloid deposition.**[20]

CURCUMIN, BISDEMETHOXYCURCUMIN, AND DEMETHOXYCURCUMIN (↑) SYNAPTIC FUNCTION
Neuroscience (United States) (2010)
Natural Products Research Unit, Department of Biological and Biomedical Sciences, The Aga Khan University Medical College, Karachi, Pakistan.

The objective of this study was to investigate the effects of curcuminoid mixture and individual constituents on spatial learning and memory in an amyloid-beta (Abeta) peptide-infused rat model of AD and on the expression of PSD-95, synaptophysin and camkIV... **These compounds salvaged PSD-95, synaptophysin and cam-kIV expression levels in the hippocampus in the rat AD model, which suggests multiple target sites with the potential of curcuminoids in spatial memory enhancing and disease modifying in AD.**[21]

CURCUMIN, BISDEMETHOXYCURCUMIN, AND DEMETHOXYCURCUMIN (↑) SYNAPTIC FUNCTION
Synapse (United States) (2011)
Natural Products Research Unit, Department of Biological and Biomedical Sciences, The Aga Khan University Medical College, Karachi, Pakistan.

Curcuminoids are vital constituent of turmeric, with therapeutic potential in the treatment of Alzheimer's disease. Electrically, stimulus train-elicited plastic changes in hippocampal CA1 excitability were used as an experimental paradigm to study the effects of curcuminoid mixture and individual components on functional failure induced by Aβ peptide in vitro. Electrical stimulation was applied on Schaffer collaterals, and population spikes (PS) were recorded from stratum pyramidale. To induce long-term potentiation (LTP) of PS, primed burst stimulation (PBs) was used… **These results showed that curcuminoids can restore susceptibility for plastic changes in CA1 excitability that is injured by exposure to Aβ peptide and rescue sinking PS LTP in Aβ-peptide-exposed hippocampal CA1 neurons.**[22]

Curcumin (↑) Synaptic Function
Toxicology in Vitro (England) (2013)

Reading School of Pharmacy, University of Reading, Whiteknights Campus, Reading, England; Laboratory of Neuroprotection and Cell Signalling, Department of Biochemistry, Federal University of Rio Grande do Sul, Porto Alegre, Brazil.

Increasing evidence demonstrates that beta-amyloid (Aβ) is toxic to synapses, resulting in the progressive dismantling of neuronal circuits. Counteract[ing] the synaptotoxic effects of Aβ could be particularly relevant for providing effective treatments for Alzheimer's disease (AD)… **Curcumin counteracted both deleterious effects of Aβ; the initial synaptic dysfunction and the later neuronal death. Curcumin-mediated attenuation of Aβ-induced synaptic dysfunction involved regulation of synaptic proteins. Our results expand the neuroprotective role of curcumin to a synaptic level.**[23]

Curcumin (↑) Synaptic Function
Neurological Research (England) (2013)

Federal University of Rio Grande do Sul, Porto Alegre, Brazil.

Organotypic hippocampal slice cultures were treated with curcumin and exposed to Abeta1–42 for 48 hours. **Synaptic dysfunction, cell death, ROS formation, neuroinflammation and beta-catenin, Akt, and GSK-3beta phosphorylation were measured to determine the effects of curcumin against Abeta toxicity. Curcumin significantly attenuated Abeta-induced cell death, loss of synaptophysin, and ROS generation.**[24]

Curcumin (↑) Synaptic Function
Advances in Experimental Medicine and Biology (United States) (2015)

Department of Neurology and Neurobiology of Aging, Kanazawa University Graduate School of Medical Sciences, Japan.

We investigated the effects of natural phenolic compounds, such as myricetin, rosmarinic acid, ferulic acid, curcumin, and nordihydroguaiaretic acid on the aggregation of amyloid β-protein (Aβ), using in vitro and in vivo models of cerebral Aβ amyloidosis. **The in vitro studies revealed that these phenolic compounds efficiently inhibit oligomerization as well as fibril formation of Aβ through differential binding, whilst reducing Aβ oligomer-induced synaptic and neuronal toxicity.** Furthermore, a transgenic mouse model fed orally with such phenolic compounds showed significant reduction of soluble Aβ oligomers as well as of insoluble Aβ deposition in the brain.[25]

Ginkgo biloba EXTRACTS (↑) SYNAPTIC FUNCTION

Ginkgo biloba EXTRACTS (GINKGOLIDES) (↑) SYNAPTIC FUNCTION
Molecular Neurodegeneration (England) (2008)
Department of Pathology and Infectious Diseases, Royal Veterinary College, North Mymms, Hertfordshire, England.
The early stages of Alzheimer's disease are closely associated with the production of the Abeta1–42 peptide, loss of synapses, and gradual cognitive decline... **These observations suggest that the ginkgolides are active components of *Ginkgo biloba* preparations and may protect against the synapse damage and the cognitive loss seen during the early stages of Alzheimer's disease.**[26]

Ginkgo biloba EXTRACTS (QUERCETIN AND BILOBALIDE) (↑) SYNAPTIC FUNCTION
Journal of Alzheimer's Disease (The Netherlands) (2009)
Department of Pharmaceutical Sciences, School of Pharmacy, University of Maryland, Baltimore.
Loss of synapses has been correlated with dementia in Alzheimer's disease as an early event during the disease progression. Hence, synaptogenesis and neurogenesis in adulthood could serve as a therapeutic target for the prevention and treatment of Alzheimer' disease... **Furthermore, both [*Ginkgo biloba* extract] constituents [quercetin and bilobalide] restored amyloid-beta oligomers-induced synaptic loss and phosphorylation of CREB. The present findings suggest that enhanced neurogenesis and synaptogenesis by bilobalide and quercetin may share a common final signaling pathway mediated by phosphorylation of CREB.**[27]

Ginkgo biloba EXTRACT 761 (↑) SYNAPTIC FUNCTION
International Psychogeriatrics (England) (2012)
Department of Pharmacology, Biocenter, Goethe University, Frankfurt, Germany.
Neuroplasticity, the ability of synapses to undergo structural adaptations in response to functional demand or dysfunctions, is increasingly impaired in aging and Alzheimer's disease. **EGb 761 has been shown in several preclinical reports to increase nearly all aspects of impaired neuroplasticity (long-term potentiation, spine density, neuritogenesis, neurogenesis).**[28]

Ginkgo biloba EXTRACT 761 (↑) SYNAPTIC FUNCTION
Brain, Behavior, and Immunity (United States) (2015)

Department of Neurology, University of the Saarland, Homburg/Saar, Germany; German Institute for Dementia Prevention (DIDP), University of the Saarland; The Institute of Neuroscience, Soochow University, Suzhou, China; Department of Pharmacy, Putuo People's Hospital, Shanghai, China; Dr. Willmar Schwabe GmbH & Co., Karlsruhe, Germany; Preclinical Research, Dr. Willmar Schwabe GmbH & Co.; Institute for Clinical and Experimental Surgery, University of the Saarland, Germany.

We gave APP-transgenic mice EGb 761 as a dietary supplement for 2 or 5 months. Plasma concentrations of EGb 761 components in mice were in the same range as such concentrations in humans taking EGb 761 at the recommended dose (240 mg daily). **Treatment with EGb 761 for 5 months significantly improved the cognitive function of the mice as measured by the Barnes Maze test. It also attenuated the loss of synaptic structure proteins, such as PSD-95 [post-synaptic density protein 95], Munc18-1 [mammalian uncoordinated 18-1 protein], and SNAP25 [synaptosomal–associated protein 25].**[29]

GREEN TEA EXTRACTS (↑) SYNAPTIC FUNCTION

GREEN TEA CATECHINS (↑) SYNAPTIC FUNCTION
Neuroscience (United States) (2009)

Department of Nutrition and Food Hygiene, School of Public Health, Peking University, Beijing, China.

[C]hronic 0.05% or 0.1% **green tea catechins consumption prevented the reductions of three representative proteins of synaptic function and synaptic structure, including brain-derived neurotrophic factor, post-synaptic density protein-95, and Ca(2+)/calmodulin-dependent protein kinase II. These results demonstrated that long-term 0.05% or 0.1% green tea catechin administration may prevent spatial learning and memory decline of SAMP8 mice by decreasing Abeta(1–42) oligomers and upregulating synaptic plasticity-related proteins in the hippocampus.**[30]

GREEN TEA EXTRACT (EGCG) AND RESVERATROL (↑) SYNAPTIC FUNCTION
Frontiers in Cellular Neuroscience (Switzerland) (2013)

Laboratory of Neuroendocrinology of Aging, Centre Hospitalier de l'Université de Montréal Research Center, Montréal, Canada; Department of Medicine, University of Montréal; Douglas Mental Health University Institute, McGill University, Montréal; Department of Psychiatry, McGill University, Montreal.

[I]mmunofluorescence data revealed that cells treated with these polyphenols **[EGCG and resveratrol] increased PKC gamma (γ) activation and** promoted neuronal interconnections. Finally, we found that the protective effects of both polyphenols [EGCG and resveratrol] on the cytoskeleton and synaptic plasticity were mediated by the PKCγ subunit.[31]

OLIVE OIL EXTRACTS (↑) SYNAPTIC FUNCTION

OLIVE OIL EXTRACT (OLEOCANTHAL) (↑) SYNAPTIC FUNCTION
Toxicology and Applied Pharmacology (United States) (2009)

Department of Neurobiology and Physiology, Northwestern University, Evanston, Illinois.
Phenolic compounds are of particular interest for their ability to disrupt Abeta oligomerization and reduce pathogenicity. This study has focused on oleocanthal, a naturally-occurring phenolic compound found in extra-virgin olive oil… **Treatment with oleocanthal improved antibody clearance of ADDLs [Abeta oligomers]. These results indicate oleocanthal is capable of altering the oligomerization state of ADDLs while protecting neurons from the synaptopathological effects of ADDLs and suggest oleocanthal as a lead compound for development in AD therapeutics.**[32]

OLIVE OIL/LEAF EXTRACT (OLEUROPEIN) (↑) SYNAPTIC FUNCTION
Neurobiology of Aging (United States) (2015)

Department of Neuroscience, Psychology, Drug Research and Child Health, Division of Pharmacology and Toxicology, University of Florence, Italy; Department of Experimental and Clinical Biomedical Sciences "Mario Serio," University of Florence, Italy; Department of Health Sciences, University of Florence, Italy; Department of NEUROFARBA, Newborn Screening, Clinical Chemistry and Pharmacology Lab, Meyer Children's University Hospital, Florence, Italy.
[T]he phenol [oleuropein] astonishingly activates neuronal autophagy even in mice at advanced stage of pathology, where it increases histone 3 and 4 acetylation, which matches both a decrease of histone deacetylase 2 expression and a significant improvement of synaptic function. The occurrence of these functional, epigenetic, and histopathologic beneficial effects even at a late stage of the pathology suggests that the phenol could be beneficial at the therapeutic, in addition to the prevention, level.[33]

Given the above research, it appears that synaptic dysfunction induced by soluble Abeta oligomers plays a key pathogenic role in the development of mild cognitive impairment due to AD. It is also apparent that certain standardized plant compounds—the curcuminoids, EGb 761, epigallocatechin-gallate (EGCG), oleocanthal and oleuropein—possess the ability to potentially reverse synaptic dysfunction and possibly prevent its associated memory impairment due to the neurotoxicity of soluble Abeta oligomers.

In their biography of Lois Alzheimer, published by Columbia University Press in 2003, Konrad and Ulrike Maurer began as follows:

> The whole world speaks of Alzheimer's, the incurable disease that afflicts so many older people. Yet Alois Alzheimer, after whom the disease is named, remains largely unknown. Alzheimer was an obsessed doctor and scientist. By day he calmly examined his patients and cared for them tenderly; deep into the night he sat at his microscope and studied slides of the brain that he had prepared himself. His contemporaries called him "the psychiatrist with a microscope" because he was convinced that mental illnesses were diseases of the brain, in stark contrast to the then-burgeoning approach of psychoanalysis, which traced psychological problems to traumatic childhood experiences. An unavoidable clash between the two sides took place at a conference in 1906. Alzheimer stood there as his contribution on the case of Auguste D. [Auguste Deter, the first known Alzheimer's disease patient] met with no interest; the minutes of the proceedings called it "inappropriate for a brief report."[34]

Auguste Deter died in April 1906. Shortly afterward, Alzheimer and his colleagues, Gaetano Perusini and Francisco Bonfiglio, examined Deter's brain, as the Maurers reported: "All three agreed that what was before them was a peculiar clinical picture. Anatomically it was characterized by an atrophy of the cerebral cortex, with cellular failure on a large scale and a distinct fibrillary disease of the nerve cells, dense growth of fibrous glia, and formation of numerous rod-like glial cells. To their surprise, they found deposits of a particular metabolic product in the form of plaques throughout the cerebral cortex, with signs of growth on the vessels."[35]

On describing "distinct fibrillary disease of the nerve cells" and "a particular metabolic product in the form of plaques throughout the cerebral cortex," the three scientists had identified, respectively, the intracellular tau neurofibrillary tangles and extracellular amyloid plaques that are still viewed today as the principal histopathological signposts of AD.

Toward the end of his slide presentation of the brain of Auguste Deter, and seemingly aware of the professional skepticism among the German psychiatrists in the lecture hall at Tübingen, Alzheimer closed his talk as follows: "Taken all in all, we clearly have a distinct disease process before us. Such processes have been discovered in great numbers in recent years. This observation suggests to us that we should not be content to locate any clinically unclear cases of illness in one of the familiar categories of disease known to us to save ourselves the effort of understanding them. There are undoubtedly far more mental illnesses than are listed in our textbooks."[36] As Alzheimer finished his presentation and stood before his silent audience of prominent psychiatrists, including Carl Jung, and as the Maurers reported, "No one answered to the chair's call for responses. Nor did a further request encourage anybody to pose a question."[37]

More than a 100 years later, and after identifying amyloid plaques as the first histopathological hallmark of AD, NIA analysts report that "the second hallmark of AD, also described by Dr. Alzheimer, is neurofibrillary tangles." The NIA booklet describes such tangles as follows: "Tangles are abnormal collections of twisted protein threads found inside nerve cells. The chief component of tangles is a protein called tau. Healthy neurons are internally supported in part by structures called microtubules, which help transport nutrients and other cellular components, such as neurotransmitter-containing vesicles, from the cell body down the axon. Tau, which usually has a certain number of phosphate molecules attached to it, binds to microtubules and appears to stabilize them. In AD, an abnormally large number of additional phosphate molecules attach to tau. As a result of this 'hyper-phosphorylation,' tau disengages from the microtubules and begins to come together with other tau threads. These tau threads form structures called paired helical filaments, which can become enmeshed with one another, forming tangles within the cell. The microtu-bules can disintegrate in the process, collapsing the neuron's internal transport network. This collapse damages the ability of neurons to communicate with each other."[4]

In a 2015 report of a major study published in *Brain* of 3600 postmortem human brains, *Science Daily* reported that "researchers at Mayo Clinic's campuses in Jacksonville, Florida and Rochester, Minnesota, have found that the progression of dysfunctional tau protein drives the cognitive decline and memory loss seen in Alzheimer's." The lead scientist in the study, Melissa Murray at the Mayo Clinic in Jacksonville, explained how hyperphosphorylated tau functions as a key mechanism in AD pathogenesis: "Tau can be compared to railroad ties that stabilize a train track that brain cells use to transport food, messages and other vital cargo throughout neurons. In Alzheimer's, changes in the tau protein cause the tracks to become unstable in neurons of the hippocampus, the center of memory. The abnormal tau builds up in neurons, which eventually leads to the death of these neurons. Evidence suggests that abnormal tau then spreads from cell to cell, disseminating pathological tau in the brain's cortex. The cortex is the outer part of the brain that is involved in higher levels of thinking, planning, behavior and attention—mirroring later behavioral changes in Alzheimer's patients."[39]

This study supported earlier reports on the role of hyperphosphorylated tau in AD and its relevance as a therapeutic target. For example, in "Hyperphosphorylation of Microtubule-Associated Protein Tau: A Promising Therapeutic Target for Alzheimer's Disease," *Current Medicinal Chemistry* reported in 2008 that "recent studies have sug-gested that abnormal hyperphosphorylation of tau in the brain plays a vital role in the molecular pathogenesis of AD and in neurodegeneration."[40]

In 2012, in "Tau as a Therapeutic Target in Neurodegenerative Disease," *Pharmacology and Therapeutics* reported: "Tau is a microtubule-associated protein thought to help modu-late the stability of neuronal microtubules. In tauopathies, including Alzheimer's disease and several frontotemporal dementias, tau is abnormally modified and misfolded resulting in its disassociation from microtubules and the generation of pathological lesions charac-teristic for each disease. A recent surge in the population of people with neurodegenerative tauopathies has highlighted the immense need for disease-modifying therapies for these conditions, and new attention has focused on tau as a potential target for intervention."[41]

In 2014, *Expert Opinion on Therapeutic Targets* reported: "Neurofibrillary pathology, which is made up from abnormally hyperphosphorylated microtubule-associated protein tau, is both a hallmark and key lesion of AD and related tauopathies… In our opinion, inhibition of abnormal hyperphosphorylation of tau is the most rational therapeutic target."[42]

In "Targeting Tau as a Treatment for Tauopathies," the Mayo Clinic reported in 2015 that "the accumulation of hyperphosphorylated tau causes the formation of neurofibrillary tangles, a pathological hallmark of tauopathies, a group of diseases which includes Alzheimer's disease," and that "therapeutics aimed at eliminating hyperphosphorylated tau are thus of considerable interest."[43] At the same time, many of the tau kinases contributing to the hyperphosphorylation of tau are critical for memory, such as GSK3, ERK, and CAMKII, and thus drugs that simply suppress activity while interfering with essential dynamic regulation are unlikely to be effective. Overall, therapeutically targeting tau hyperphosphorylation by correcting dysregulation of tau kinases without interfering with their dynamic role in memory is a potentially important strategy for treating clinical AD.

Downstream from the hyperphosphorylation of tau monomers in AD-related tau pathology is the formation of tau oligomers, which function as an additional pathogenic element. With respect to tau oligomers, scientists from the Department of Neurology at the University of Texas Medical Branch at Galveston reported in 2011 in *Current Alzheimer Research*: "The aggregation and accumulation of the microtubule-associated protein tau is a pathological hallmark of Alzheimer disease and many neurodegenerative diseases. For a long time, research has focused on neurofibrillary tangles (NFTs) and other large meta-stable inclusions composed of aggregated hyperphosphorylated tau protein. The correlation between these structures and disease progression produced conflicting results; moreover, the mechanism of their formation remains poorly understood. Lately, the significance and toxicity of NFTs have been challenged and a new aggregated tau entity has emerged as the true pathogenic species in tauopathies and a possible mediator of Aβ toxicity in AD; specifically, aggregates of a size intermediate between [tau] monomers and [tau] NFTs, the so-called tau oligomers."[44] In 2012, in *Biochemical Society Transactions*, scientists from the Department of Cell and Molecular Biology at Northwestern University reported "findings [that] support the hypothesis that tau oligomers are the toxic form of tau in neurodegenerative disease."[45]

Likewise, in 2012 scientists at the University of Southampton in England reported in *Biochemical Society Transactions*: "Recent findings from both Drosophila and rodent models of tauopathy suggest that large insoluble aggregates such as tau filaments and tangles may not be the key toxic species in these diseases. Thus, some investigators have shifted their focus to study pre-filament tau species such as tau oligomers and hyperphosphorylated tau monomers." However, these same researchers reported: "Interestingly, tau oligomers can exist in a variety of states including hyperphosphorylated and unphosphorylated forms, which can be both soluble and insoluble. It remains to be determined which of these oligomeric states of tau are causally involved in neurodegeneration and which signal the beginning of the formation of inert/protective filaments. It will be important to better understand this so that tau-based therapeutic interventions can target the most toxic tau species."[46]

In 2013, scientists at Arizona State University reported in the *International Journal of Cell Biology*: "In Alzheimer's disease, tau aggregates into fibrils and higher order neurofibrillary tangles, a key histopathological feature of AD. However, soluble oligomeric tau species may play a more critical role in AD progression since these tau species correlate better with neuronal loss and cognitive dysfunction. Recent studies show that extracellular oligomeric tau can inhibit memory formation and synaptic function and also transmit pathology to neighboring neurons. However, the specific forms of oligomeric tau involved in toxicity are still unknown."[47]

Given these reports, and while the neurotoxicity of tau oligomers appears to have been established, there is no certainty as of this writing about which structure of tau oligomers is neurotoxic and thus which tau oligomers to therapeutically target. However, in 2013 researchers at the Department of Neurochemistry at the New York State Institute for Basic Research in Developmental Disabilities reported in *Frontiers in Neurology*, in a study titled, "Hyperphosphorylation-Induced Tau Oligomers," that tau hyperphosphorylation precedes the formation of tau oligomers: "The normal brain tau interacts with tubulin and promotes its assembly into microtubules and stabilizes these fibrils. In Alzheimer disease, brain tau is three- to four-fold hyperphosphorylated. The abnormally hyperphosphorylated tau binds to normal tau instead of the tubulin and this binding leads to the formation of tau oligomers… Unlike Aβ and prion protein oligomers, tau oligomerization in AD and related tauopathies is hyperphosphorylation-dependent."[48] The same researchers also indicated in 2008 that "abnormally hyperphosphorylated tau" follows "normal tau" and that "inhibiting tau hyperphosphorylation" should function as an important tau-related target in AD. The researchers also reported: "Alternatively, a 'cocktail' strategy of multitarget therapy, targeting amyloid-beta, tau, acetylcholinesterase, inflammation, oxidative stress, and cognitive symptoms, could be more efficacious than monotherapy."[40]

Given the information presented above, we can identify three key tau-related targets in AD pathophysiology: (1) inhibition of tau hyperphosphorylation, (2) inhibition of the assembly of tau oligomers from phosphorylated tau monomers (peptides), and (3) enhanced clearance from the brain of tau peptides and oligomers. These targets can be discerned below amid the reported therapeutic impacts on tau by our group of plant polyphenols as reported in numerous studies.

CURCUMINOIDS (↓) TAU HYPERPHOSPHORYLATION

CURCUMIN (↓) TAU HYPERPHOSPHORYLATION
Food and Chemical Toxicology (England) (2008)
Environmental Toxico-Genomic and Proteomic Center, College of Medicine, Korea University, Seoul, South Korea.
Taken together, these data indicate that curcumin protected PC12 cells against Abeta-induced neurotoxicity through the inhibition of oxidative damage, intracellular calcium influx, and tau hyperphosphorylation.[49]

CURCUMIN AND DHA (↓) TAU HYPERPHOSPHORYLATION
The Journal of Neuroscience (United States) (2009)
Department of Medicine, University of California, Los Angeles.

Treatment of the 3xTg-AD mice on high-fat diet with fish oil or curcumin or a combination of both for 4 months reduced phosphorylated JNK, IRS-1, and tau and prevented the degradation of total IRS-1. This was accompanied by improvement in Y-maze performance. **Mice fed with fish oil and curcumin for 1 month had more significant effects on Y-maze, and the combination showed more significant inhibition of JNK, IRS-1, and tau phosphorylation.**[50]

CURCUMIN (↓) TAU HYPERPHOSPHORYLATION
Journal of Receptors and Signal Transduction Research (England) (2014)
Beijing Key Laboratory of Bioactive Substances and Functional Foods, Beijing Union University, China.

Increasing evidence suggests that Aβ induces tau hyperphosphorylation in AD pathology, but the signaling pathway is not completely understood. Inhibiting Aβ-induced cellular signaling is benefic[ial] to AD treatment. In this study, cellular signaling of tau phosphorylation induced by Aβ and the inhibiting effects of curcumin on this signaling were investigated on human neuroblastoma SH-SY5Y cells. Curcumin depresses Aβ-induced up-regulation of PTEN (phosphatase and tensin homolog) induced by Aβ. **These results imply that curcumin inhibits Aβ-induced tau hyperphosphorylation involving PTEN/Akt/GSK-3β pathway.**[51]

Ginkgo biloba EXTRACTS (↓) TAU HYPERPHOSPHORYLATION

Ginkgo biloba EXTRACT (GINKGOLIDE A) (↓) TAU HYPERPHOSPHORYLATION
Planta Medica (Germany) (2012)
School of Life Science, Anhui University, China.

The results showed that ginkgolide A could increase cell viability and suppress the phosphorylation level of tau in cell lysates, meanwhile, GSK3β was inhibited with phosphorylation at Ser9.[52]

Ginkgo biloba EXTRACT 761 (↓) TAU HYPERPHOSPHORYLATION
Food and Function (England) (2015)
Department of Neuroscience, Center for Neuroscience Research, Institute of Biomedical Science and Technology, Konkuk University School of Medicine, Seoul, South Korea.

Interestingly, **EGb761 treatment attenuated the zinc-induced tau hyperphosphorylation at Ser262 in a concentration-dependent manner while the**

antioxidant *N*-acetylcysteine showed a similar effect... Therefore, EGb 761 may be a candidate for the treatment of tauopathy present in neurological disorders such as Alzheimer's disease.[53]

GRAPE SEED EXTRACT (↓) TAU HYPERPHOSPHORYLATION

GRAPE SEED EXTRACT (↓) TAU HYPERPHOSPHORYLATION
Neurobiology of Aging (United States) (2012)
Department of Neurology, Mount Sinai School of Medicine, New York.
We found that GSPE [grape seed polyphenolic extract] treatment significantly reduced the number of motor neurons immuno-reactive for hyperphosphorylated and conformationally modified tau in the ventral horns of the spinal cord identified using AT100, PHF-1, AT8, and Alz50 tau antibodies... **Furthermore, the reduction of tau pathology was accompanied by an improvement in the motor function** assessed by a wire hang test. Collectively, our results suggest that GSPE can interfere with tau-mediated neurodegenerative mechanisms and ameliorate neurodegenerative phenotype in an animal model of tauopathy.[54]

GREEN TEA EXTRACTS (↓) TAU HYPERPHOSPHORYLATION

GREEN TEA EXTRACT (EGCG) (↓) TAU HYPERPHOSPHORYLATION
Brain Research (The Netherlands) (2008)
Rashid Laboratory for Developmental Neurobiology, Silver Child Development Center, Department of Psychiatry and Behavioral Medicine, and Department of Neurosurgery, University of South Florida, Tampa, Florida.
In the present study, we also investigated the effect EGCG administration had on tau pathology and cognition in Tg mice. Both i.p. [intraperitoneal injection] and orally-treated Tg animals were found to have modulated tau profiles, with markedly suppressed sarkosyl-soluble phosphorylated tau isoforms.[55]

RESVERATROL (↓) TAU HYPERPHOSPHORYLATION

RESVERATROL (↓) TAU HYPERPHOSPHORYLATION
Age (The Netherlands) (2013)
Unitat de Farmacologia i Farmacognòsia, Facultat de Farmàcia, Institut de Biomedicina (IBUB), Centros de Investigación Biomédica en Red de Enfermedades Neurodegenerativas (CIBERNED), Barcelona, Spain.
We found that long-term dietary resveratrol activates AMPK pathways and pro-survival routes such as SIRT1 in vivo. **[Resveratrol] also reduces cognitive**

impairment and has a neuroprotective role, decreasing the amyloid burden and reducing tau hyperphosphorylation.[56]

RESVERATROL (↓) TAU HYPERPHOSPHORYLATION
Age (The Netherlands) (2014)
Department of Pathophysiology, Key Laboratory of Neurological Diseases of Education Ministry of China, Tongji Medical College, Huazhong University of Science and Technology, Wuhan, China.
Resveratrol, a specific activator of SIRT1, reversed the streptozotocin-induced decrease in SIRT1 activity and reversed the increases in ERK1/2 phosphorylation, tau phosphorylation, and impairment of cognitive capability. SIRT1 protects hippocampus neurons from tau hyperphosphorylation and prevents cognitive impairment induced by streptozotocin brain insulin resistance with decreased hippocampus ERK1/2 activity.[57]

CURCUMINOIDS (↓) TAU AGGREGATION

CURCUMINOIDS (↓) TAU AGGREGATION
American Chemical Neuroscience (United States) (2011)
Department of Chemistry, The City University of New York at College of Staten Island, New York.
Curcumin inhibits amyloid-β and tau peptide aggregation at micromolar concentrations. The sugar-curcumin conjugate inhibits Aβ and tau peptide aggregation at concentrations as low as 8 nM and 0.1 nM, respectively.[58]

GRAPE SEED EXTRACT (↓) TAU AGGREGATION

GRAPE SEED EXTRACT (↓) TAU AGGREGATION
Journal of Alzheimer's Disease (The Netherlands) (2009)
Department of Psychiatry, Mount Sinai School of Medicine, New York.
We demonstrate that this GSPE [grape seed polyphenolic extract] is capable of inhibiting tau peptide aggregations, as well as dissociating preformed tau peptide aggregates. Results from this study suggest that this GSPE might provide beneficial disease-modifying bioactivities in tau-associated neurodegenerative disorders by modulating tau-mediated neuropathologic mechanisms.[59]

GRAPE SEED EXTRACT (↓) TAU AGGREGATION
Journal of Alzheimer's Disease (The Netherlands) (2010)
Department of Neurology, Mount Sinai School of Medicine, New York.
Tau is abnormally hyperphosphorylated in AD and aberrant tau phosphorylation contributes to the neuropathology of AD and other tauopathies. Anti-aggregation and anti-phosphorylation are main approaches for tau-based therapy. **In this study,**

we report that a select grape-seed polyphenol extract could potently interfere with the assembly of tau peptides into neurotoxic aggregates.[60]

GRAPE SEED EXTRACT (↓) TAU AGGREGATION
Neurobiology of Aging (United States) (2012)
Department of Neurology, Mount Sinai School of Medicine, New York.
Collectively, our results suggest that grape seed polyphenol extract has a significant potential for therapeutic development by neutralizing phospho-epitopes and disrupting [tau] fibrillary conformation leading to disintegration of paired [tau] helical filaments.[61]

GREEN TEA EXTRACTS (↓) TAU AGGREGATION

GREEN TEA EXTRACT (EGCG) (↓) TAU AGGREGATION
FEBS Letters (The Netherlands) (2015)
AstraZeneca-Tufts Lab for Basic and Translational Medicine, Boston, Massachusetts; Max Delbrück Center for Molecular Medicine, Berlin, Germany; Department of Neurology and Neurotherapeutics, University of Texas, Southwestern Medical Center, Dallas.
We generated a fragment of tau (His-K18ΔK280) that forms stable, toxic, oligomeric tau aggregates in vitro. We show that epigallocatechin gallate (EGCG), a green tea polyphenol that was previously found to reduce Aβ aggregation, inhibits the aggregation of tau K18ΔK280 into toxic oligomers at ten- to hundred-fold substoichiometric concentrations, thereby rescuing toxicity in neuronal model cells.[62]

OLIVE OIL EXTRACTS (↓) TAU AGGREGATION

OLIVE OIL EXTRACTS (OLEUROPEIN, OLEUROPEIN AGLYCONE, AND HYDROXYTYROSOL) (↓) TAU AGGREGATION
Neurochemistry International (England) (2011)
University of Lille Nord de France, Lille, France.
We report herein the ability of three natural phenolic derivatives obtained from olives and derived food products to prevent such tau fibrillization in vitro, namely hydroxytyrosol, oleuropein, and oleuropein aglycone... These findings might provide further experimental support for the beneficial nutritional properties of olives and olive oil as well as a chemical scaffold for the development of new drugs aiming at neurodegenerative tauopathies.[63]

OLIVE OIL EXTRACT (OLEOCANTHAL) (↓) TAU AGGREGATION
Food and Function (England) (2011)
Dipartimento di Scienze Farmaceutiche e Biomediche, Università degli Studi di Salerno, Via Ponte don Melillo, Fisciano, Italy.

Oleocanthal is a phenolic component of extra-virgin olive oil, recently supposed to be involved in the modulation of some human diseases, such as inflammation and Alzheimer. In particular, **oleocanthal has been shown to abrogate fibrillization of tau protein**, one of the main causes of Alzheimer neurodegeneration… **These data give new insights on the mechanism of inhibition of tau fibrillization mediated by oleocanthal.**[64]

OLIVE OIL EXTRACT (OLEOCANTHAL) (↓) TAU AGGREGATION
Journal of Natural Products (United States) (2012)
Dipartimento di Scienze Farmaceutiche e Biomediche, Università degli Studi di Salerno, Via Ponte don Melillo, Fisciano, Italy.
In the present study… oleocanthal has been found to interact with tau-441, inducing stable conformational modifications of the protein secondary structure and also interfering with tau aggregation. These findings provide experimental support for the potential reduced risk of AD and related neurodegenerative diseases associated with olive oil consumption and may offer a new chemical scaffold for the development of AD-modulating agents.[65]

CURCUMINOIDS (↑) TAU CLEARANCE

CURCUMIN (↑) SOLUBLE TAU AGGREGATE CLEARANCE
The Journal of Biological Chemistry (United States) (2013)
Department of Neurology, David Geffen School of Medicine at the University of California at Los Angeles; Geriatric, Research, and Clinical Center, Greater Los Angeles Veterans Affairs Healthcare System, Los Angeles.
In summary, curcumin reduced soluble tau and elevated HSPs [heat shock proteins] involved in tau clearance, showing that even after tangles have formed, tau-dependent behavioral and synaptic deficits can be corrected.[66]

CURCUMIN (↑) TAU CLEARANCE
Neuroscience Letters (Ireland) (2013)
Cellular and Biomolecular Laboratory, Department of Chemical Engineering and Material Science, Michigan State University, East Lansing.
The present results demonstrate novel activity of polyphenol curcumin in up-regulating an anti-tau cochaperone BAG2 [involved in clearance of tau from neurons] and thus suggest probable benefit of curcumin against AD-associated tauopathy.[67]

GREEN TEA EXTRACTS (↑) TAU CLEARANCE

GREEN TEA EXTRACT (EGCG) (↑) TAU CLEARANCE
Nutritional Neuroscience (England) (2015)

Neuroscience Graduate Program, University of Rochester, New York

This study aimed to explore a novel mechanism for enhancing the clearance of these pathological tau species using the green tea flavonoid epigallocatechin-3-gallate (EGCG)… Taken together, these results demonstrate that **EGCG has the ability to clear phosphorylated tau species in a highly specific manner**, likely through increasing adaptor protein expression.[68]

Although Alois Alzheimer would originally find and describe the Abeta plaques and tau neurofibrillary tangles in the brain of Auguste Deter in 1906, it would take a 100 years for scientists to begin to understand the relationship between the two major histopathological manifestations of AD. For example, in 2006 scientists in the Department of Neurobiology at the University of California at Irvine reported as follows in *Current Alzheimer Research*: "Since the initial description one hundred years ago by Dr. Alois Alzheimer, the disorder that bears his name has been characterized by the occurrence of two brain lesions: amyloid plaques and [tau] neurofibrillary tangles. Yet the precise relationship between beta-amyloid (Abeta) and tau, the two proteins that accumulate within these lesions, has proven elusive. Today, a growing body of work supports the notion that Abeta may directly or indirectly interact with tau to accelerate neurofibrillary tangle formation. Here we review recent evidence that Abeta can adversely affect distinct molecular and cellular pathways, thereby facilitating tau phosphorylation, aggregation, mislocalization, and accumulation."[69] Also in 2006, scientists at the University of Virginia at Charlottesville reported in *The Journal of Cell Biology* that "a seminal cell biological event in AD pathogenesis is acute, tau-dependent loss of microtubule integrity caused by exposure of neurons to readily diffusible Abeta."[70]

Evidence that tau-related AD pathology interacts with the neurotoxic effects of Abeta oligomers was published in 2001 in *Science*, in which researchers reported: "Injection of beta-amyloid Abeta42 fibrils into the brains of P301L mutant tau transgenic mice caused fivefold increases in the numbers of NFTs [neurofibrillary tangles] in cell bodies within the amygdala from where neurons project to the injection sites. Gallyas silver impregnation identified NFTs that contained tau phosphorylated at serine 212/ threonine 214 and serine 422. NFTs were composed of twisted filaments and occurred in 6-month-old mice as early as 18 days after Abeta42 injections. Our data support the hypothesis that Abeta42 fibrils can accelerate NFT formation in vivo."[71]

In the same 2001 issue of *Science*, another group of scientists drew a link between Abeta and tau as follows: "JNPL3 transgenic mice expressing a mutant tau protein, which develop neurofibrillary tangles and progressive motor disturbance, were crossed with Tg2576 transgenic mice expressing mutant beta-amyloid precursor protein (APP), thus modulating the APP-Abeta (beta-amyloid peptide) environment. The resulting double mutant (tau/APP) progeny and the Tg2576 parental strain developed Abeta deposits at the same age; however, relative to JNPL3 mice, the double mutants exhibited neurofibrillary tangle pathology that was substantially enhanced in the limbic system and olfactory cortex. These results indicate that either APP or Abeta influences the formation of neurofibrillary tangles. The interaction between Abeta and tau pathologies in these mice supports the hypothesis that a similar interaction occurs in Alzheimer's disease."[72]

In 2008, researchers at Northwestern University reported in *The Journal of Nutrition, Health and Aging* that "the attack on synapses" by "neurologically active Abeta oligomers" provides "a plausible mechanism unifying memory dysfunction with major features of AD neuropathology," and that "recent findings show that ADDL (Abeta oligomer) binding instigates synapse loss, oxidative damage, and AD-type tau hyperphosphorylation."[73] In 2010 scientists at the Department of Neurology at the University of Texas at Galveston reported in *Biochemistry* that "both Aβ and α-synuclein oligomers induce tau aggregation and the formation of β-sheet-rich neurotoxic tau oligomers."[74] These two reports thus provide evidence that Abeta oligomers induce both microtubule-destabilizing tau hyperphosphorylation and the subsequent aggregation of hyperphosphorylated tau peptides to form neurotoxic tau oligomers.

In his 2011 interview with *Science Watch*, "On the Amyloid Hypothesis of Alzheimer's Disease," Dennis Selkoe cited additional evidence of the influence of Abeta oligomers on tau pathology:

Tau is a subunit of the Alzheimer's neurofibrillary tangles. And [Lennart] Mucke [at the University of California at San Francisco] and his colleagues were able to show that if mice don't have tau, they still get the amyloid pathology of Alzheimer's, but they don't get all that much behavioral trouble. They have behavioral symptoms but much less. I thought that was really cool, and I would have enjoyed coming up with that myself.

We did follow up on it in our laboratory [at Brigham and Women's Hospital in Boston] with a paper that's almost in press at PNAS [Proceedings at the National Academy of Sciences]. I'm very excited about it. We said, "Let's reduce Mucke's discovery to an even simpler form." We took some primary cultured neurons from rats and put them in a dish and then put on top of them some soluble amyloid beta dimers that we isolated from the brains of Alzheimer's patients after they died. The idea was to see if those amyloid beta dimers are themselves necessary and sufficient to induce alterations of tau. And we showed it quite nicely. When we put on these amyloid beta dimers, even in exquisitely small amounts, they were very potent.

First we induced an increased phosphorylation of the tau protein in these healthy neurons. Then we saw the microtubule cytoskeleton begin to collapse, and then the nerve endings degenerate into what we call neuritic dystrophy. So, in a test tube, a culture dish, we can show how AD brain amyloid beta protein dimers directly induce these tau alterations, the abnormal phosphorylation of the tau protein, just about the same as happens in the Alzheimer's brain… I like to think that this experiment shows the bridge between the two classical lesions that Alzheimer first described in 1906—plaques and tangles.[75]

In the study published in the *Proceedings of the National Academy of Sciences* that Selkoe mentioned, he and his research colleagues reported in 2011: "Here, we isolated Aβ dimers, the most abundant form of soluble oligomer detectable in the human brain, from the cortices of typical AD subjects and found that at sub-nanomolar concentrations, they first induced hyperphosphorylation of tau at AD-relevant epitopes in hippocampal neurons and then disrupted the microtubule cytoskeleton and caused neuritic degeneration, all in the absence of amyloid fibrils. Application of pure, synthetic dimers confirmed the effects of the natural AD dimers, although the former was far less potent. Knocking down endogenous tau fully prevented the neuritic changes, whereas overexpressing human tau accelerated them. Co-administering Aβ N-terminal antibodies neutralized the cytoskeletal disruption." They concluded that "natural [Abeta] dimers isolated from the AD brain are sufficient to potently induce AD-type tau phosphorylation and then neuritic dystrophy."[76]

In 2012, scientists at the University of California at Irvine extended Selkoe's study to an in vivo experiment by inhibiting BACE activity in a mouse model of AD, as they reported in *The Journal of Neuroscience*: "Growing evidence suggests that soluble Aβ species can drive Alzheimer disease pathogenesis by inducing a cascade of events including tau hyperphosphorylation, proteasome impairment, and synaptic dysfunction. However, these studies have relied largely on in vitro approaches to examine the role of soluble Aβ in AD. In particular, it remains unknown whether soluble Aβ oligomers can facilitate the development of human wild-type tau pathology in vivo. To address this question, we developed a novel transgenic model that expresses low levels of APP with the Arctic familial AD mutation to enhance soluble Aβ oligomer formation in conjunction with wild-type human tau. Using a genetic approach, we show that reduction of β-site APP cleaving enzyme (BACE) in these ArcTau mice decreases soluble Aβ oligomers, rescues cognition, and, more importantly, reduces tau accumulation and phosphorylation." The researchers concluded that "these studies provide critical in vivo evidence for a strong mechanistic link between soluble Aβ, wild-type tau, and synaptic pathology."[77]

In 2014, in a multicenter study conducted in China, Sweden, and the United States and titled, "A Lifespan Observation of a Novel Mouse Model: In Vivo Evidence Supports Aβ Oligomer Hypothesis," the researchers reported in *PLoS One*: "Recently, Aβ oligomers have been identified as more neurotoxic than Aβ plaques. However, no ideal transgenic mouse model directly supports Aβ oligomers as a neurotoxic species due to the puzzling effects of amyloid plaques in the more widely-used models. Here, we constructed a single-mutant transgenic (Tg) model harboring the PS1V97L mutation and used non-Tg littermates as a control group." The researchers concluded that "following Aβ oligomers, we detected synaptic alteration, tau hyperphosphorylation and glial activation" and that "this model supports an initial role for Aβ oligomers in the onset of AD."[78]

In 2014, George Bloom, who directs the Alzheimer's lab at the University of Virginia at Charlottesville, in a review article published in *JAMA Neurology* titled "Amyloid-β and Tau: The Trigger and Bullet in Alzheimer's Disease Pathogenesis," observed: "Aβ is upstream of tau in AD pathogenesis and triggers the conversion of tau from a normal to a toxic state, but there is also evidence that toxic tau enhances Aβ toxicity via a feedback loop. Because soluble toxic aggregates of both Aβ and tau can self-propagate and spread throughout the brain by prion-like mechanisms, successful therapeutic intervention for AD would benefit from detecting these species before plaques, tangles, and cognitive impairment become evident and from interfering with the destructive biochemical pathways that they initiate."[79] Support for an Abeta-tau feedback loop was published a year later in *Neurobiology of Aging*, which reported that "Aβ drives the disease pathway through Tau, with eTau [extracellular tau] further increasing Aβ levels, perpetuating a destructive cycle."[80]

In 2000, a large number of scientists, the Neuroinflammation Working Group, in a seminal review of inflammation and AD published in *Neurobiology of Aging*, reported: "Inflammation clearly occurs in pathologically vulnerable regions of the Alzheimer's disease (AD) brain, and it does so with the full complexity of local peripheral inflammatory responses. In the periphery, degenerating tissue and the deposition of highly insoluble abnormal materials are classical stimulants of inflammation. Likewise, in the AD brain damaged neurons and neurites and highly insoluble amyloid beta peptide deposits and neurofibrillary tangles provide obvious stimuli for inflammation. Because these stimuli are discrete, microlocalized, and present from early preclinical to terminal stages of AD, local upregulation of complement, cytokines, acute phase reactants, and other inflammatory mediators is also discrete, microlocalized, and chronic. Cumulated over many years, direct and bystander damage from AD inflammatory mechanisms is likely to significantly exacerbate the very pathogenic processes that gave rise to it. Thus, animal models and clinical studies, although still in their infancy, strongly suggest that AD inflammation significantly contributes to AD pathogenesis."[81]

In a 2005 review, titled "Clinical Aspects of Inflammation in Alzheimer's Disease," the *International Review of Psychiatry* reported: "In Alzheimer's disease there is increasing evidence that neurotoxicity is mediated by CNS inflammatory processes. These processes involve activation of microglia by amyloid-beta leading to release of pro-inflammatory cytokines including IL-1beta, IL-6, and TNF-alpha among others. Neurotoxic processes mediated by these cytokines may include direct neuronal death by enhancement of apoptosis, decreased synaptic function as evidenced by inhibition of long-term potentiation, and inhibition of hippocampal neurogenesis. Central nervous system inflammation may predate the development of [beta-amyloid] senile plaques and [tau] neurofibrillary tangles in AD and may prove to be a more sensitive marker of prodromal AD."[82]

In 2010, scientists at the University of Arkansas, in "Neuroinflammatory Cytokines—The Common Thread in Alzheimer's Pathogenesis," wrote in *US Neurology*: "Based on the discovery of cytokine overexpression as an accompaniment to the dementia-related glial activation, the cytokine hypothesis was proposed. This states that in response to the negative impact on neurons of known and unknown risk factors—which include genetic inheritance, comorbid and environmental factors—microglia and astrocytes become activated and produce excess amounts of the immune-modulating cytokine interleukin-1 (IL-1) and the neuritogenic cytokine S100B, respectively. Finding that these glial events occur in fetuses and neonates with Down syndrome provided the first evidence that productive immune responses by activated glia precede rather than follow overt AD-related pathology. This finding can be added to the demonstration of IL-1 induction of amyloid β (Aβ) precursor protein and astrocyte activation with excess production of neuritogenic factor S100B. This combination suggests that IL-1 and S100B overexpression would favor the Aβ production and dystrophic neurite growth necessary for laying down neuritic Aβ plaques. This, together with demonstration of IL-1 induction of excessive production of the precursors of other features common in AD prompted a corollary to the cytokine hypothesis. The corollary states that regardless of the primary cause of the neuronal insult, the result will be chronic glial activation, which in turn will result in further neuronal injury, still more glial activation with excess cytokine expression and so on."[83]

Also in 2010, Patrick and Edith McGeer, neuroscientists at the University of British Columbia in Vancouver, in "Neuroinflammation in Alzheimer's Disease and Mild Cognitive Impairment: A Field in its Infancy," reported as follows in the *Journal of Alzheimer's Disease*: "Neuroinflammation is a prominent feature of Alzheimer disease (AD) and other chronic neurodegenerative disorders. It exacerbates the fundamental pathology by generating a plethora of inflammatory mediators and neurotoxic compounds. Inflammatory cytokines, complement components, and toxic free radicals are among the many species that are generated. Microglia attack the pathological entities and may inadvertently injure host neurons. Recent evidence indicates that microglia can be stimulated to assume an antiinflammatory state rather than a proinflammatory state which may have therapeutic potential. Proinflammatory cytokines include IL-1, IL-6 and TNF, while antiinflammatory cytokines include IL-4 and IL-10."[84]

In 2011, AD scientists reported in the *Journal of Immunology*: "Inflammation is a key pathological hallmark of Alzheimer's disease, although its impact on disease progression and neurodegeneration remains an area of active investigation. Among numerous inflammatory cytokines associated with AD, IL-1β in particular has been implicated in playing a pathogenic role."[85] Inflammation also drives tau kinase activation, thus increasing phosphorylated tau even in conditions of amyloid clearance.[86]

Another review on AD-related inflammation published in 2012 reported: "Neuropathological hallmarks [of Alzheimer's disease] are β-amyloid (Aβ) plaques and neurofibrillary tangles, but the inflammatory process has a fundamental role in the pathogenesis of AD. Inflammatory components related to AD neuroinflammation include brain cells such as microglia and astrocytes, the complement system, as well as cytokines and chemokines. Cytokines play a key role in inflammatory and anti-inflammatory processes in AD. An important factor in the onset of inflammatory process is the overexpression of interleukin (IL)-1, which produces many reactions in a vicious circle that cause dysfunction and neuronal death. Other important cytokines in neuroinflammation are IL-6 and tumor necrosis factor (TNF)-α... Proinflammatory responses may be countered through polyphenols. Supplementation of these natural compounds may provide a new therapeutic line of approach to this brain disorder."[87]

In 2015, Patrick and Edith McGeer and their colleague Moonhee Lee at the University of British Columbia reported in *Neurobiology of Aging*: "Neuroinflammation is hypothesized to be a major driving force behind Alzheimer's disease pathogenesis. This hypothesis predicts that activated microglial cells can stimulate neurons to produce excessive amounts of β-amyloid protein ($A\beta_{1-42}$) and tau. The excess $A\beta_{1-42}$ forms extracellular deposits which stimulate further microglial activation. The excess tau is partially released but also becomes phosphorylated forming intracellular neurofibrillary deposits. The end result is a positive feedback mechanism which drives the disease development... Our data indicate that reactive microglia play an important role in governing the expression of Aβ and tau, and therefore the progression of AD."[88]

Also in 2015, the original review published in 2000 by the Neuroinflammation Working Group was updated by more than 30 scientists from research centers worldwide, and published in *Lancet Neurology* in part as follows: "Increasing evidence suggests that Alzheimer's disease pathogenesis is not restricted to the neuronal compartment, but includes strong interactions with immunological mechanisms in the brain. Misfolded and aggregated proteins bind to pattern recognition receptors on microglia and astroglia, and trigger an innate immune response characterised by release of inflammatory mediators, which contribute to disease progression and severity. Genome-wide analysis suggests that several genes that increase the risk for sporadic Alzheimer's disease encode factors that regulate glial clearance of misfolded proteins and the inflammatory reaction. External factors, including systemic inflammation and obesity, are likely to interfere with immunological processes of the brain and further promote disease progression. Modulation of risk factors and targeting of these immune mechanisms could lead to future therapeutic or preventive strategies for Alzheimer's disease."[89]

Like neuroinflammation, oxidative stress plays a diverse role in AD pathogenesis. In 2000, researchers in the Department of Neurology at the University of Tokyo reported in *Biochemistry* that "several lines of evidence suggest that enhanced oxidative stress is involved in the pathogenesis and/or progression of Alzheimer's disease." A positive feedback loop likely exists whereby amyloid accumulation induces oxidative stress, which in turn increases amyloid production. This includes indications that "oxidative stress promotes intracellular accumulation of Abeta through enhancing the amyloidogenic pathway."[90] Furthermore, in 2001, scientists reported in *The American Journal of Pathology* that lipid peroxidation of DHA and arachidonic acid to neuroprostanes and isoprostanes is greatly increased in Alzheimer brains and closely corresponds to loss of synapses.[91]

In 2008, investigators at the University of Torino in Italy reported in the *Journal of Neurochemistry* that "oxidative stress stimulates BACE1 expression" and "BACE1 levels are increased in response to oxidative stress in normal cells."[92] Similarly, in 2009 scientists at the National Autonomous University of Mexico studied the effect of two "pro-oxidant molecules [hydrogen peroxide and iron chloride] on the levels of processing of human APP by alpha-, beta- and gamma-secretase" and reported evidence in *Neurochemistry International* that this effect influenced the undesirable "down-regulation of alpha-secretase and up-regulation of gamma-secretase and particularly of beta-secretase."[93] In 2015, scientists at the neurodegenerative disease section of the Department of Neurology at Massachusetts General Hospital and Harvard Medical School, in a study titled, "Oxidative Stress and Lipid Peroxidation Are Upstream of Amyloid Pathology," reported in *Neurobiology of Disease* that "the present study demonstrates a direct cause-and-effect correlation between oxidative stress and altered amyloid-β production, and provides a molecular mechanism by which [the] naturally occurring product of lipid peroxidation may trigger generation of toxic amyloid-β42 species."[94]

Also, it has long been known that amyloid induces oxidative stress that mediates neuron death.[95] In 2009, scientists in the Department of Chemistry at the University of Kentucky reported in the *Journal of Bioenergetics and Biomembranes* that Abeta oligomers are themselves "capable of inducing oxidative stress under both in vitro and in vivo conditions" and that this and other evidence reported by the scientists suggests "the possible role of Abeta in the initiation of the free-radical mediated process and consequently to the buildup of oxidative stress and AD pathogenesis."[96] And in 2011 scientists at the University of Torino reported in the *Journal of Alzheimer's Disease* that "expression of BACE1 is regulated by Aβ42."[97] Thus, in addition to the Abeta-tau and the neuroinflammation-based feedback systems noted above, the circular interactions between AD-related oxidative stress and Abeta production may function as yet another feedback process driving AD pathogenesis.

Perhaps unsurprisingly at this point, there is significant evidence that the plant polyphenols featured in this volume possess the capabilities to inhibit these dynamic processes involving neuroinflammation and AD-related oxidative stress, as indicated below.

CURCUMINOIDS (↓) NEUROINFLAMMATION

CURCUMIN (↓) NEUROINFLAMMATION
Journal of Alzheimer's Disease (The Netherlands) (2010)
Department of Neurology and Institute of Neurology, Ruijin Hospital, Shanghai Jiao Tong University School of Medicine, China.

Accumulating data indicate that astrocytes play an important role in the neuroinflammation related to the pathogenesis of AD. It has been shown that microglia and astrocytes are activated in AD brain and amyloid-beta can increase the expression of cyclooxygenase 2 (COX-2), interleukin-1, and interleukin-6. Suppressing the inflammatory response caused by activated astrocytes may help to inhibit the development of AD... **Curcumin has been shown to suppress activated astroglia in amyloid-beta protein precursor transgenic mice... These results show that curcumin might act as a**

PPARgamma [peroxisome proliferator-activated receptor gamma]–agonist to inhibit the inflammation in Abeta-treated astrocytes.[98]

CURCUMIN (↓) NEUROINFLAMMATION
Experimental Neurology (United States) (2014)
Department of Neurology, Johns Hopkins University School of Medicine, Baltimore, Maryland. Axon degeneration is a hallmark of several central nervous system (CNS) disorders, including multiple sclerosis (MS), Alzheimer's disease (AD), and Parkinson's disease (PD). Previous neuroprotective approaches have mainly focused on reversal or prevention of neuronal cell body degeneration or death. However, experimental evidence suggests that mechanisms of axon degeneration may differ from cell death mechanisms, and that therapeutic agents that protect cell bodies may not protect axons. Moreover, axon degeneration underlies neurologic disability and may, in some cases, represent an important initial step that leads to neuronal death… **Overall, our platform provides mechanistic insights into local axon degeneration, identifies curcumin as a novel axon protectant in the setting of neuroinflammation, and allows for ready screening of axon protective drugs.**[99]

CURCUMIN (↓) NEUROINFLAMMATION
Journal of Medicinal Food (United States) (2015)
Bioactive Botanical Research Laboratory, Department of Biomedical and Pharmaceutical Sciences, College of Pharmacy, University of Rhode Island, Kingston.
Inflammation and the presence of pro-inflammatory cytokines are associated with numerous chronic diseases such as type-2 diabetes mellitus, cardiovascular disease, Alzheimer's disease, and cancer. An overwhelming amount of data indicates that curcumin, a polyphenol obtained from the Indian spice turmeric, *Curcuma longa*, is a potential chemopreventive agent for treating certain cancers and other chronic inflammatory diseases. However, the low bioavailability of curcumin, partly due to its low solubility and stability in the digestive tract, limits its therapeutic applications. Recent studies have demonstrated increased bioavailability and health-promoting effects of a novel solid lipid particle formulation of curcumin (Curcumin SLCP, Longvida)… Taken together, these results show that in RAW 264.7 murine macrophages, **SLCPs [solid lipid curcumin particles] have improved solubility over unformulated curcumin, and significantly decrease the LPS-induced pro-inflammatory mediators NO, PGE2, and IL-6 by inhibiting the activation of NF-κB.**[100]

CURCUMIN (↓) NEUROINFLAMMATION
Neuroscience Letters (Ireland) (2015)
Guangdong Key Laboratory for Diagnosis and Treatment of Major Neurological Diseases, Department of Neurology, The First Affiliated Hospital, Sun Yat-sen University, Guangzhou, China; Department of Neurology, Zengcheng People's Hospital, Zengcheng, China.

Curcumin, the major yellow pigment in turmeric (*Curcuma longa*), is proposed for its anti-inflammatory properties… **These results show that curcumin suppresses ERK1/2 and p38 signaling, thus attenuating inflammatory responses of brain microglia.**[101]

GRAPE SEED EXTRACT (↓) NEUROINFLAMMATION

GRAPE SEED EXTRACT (↓) NEUROINFLAMMATION
Neurotoxicity Research (United States) (2009)
Department of Human Physiology and Centre for Neuroscience, Flinders University, Adelaide, Australia.
Conclusively, polyphenol-rich grape seed extract prevents the Abeta deposition and attenuates the inflammation in the brain of a transgenic mouse model, and thus is promising in delaying development of AD.[102]

GREEN TEA EXTRACTS (↓) NEUROINFLAMMATION

GREEN TEA EXTRACT (EGCG) (↓) NEUROINFLAMMATION
The Journal of Nutritional Biochemistry (United States) (2013)
College of Pharmacy, Chungbuk National University 12, Chungbuk, South Korea.
Neuroinflammation has been known to play a critical role in the pathogenesis of Alzheimer's disease (AD) through amyloidogenesis… **[T]his study shows that EGCG prevents memory impairment as well as amyloidogenesis via inhibition of neuroinflammatory-related cytokines released from astrocytes and suggests that EGCG might be a useful intervention for neuroinflammation-associated AD.**[103]

RESVERATROL (↓) NEUROINFLAMMATION

RESVERATROL (↓) NEUROINFLAMMATION
Journal of Neuroinflammation (England) (2007)
Neurochemistry Research Group, Department of Psychiatry, University of Freiburg Medical School, Germany.
Attenuation of microglial activation has been shown to confer protection against different types of brain injury… **These findings suggest that the naturally occurring polyphenol resveratrol is able to reduce microglial activation, an effect that might help to explain its neuroprotective effects in several in vivo models of brain injury.**[104]

RESVERATROL (↓) NEUROINFLAMMATION
European Journal of Pharmacology (The Netherlands) (2010)

Shanghai University of Traditional Chinese Medicine, China.
This mini-review summarized the anti-inflammatory activities of resveratrol in the brain from both in vivo and in vitro studies, and highlighted the inhibition of activated microglia as a potential mechanism of neuroprotection… **Taken together, microglia are an important target for anti-inflammatory activities of resveratrol in the brain.**[105]

RESVERATROL (↓) NEUROINFLAMMATION
Journal of Neuroinflammation (England) (2010)
Key Laboratory of Nutrition and Metabolism, Institute for Nutritional Sciences, Shanghai Institutes for Biological Sciences, Chinese Academy of Sciences, China.
We investigated the capacity of resveratrol to protect microglia and astrocytes from inflammatory insults and explored mechanisms underlying different inhibitory effects of resveratrol on microglia and astrocytes… **Resveratrol inhibited LPS-induced expression and release of TNF-alpha, IL-6, MCP-1, and iNOS/ NO in both cell types [microglia and astrocytes] with more potency in microglia, and inhibited LPS-induced expression of IL-1beta in microglia but not astrocytes.**[106]

RESVERATROL (↓) NEUROINFLAMMATION
Journal of Neurochemistry (England) (2012)
The Litwin-Zucker Research Center for the Study of Alzheimer's Disease, The Feinstein Institute for Medical Research, Manhasset, New York.
Activation of microglia, the resident macrophages of the brain, around the amyloid plaques is a key hallmark of Alzheimer's disease (AD). Recent evidence in mouse models indicates that microglia are required for the neurodegenerative process of AD. Amyloid-β (Aβ) peptides, the core components of the amyloid plaques, can trigger microglial activation… **Together this work provides strong evidence that resveratrol has in vitro and in vivo anti-inflammatory effects against Aβ-triggered microglial activation.**[107]

RESVERATROL (↓) NEUROINFLAMMATION
Phytotherapy Research (England) (2013)
Department of Pharmacology and Key Lab of Basic Pharmacology of Guizhou, Zunyi Medical College, China.
Resveratrol significantly inhibited LPS-induced microglial activation and subsequent production of multiple pro-inflammatory and cytotoxic factors such as tumor necrosis factor-α, nitric oxide, and IL-1β. Collectively, resveratrol produced neuroprotection against microglia-induced neurotoxicity. Thus, resveratrol might represent a potential benefit for the treatment of inflammation-related neurological disorders.[108]

RESVERATROL (↓) NEUROINFLAMMATION
Journal of Biomedical Nanotechnology (United States) (2013)
Programa de Pós-Graduação em Bioquímica, Universidade Federal do Rio Grande do Sul (UFRGS), Porto Alegre, Brazil.
Our findings suggest that modulation of neuroinflammation through a combination of resveratrol and a lipid-core nanocapsule-based delivery system might represent a promising approach for preventing or delaying the neurodegenerative process triggered by Abeta.[109]

CURCUMINOIDS (↓) OXIDATIVE STRESS

CURCUMIN, BISDEMETHOXYCURCUMIN, DEMETHOXYCURCUMIN (↓) OXIDATIVE STRESS
Neuroscience Letters (Ireland) (2001)
The Program for Collaborative Research in Pharmaceutical Science, College of Pharmacy, University of Illinois at Chicago.
Beta-Amyloid (betaA) induced oxidative stress is a well-established pathway of neuronal cell death in Alzheimer's disease. **From turmeric, *Curcuma longa* L. (Zingiberaceae), three curcuminoids, curcumin, demethoxycurcumin, and bisdemethoxycurcumin, were found to protect PC12 rat pheochromocytoma and normal human umbilical vein endothelial (HUVEC) cells from betaA(1-42) insult... The results suggest that these compounds [curcuminoids] may be protecting the cells from betaA(1-42) insult through antioxidant pathway.**[110]

CURCUMIN (↓) OXIDATIVE STRESS
Amino Acids (Austria) (2003)
Department of Chemistry, Section of Biochemistry and Molecular Biology, Faculty of Medicine, University of Catania, Italy.
Given the broad cytoprotective properties of the heat shock response, there is now strong interest in discovering and developing pharmacological agents capable of inducing the heat shock response... [C]urcumin, a powerful antioxidant derived from the curry spice turmeric, has emerged as a strong inducer of the heat shock response. **In light of this finding, curcumin supplementation has been recently considered as an alternative, nutritional approach to reduce oxidative damage and amyloid pathology associated with AD.**[111]

CURCUMIN (↓) OXIDATIVE STRESS
Journal of Alzheimer's Disease (The Netherlands) (2004)
Department of Medicine and Therapeutics, Chinese University of Hong Kong.
Because of its anti-oxidant and anti-inflammatory effects, [curcumin] was tested in animal models of Alzheimer's disease, reducing levels of amyloid and oxidized proteins and preventing cognitive deficits.[112]

CURCUMIN (↓) OXIDATIVE STRESS

Proceedings of the National Academy of Sciences (United States) (2006)

Nutrition and Metabolism Center, Children's Hospital Oakland Research Institute (CHORI), California.

The Abeta-heme complex is shown to be a peroxidase, which catalyzes the oxidation of serotonin and 3,4-dihydroxyphenylalanine by hydrogen peroxide. **Curcumin, which lowers oxidative damage in the brain in a mouse model for AD, inhibits this peroxidase.** The binding of Abeta to heme supports a unifying mechanism by which excessive Abeta induces heme deficiency, causes oxidative damage to macromolecules, and depletes specific neurotransmitters.[113]

CURCUMIN AND FERULIC ACID (↓) OXIDATIVE STRESS

Frontiers in Bioscience (United States) (2007)

Institute of Pharmacology, Catholic University School of Medicine, Rome, Italy.

Food supplementation with curcumin and ferulic acid is considered a nutritional approach to reduce oxidative damage and amyloid pathology in Alzheimer's disease.[114]

CURCUMIN AND GREEN TEA EXTRACT (EGCG) (↓) OXIDATIVE STRESS

Neuroreport (England) (2008)

Department of Neuroscience for Drug Discovery, Graduate School of Pharmaceutical Sciences, Kyoto University, Japan.

EGCG or curcumin significantly attenuated beta amyloid-induced radical oxygen species production and beta-sheet structure formation. These two compounds (EGCG and curcumin) have novel pharmacological effects that may be beneficial for Alzheimer's disease treatment.[115]

CURCUMIN (↓) OXIDATIVE STRESS

Behavioral Brain Research (The Netherlands) (2008)

Pharmacology Division, University Institute of Pharmaceutical Sciences, Panjab University, Chandigarh, India.

Aluminum is a potent neurotoxin and has been associated with Alzheimer's disease (AD) causality for decades… **Chronic administration of curcumin significantly improved memory retention and attenuated oxidative damage … in aluminum-treated rats. Curcumin has neuroprotective effects against aluminum-induced cognitive dysfunction and oxidative damage.**[116]

CURCUMIN (↓) OXIDATIVE STRESS

Pharmacological Research (England) (2010)

Division of Pharmacology, Central Drug Research Institute, Council of Scientific and Industrial Research, Lucknow, India.

[S]treptozotocin resulted in enhanced acetylcholinesterase activity in the hippocampus and cerebral cortex, which was normalized by curcumin pre- and post-treatment. **An increase in malondialdehyde level and a decrease in glutathione level were obtained in both hippocampus and cerebral cortex in the streptozotocin-treated group, indicating the state of oxidative stress, which also was attenuated by pre- and post-treatment of curcumin.**[117]

CURCUMIN, RESVERATROL, AND GREEN TEA EXTRACT (EGCG) (↓) OXIDATIVE STRESS
Journal of Neurochemistry (England) (2010)
Department of Agricultural Biotechnology, Research Institute for Agriculture and Life Sciences, Seoul National University, South Korea.
Certain naturally occurring dietary polyphenolic phytochemicals have received considerable recent attention as alternative candidates for AD therapy. In particular, **curcumin, resveratrol, and green tea catechins have been suggested to have the potential to prevent AD because of their anti-amyloidogenic, anti-oxidative, and anti-inflammatory properties.**[118]

CURCUMIN (↓) OXIDATIVE STRESS
Neurochemical Research (United States) (2012)
College of Life Science and Technology, Beijing University of Chemical Technology, China.
Curcumin depresses Aβ-induced up-regulation of neuronal oxidative stress [in rats]… Curcumin shows a more protective effect on neuronal oxidative damage when curcumin was added into cultured neurons not later than Aβ, especially prior to Aβ. The abilities of curcumin to scavenge free radicals and to inhibit the formation of β-sheeted aggregation are both beneficial to depress Aβ-induced oxidative damage. **Curcumin prevents neurons from Aβ-induced oxidative damage, implying the therapeutic usage for the treatment of Alzheimer's disease patients.**[119]

CURCUMIN (↓) OXIDATIVE STRESS
Journal of Alzheimer's Disease (The Netherlands) (2012)
Institut National de la Recherche Scientifique, Institut Armand-Frappier, Laval, Canada.
Taken together, these results suggest that Nps-Cur [a nanoparticulate formulation of curcumin] could be a promising drug delivery strategy to protect neurons against oxidative damage as observed in Alzheimer's disease.[120]

Ginkgo biloba EXTRACTS (↓) OXIDATIVE STRESS

Ginkgo biloba EXTRACT 761 (↓) OXIDATIVE STRESS
The Journal of Pharmacology and Experimental Therapeutics (United States) (2000)

Department of Pharmacology, Shanghai Institute of Materia Medica, Chinese Academy of Sciences.

Although clinical studies have demonstrated that EGb 761, a standard extract of *Ginkgo biloba*, was effective in mild-to-moderate dementia of the Alzheimer's disease patients, the mechanism underlying its neuroprotective effect remains unclear… **Our results provide the first direct evidence that [Ginkgo] bilobalide can protect neurons against oxidative stress.**[121]

Ginkgo biloba EXTRACT 761 (↓) OXIDATIVE STRESS
Journal of Neurochemistry (England) (2000)

Douglas Hospital Research Centre, Department of Psychiatry, McGill University, Verdun, Québec, Canada.

An excess of the free radical nitric oxide (NO) is viewed as a deleterious factor involved in various CNS disorders. Numerous studies have shown that the *Ginkgo biloba* extract EGb 761 is a NO scavenger with neuroprotective properties… **These data suggest that the protective and rescuing abilities of EGb 761 are not only attributable to the antioxidant properties of its flavonoid constituents but also via their ability to inhibit NO-stimulated PKC activity.**[122]

Ginkgo biloba EXTRACT 761 (↓) OXIDATIVE STRESS
Phytotherapy Research (England) (2001)

Programa de Pós-Graduação em Ciências Farmacêuticas, Universidade Federal do Rio Grande do Sul, Porto Alegre, Brazil.

The standardized extract of *Ginkgo biloba* (EGb 761) has been widely employed for its significant benefit in neurodegenerative disorders… **An increase in the catalase and superoxide dismutase activities in the hippocampus, striatum, and substantia nigra, and a decrease of the lipid peroxidation in the hippocampus [among rats treated with EGb 761] were observed.**[123]

Ginkgo biloba EXTRACT 761 (↓) OXIDATIVE STRESS
Journal of Alzheimer's Disease (The Netherlands) (2003)

Laboratory of Cellular and Molecular Neuroscience, Department of Biological Sciences, The University of Southern Mississippi, Hattiesburg.

A rise in levels of hydrogen peroxide was observed in both in vitro and in vivo AD-associated transgenic models expressing the Abeta peptide compared with the wild type controls. **Treatment of the cells or C. elegans with *Ginkgo biloba* extract EGb 761 significantly attenuated the basal as well as the induced levels of hydrogen peroxide-related reactive oxygen species.**[124]

Ginkgo biloba EXTRACT (BILOBALIDE) AND VITAMIN E (↓) OXIDATIVE STRESS
Journal of Neuropathology and Experimental Neurology (United States) (2006)
Department of Neurology/Alzheimer's Disease Research Laboratory, Massachusetts General Hospital, Charlestown, Massachusetts.
Both *Ginkgo biloba* extract and vitamin E reduced the oxidative stress resulting from senile [Abeta] plaques in vivo as monitored with intracranial imaging. Both treatments also lead to a progressive reversal of the structural changes in dystrophic neurites [axons and dendrites] associated with senile plaques. These results suggest a causal relationship between plaque-associated oxidative stress and neuritic alterations and demonstrate for the first time that the focal neurotoxicity associated with the senile plaques of AD is partially reversible with antioxidant therapies.[125]

GRAPE SEED EXTRACT (↓) OXIDATIVE STRESS

BLACK GRAPE SKIN EXTRACT (↓) OXIDATIVE STRESS
Life Sciences (The Netherlands) (2003)
Department of Biological Chemistry, Medical Chemistry and Molecular Biology, University of Catania, Italy.
The results demonstrate that black grape skin extract reduces the ROS production, protects the cellular membrane from oxidative damage, and consequently prevents DNA fragmentation. The experimental results suggest that this natural compound may be used to ameliorate the progression of pathology in Alzheimer disease therapy.[126]

GRAPE SEED EXTRACT (↓) OXIDATIVE STRESS
Annals of the New York Academy of Sciences (United States) (2004)
Research Institute of Pharmaceutical Sciences, College of Pharmacy, Seoul National University, South Korea.
There is compelling evidence supporting the notion that Abeta-induced cytotoxicity is mediated through the generation of reactive oxygen species (ROS)… **Oligonol [a grape seed extract product] attenuated Abeta-induced cytotoxicity, apoptotic features, intracellular ROS accumulation, and lipid peroxidation, and increased the cellular glutathione pool.** Moreover, Abeta transiently induced the activation of nuclear factor kappaB in PC12 cells, which was suppressed by pretreatment with Oligonol.[127]

GREEN TEA EXTRACTS (↓) OXIDATIVE STRESS

GREEN TEA EXTRACT (EGCG) (↓) OXIDATIVE STRESS
Life Sciences (The Netherlands) (2001)

Department of Psychiatry, Keimyung University School of Medicine, Taegu, South Korea.
Taken together, the results suggest that **EGCG has protective effects against betaA-induced neuronal apoptosis through scavenging reactive oxygen species, which may be beneficial for the prevention of Alzheimer's disease.**[128]

GREEN TEA EXTRACT (EGCG) (↓) OXIDATIVE STRESS
Archives of Pharmacal Research (South Korea) (2009)
College of Oriental Medicine, Daegu Haany University, Daegu, South Korea.
These results suggest that EGCG may have preventive and/or therapeutic potential in Alzheimer's disease patients by augmenting cellular antioxidant defense capacity and attenuating Abeta-mediated oxidative and/or nitrosative cell death.[129]

GREEN TEA EXTRACT (EGCG) (↓) OXIDATIVE STRESS
Chonnam Medical Journal (South Korea) (2014)
Department of Neurology, Chonnam National University Medical School, Gwangju, South Korea; Department of Pharmacology, Chonnam National University Medical School, Gwangju, South Korea.
Epicatechin, EGCG, luteolin, and myricetin showed more potent and persistent neuroprotective action than did the other compounds. **These results demonstrated that oxidative stress was involved in Aβ-induced neuronal death, and antioxidative flavonoid compounds, especially epicatechin, EGCG, luteolin, and myricetin, could inhibit neuronal death.**[130]

RESVERATROL (↓) OXIDATIVE STRESS

RESVERATROL (↓) OXIDATIVE STRESS
GERONTOLOGY (SWITZERLAND) (2003)
Psychiatric Clinic, University of Basel, Switzerland.
The present study aimed at elucidating the possible neuroprotective effects of resveratrol against Abeta-induced neurotoxicity. **Resveratrol maintains cell viability and exerts an anti-oxidative action by enhancing the intracellular free-radical scavenger glutathione.** Our findings suggest that red wine may be neuroprotective through the actions of resveratrol.[131]

RESVERATROL (↓) OXIDATIVE STRESS
Pharmacology (Switzerland) (2007)
Pharmacology Division, University Institute of Pharmaceutical Sciences, Panjab University, Chandigarh, India.

Alzheimer's disease is a complex and multifactorial neurodegenerative disease. Central administration of colchicine, a microtubule-disrupting agent, causes loss of cholinergic neurons and cognitive dysfunction that is associated with excessive free radical generation... **Results of the present study indicated that trans-resveratrol has a neuroprotective role against colchicine-induced cognitive impairment and associated oxidative stress.**[132]

RESVERATROL (↓) OXIDATIVE STRESS
Current Medicinal Chemistry (The Netherlands) (2008)
Department of Neurology, Leonard M. Miller School of Medicine, University of Miami, Florida.
Current research suggests resveratrol may enhance prognosis of neurological disorders, such as Parkinson's, Huntington's, Alzheimer's diseases and stroke. **The beneficial effects of resveratrol include: antioxidation, free radical scavenger, and modulation of neuronal energy homeostasis and glutamatergic receptors/ion channels.**[133]

RESVERATROL AND MELATONIN (↓) OXIDATIVE STRESS
Journal of Clinical Neurology (South Korea) (2010)
Department of Neurology, Center for Geriatric Neuroscience Research, Institute of Biomedical Science and Technology, School of Medicine, Konkuk University, Seoul, South Korea.
Recent studies have demonstrated that resveratrol reduces the incidence of age-related macular degeneration, Alzheimer's disease (AD), and stroke, while melatonin supplementation reduces the progression of the cognitive impairment in AD patients... **Together, our results suggest that co-administration of melatonin and resveratrol acts as an effective treatment for AD by attenuating Aβ1-42-induced oxidative stress and the AMPK-dependent pathway.**[134]

RESVERATROL AND MELATONIN (↓) OXIDATIVE STRESS
Journal of Pineal Research (England) (2011)
Departments of Neurology Pharmacology Rehabilitation, Center for Geriatric Neuroscience Research, Institute of Biomedical Science and Technology, School of Medicine, Konkuk University, Seoul, South Korea.
Recent studies have demonstrated that resveratrol reduced the incidence of Alzheimer's disease and stroke, while melatonin supplementation was found to reduce the progression of the cognitive impairment in AD... **These data suggest that melatonin potentiates the neuroprotective effect of resveratrol against oxidative injury by enhancing HO-1 [heme oxygenase-1] induction through inhibiting ubiquitination-dependent proteasome pathway, which may provide an effective means to treat neurodegenerative disorders.**[135]

RESVERATROL (↓) OXIDATIVE STRESS
Neural Regeneration Research (India) (2013)

Department of Neurology, the First Affiliated Hospital, Zhengzhou University, Zhengzhou, Henan Province, China; Department of Neurology, Henan Provincial People's Hospital, Zhengzhou, Henan Province, China.

Recently, resveratrol has been shown to exhibit neuroprotective effects in models of Parkinson's disease, cerebral ischemia, and Alzheimer's disease… **These results confirmed that the neuroprotective effects of resveratrol on vascular dementia were associated with its anti-oxidant properties.**[136]

RESVERATROL (↓) OXIDATIVE STRESS
Current Alzheimer Research (United Arab Emirates) (2015)
Department of Nutrition, Dietetics, and Hospitality Management, Auburn University, Alabama.
The present study evaluates the neuroprotective action of resveratrol on Aβ-induced oxidative stress and memory loss… **Resveratrol treatment attenuated the accumulation of lipid peroxide levels, up-regulated the antioxidant activities, and improved the expression of memory-associated proteins in Aβ treated H19-7 cells. These findings highlight the neuroprotective effect of resveratrol in preventing Aβ-induced oxidative damage and memory loss in vitro.**[137]

In summary, the multidimensionality of AD, featuring a disease process of multiple interactive elements and destructive cycles, would appear to present significant challenges to the mono-therapeutic "right drug, right target" approach, given the difficulty, for one thing, of identifying a "right target" amidst manifold dynamic targets. One such challenge includes the apparent incongruity between largely mono-action pharmaceuticals and the pathophysiological swarm orchestrated by amyloid- and tau-based actors amidst excessive neuroinflammation and oxidative stress. Furthermore, although oxidative stress and Abeta may turn each other on, it is not precisely clear which one initiates the interaction, and thus perhaps AD itself.

Given the attributes of the plant polyphenol pleiotropism that has been shown so far, questions about which particular pathogenic element to target become somewhat less pressing, given the totality of such elements that the polyphenols seem able to address. This is apparent in the tables presented below, which bundle the findings from the abstracts presented in Chapters 3 through 5, and thus conveniently exhibit the far-reaching therapeutic range of the plant polyphenols (see Table 5.1). Even simple two-polyphenol pairings may provide total coverage of the 12 pathogenic targets listed in the tables (see Tables 5.1A through 5.1H). It is doubtful that any pharma-based approach to the prevention and treatment of AD could duplicate these detailed portraits of therapeutic pleiotropism.

Tables 5.1A through 5.1H represent eight simple formulations of plant polyphenols, each formula being somewhat functionally comparable to the others with respect to their impact on the pathogenic categories listed in the tables. Thus, any one of the eight formulas, to one degree or another, would potentially (1) increase alpha-secretase activity while also generating the neuroprotective cleavage product, sAPPα, (2) decrease

Table 5.1 Plant polyphenol pleiotropism and Alzheimer's disease pathogenic hallmarks[a]

Plant Polyphenols and AD Hallmarks	Curcuminoids	Ginkgo 761	Grape Seed Extract	Green Tea (EGCG)	Olive Oil Oleuropein/ Oleocanthal	Resveratrol
(↑) Alpha-secretase/sAPPα	(5)	(2)		(8)	(1)	
(↓) Beta-secretase	(2)	(2)		(2)		
(↓) Gamma-secretase	(7)			(1)		
(↓) Abeta aggregation	(9)	(3)	(5)	(6)	(1)	
(↓) Abeta toxicity	(13)	(9)	(2)	(3)	(3)	(10)
(↑) Abeta clearance	(7)	(1)		(1)	(4)	(3)
(↑) Synaptic function	(3)	(4)		(2)	(2)	(1)
(↓) Tau hyperphosphorylation	(1)	(2)	(1)	(1)		(2)
(↓) Tau Aggregation	(2)		(3)	(1)	(3)	
(↑) Tau clearance				(1)		
(↓) Aberrant neuroinflammation	(4)		(1)	(1)		(6)
(↓) AD-related oxidative stress	(11)	(5)	(1)	(5)		(8)

sAPPα, soluble amyloid precursor protein alpha.

[a]Numbers in parentheses in this table and the tables that follow reflect a close accounting of the number of studies reported in this volume that support the therapeutic function indicated in the left-side column.

Table 5.1A Pleiotropism and Alzheimer's disease pathogenic hallmarks (curcuminoids/EGCG)

Curcuminoids/EGCG	Curcuminoids	EGCG	Combined
(↑) Alpha-secretase/sAPPα		(8)	(8)
(↓) Beta-secretase	(5)	(2)	(7)
(↓) Gamma-secretase	(2)	(1)	(3)
(↓) Abeta aggregation	(7)	(6)	(13)
(↓) Abeta toxicity	(9)	(3)	(12)
(↑) Abeta clearance	(13)	(1)	(14)
(↑) Synaptic function	(7)	(2)	(9)
(↓) Tau hyperphosphorylation	(3)	(1)	(4)
(↓) Tau aggregation	(1)	(1)	(2)
(↑) Tau clearance	(2)	(1)	(3)
(↓) Aberrant neuroinflammation	(4)	(1)	(5)
(↓) AD-related oxidative stress	(11)	(5)	(16)

sAPPα, soluble amyloid precursor protein alpha.

beta-secretase activity, (3) decrease gamma-secretase activity, (4) decrease Abeta aggregation, (5) reduce Abeta toxicity, (6) enhance Abeta clearance, (7) protect and repair synaptic function, (8) reduce tau hyperphosphorylation, (9) reduce tau aggregation, (10) increase tau clearance, (11) reduce aberrant neuroinflammation, and (12) reduce AD-related oxidative stress.

Table 5.1B Pleiotropism and Alzheimer's disease pathogenic hallmarks (curcuminoids/*Ginkgo biloba* extract 761)

Curcuminoids/EGb 761	Curcuminoids	EGb 761	Combined
(↑) Alpha-secretase/sAPPα		(2)	(2)
(↓) Beta-secretase	(5)	(2)	(7)
(↓) Gamma-secretase	(2)		(2)
(↓) Abeta aggregation	(7)	(3)	(10)
(↓) Abeta toxicity	(9)	(9)	(18)
(↑) Abeta clearance	(13)	(1)	(14)
(↑) Synaptic function	(7)	(4)	(11)
(↓) Tau hyperphosphorylation	(3)	(2)	(5)
(↓) Tau aggregation	(1)		(1)
(↑) Tau clearance	(2)		(2)
(↓) Aberrant neuroinflammation	(4)		(4)
(↓) AD-related oxidative stress	(11)	(5)	(16)

sAPPα, soluble amyloid precursor protein alpha.

These disease-wide processes—proceeding variously as linear, in parallel, or circular phenomena—would thus be degraded or disrupted by any given pair of plant polyphenols. This gauntlet of anti-AD effects at key junctures of AD pathogenesis represents a potentially powerful approach to the prevention and treatment of AD.

In general, the broad therapeutic coverage of the nontoxic plant polyphenols featured in Tables 5.1A–5.1H—which depict in vitro and in vivo evidence that has been published

Table 5.1C Pleiotropism and Alzheimer's disease pathogenic hallmarks [curcuminoids/olive oil extracts (oleuropein and oleocanthal)]

Curcuminoids/Olive Oil Extracts	Curcuminoids	Olive Oil Extracts	Combined
(↑) Alpha-secretase/sAPPα		(1)	(1)
(↓) Beta-secretase	(5)		(5)
(↓) Gamma-secretase	(2)		(2)
(↓) Abeta aggregation	(7)	(1)	(8)
(↓) Abeta toxicity	(9)	(3)	(12)
(↑) Abeta clearance	(13)	(4)	(17)
(↑) Synaptic function	(7)	(2)	(9)
(↓) Tau hyper-phosphorylation	(3)		(3)
(↓) Tau aggregation	(1)	(3)	(4)
(↑) Tau clearance	(2)		(2)
(↓) Aberrant neuroinflammation	(4)		(4)
(↓) AD-related oxidative stress	(11)		(11)

sAPPα, soluble amyloid precursor protein alpha.

Table 5.1D Pleiotropism and Alzheimer's disease pathogenic hallmarks (EGCG/*Ginkgo biloba* extract 761)

EGCG/EGb 761	EGCG	EGb 761	Combined
(↑) Alpha-secretase/sAPPα	(8)	(2)	(10)
(↓) Beta-secretase	(2)	(2)	(4)
(↓) Gamma-secretase	(1)		(1)
(↓) Abeta aggregation	(6)	(3)	(9)
(↓) Abeta toxicity	(3)	(9)	(12)
(↑) Abeta clearance	(1)	(1)	(2)
(↑) Synaptic function	(2)	(4)	(6)
(↓) Tau hyperphosphorylation	(1)	(2)	(3)
(↓) Tau aggregation	(1)		(1)
(↑) Tau clearance	(1)		(1)
(↓) Aberrant neuroinflammation	(1)		(1)
(↓) AD-related oxidative stress	(5)	(5)	(10)

sAPPα, soluble amyloid precursor protein alpha.

worldwide in established science and medical journals with results repeated in many studies over several years—suggest a low-risk presumption of safety and efficacy, pending more human trials. At a minimum, the published evidence on plant polyphenols and AD seems more promising than the monotargeted, pharma-based products that have emerged to date for the primary and secondary prevention of AD and the treatment of clinical AD.

Table 5.1E Pleiotropism and Alzheimer's disease pathogenic hallmarks (EGCG/grape seed extract)

EGCG/Grape Seed Extract	EGCG	Grape Seed	Combined
(↑) **Alpha-secretase/sAPPα**	(8)		(8)
(↓) **Beta-secretase**	(2)		(2)
(↓) **Gamma-secretase**	(1)		(1)
(↓) **Abeta aggregation**	(6)	(5)	(11)
(↓) **Abeta toxicity**	(3)		(3)
(↑) **Abeta clearance**	(1)	(2)	(3)
(↑) **Synaptic function**	(2)		(2)
(↓) **Tau hyper-phosphorylation**	(1)	(1)	(2)
(↓) **Tau aggregation**	(1)	(3)	(4)
(↑) **Tau clearance**	(1)		(1)
(↓) **Aberrant neuroinflammation**	(1)	(1)	(2)
(↓) **AD-related oxidative stress**	(5)	(1)	(6)

sAPPα, soluble amyloid precursor protein alpha.

Before concluding this chapter, there is one additional prevention- and treatment-related issue that needs to be addressed. In a publication current as of December 2015 titled, "Alzheimer's Disease Genetics Fact Sheet," the National Institute on Aging observed that "there are two types of Alzheimer's—early-onset and late-onset," and that "both types have a genetic component." In early onset AD, the "disease occurs in people age 30

Table 5.1F Pleiotropism and Alzheimer's disease pathogenic hallmarks [EGCG/olive oil extracts (oleuropein and oleocanthal)]

EGCG/Olive Oil Extracts	EGCG	Olive Oil Extracts	Combined
(↑) Alpha-secretase/sAPPα	(8)	(1)	(9)
(↓) Beta-secretase	(2)		(2)
(↓) Gamma-secretase	(1)		(1)
(↓) Abeta aggregation	(6)	(1)	(7)
(↓) Abeta toxicity	(3)	(3)	(6)
(↑) Abeta clearance	(1)	(4)	(5)
(↑) Synaptic function	(2)	(2)	(4)
(↓) Tau hyper-phosphorylation	(1)		(1)
(↓) Tau aggregation	(1)	(3)	(4)
(↑) Tau clearance	(1)		(1)
(↓) Aberrant neuroinflammation	(1)		(1)
(↓) AD-related oxidative stress	(5)		(5)

sAPPα, soluble amyloid precursor protein alpha.

to 60 and represents less than 5 percent of all people with Alzheimer's." In most cases, early onset AD is "caused by an inherited change in one of three genes, resulting in a type known as early-onset familial Alzheimer's disease, or FAD." The fact sheet reported: "A child whose biological mother or father carries a genetic mutation for early-onset

Table 5.1G Pleiotropism and Alzheimer's disease pathogenic hallmarks (EGCG/resveratrol)

EGCG/Resveratrol	EGCG	Resveratrol	Combined
(↑) Alpha-secretase/sAPPα	(8)		(8)
(↓) Beta-secretase	(2)		(2)
(↓) Gamma-secretase	(1)		(1)
(↓) Abeta aggregation	(6)		(6)
(↓) Abeta toxicity	(3)	(10)	(13)
(↑) Abeta clearance	(1)	(3)	(4)
(↑) Synaptic function	(2)	(1)	(3)
(↓) Tau hyper-phosphorylation	(1)	(2)	(3)
(↓) Tau aggregation	(1)		(1)
(↑) Tau clearance	(1)		(1)
(↓) Aberrant neuroinflammation	(1)	(6)	(7)
(↓) AD-related oxidative stress	(5)	(8)	(13)

sAPPα, soluble amyloid precursor protein alpha.

FAD has a 50/50 chance of inheriting that mutation. If the mutation is in fact inherited, the child has a very strong probability of developing early-onset FAD." The NIA reported further: "Early-onset FAD is caused by any one of a number of different single-gene mutations on chromosomes 21, 14, and 1. Each of these mutations causes abnormal proteins to be formed. Mutations on chromosome 21 cause the formation of abnormal amyloid precursor protein (APP). A mutation on chromosome 14 causes abnormal

Table 5.1H Alzheimer's disease-related plant polyphenol pleiotropism (curcuminoids/EGCG/resveratrol)

Curcuminoids, EGCG, Resveratrol	Curcuminoids	EGCG	Resveratrol	Combined
(↑) Alpha–secretase, sAPPα		(8)		(8)
(↓) Beta–secretase	(5)	(2)		(7)
(↓) Gamma–secretase	(2)	(1)		(3)
(↓) Abeta aggregation	(7)	(6)		(13)
(↓) Abeta toxicity	(9)	(3)	(10)	(22)
(↑) Abeta clearance	(13)	(1)	(3)	(17)
(↑) Synaptic function	(7)	(2)	(1)	(10)
(↓) Tau hyper-phosphorylation	(3)	(1)	(2)	(6)
(↓) Tau aggregation	(1)	(1)		(2)
(↑) Tau clearance	(2)	(1)		(3)
(↓) Aberrant neuroinflammation	(4)	(1)	(6)	(11)
(↓) AD-related oxidative stress	(11)	(5)	(8)	(24)

sAPPα, soluble amyloid precursor protein alpha.

presenilin 1 to be made, and a mutation on chromosome 1 leads to abnormal presenilin 2. Each of these mutations plays a role in the breakdown of APP, a protein whose precise function is not yet fully understood. This breakdown is part of a process that generates harmful forms of amyloid plaques, a hallmark of the disease."[138]

In 1993, it was discovered that apolipoprotein E epsilon 4 allele, which is often expressed more economically as APOE ε4, or simply APOE4, is the gene associated with late-onset AD.[139] According to the NIA, APOE4 "increases the risk of [late-onset] Alzheimer's" and that "most people with Alzheimer's have the late-onset form of the disease, in which symptoms become apparent in the mid-60s and later." The NIA reports further that "APOE ε4 is called a risk-factor gene because it increases a person's risk of developing the disease," but "does not mean that a person will definitely develop Alzheimer's." Some people "with an APOE ε4 allele never get the disease, and others who develop Alzheimer's do not have any APOE ε4 alleles."[138]

In the mid-1990s, researchers in the Department of Neurology at the University of Massachusetts Medical Center in Worcester reported in *Archives of Neurology* that "even with an APOE epsilon 4 allele, the lifetime risk [of acquiring AD] remains below 30%."[140] It is well established that APOE4 increases amyloid deposition, but recent work indicates that it dysregulates inflammation even in the absence of amyloid.[141] A 2016 study showed that this is possible by affecting a master regulator of brain inflammation called miRNA146.[142] It remains controversial why female carriers of APOE4 are twice as much at risk of AD as male carriers.

In 2013, a study sponsored by the Buck Institute for Research on Aging, and published in the *Proceedings of the National Academy of Sciences*, reported "insight for new [Alzheimer's disease] therapeutics that target the interaction between ApoE4 and a Sirtuin protein" called Sirt1. What this means is explained briefly below, first with brief background on the Sirt1 gene, then back to the published study on APOE4 and Sirt1 from the Buck Institute.

In an article in *Scientific American* in 2006, David A. Sinclair of Harvard University and Leonard Guarente of MIT described the health- and longevity-related implications of their discovery of a family of longevity genes, featuring Sirt1, in mammals, including humans:

You can assume quite a bit about the state of a used car just from its mileage and model year. The wear and tear of heavy driving and the passage of time will have taken an inevitable toll. The same appears to be true of aging in people, but the analogy is flawed because of a crucial difference between inanimate machines and living creatures: deterioration is not inexorable in biological systems, which can respond to their environments and use their own energy to defend and repair themselves.

At one time, scientists believed aging to be not just deterioration but an active continuation of an organism's genetically programmed development. Once an individual achieved maturity, "aging genes" began to direct its progress toward the grave. This idea has been discredited, and conventional wisdom now holds that aging really is just wearing out over time because the body's normal maintenance and repair mechanisms simply wane. Evolutionary natural selection, the logic goes, has no reason to keep them working once an organism has passed its reproductive age.

Yet we and other researchers have found that a family of genes involved in an organism's ability to withstand a stressful environment, such as excessive heat or scarcity of food or water, have the power to keep its natural defense and repair activities going strong regardless of age. By optimizing the body's functioning for survival, these genes maximize the individual's chances of getting through the crisis. And if they remain activated long enough, they can also dramatically enhance the organism's health and extend its life span. In essence, they represent the opposite of aging genes—longevity genes.

While describing ways in which Sirt1 works to enhance longevity in mammals, Sinclair and Guarente included these examples:

Increased Sirt1 in mice and rats, for example, allows some of the animals' cells to survive in the face of stress that would normally trigger their programmed suicide. Sirt1 does this by regulating the activity of several other key cellular proteins, such as p53, FoxO and Ku70, that are involved either in setting a threshold for apoptosis or in prompting cell repair. Sirt1 thus enhances cellular repair mechanisms while buying time for them to work...

A striking example of Sirt1's ability to foster survival in mammalian cells can be seen in the Wallerian mutant strain of mouse. In these mice, a single gene is duplicated, and the mutation renders their neurons highly resistant to stress, which protects them against stroke, chemotherapy-induced toxicity and neurodegenerative diseases.

In 2004 Jeffrey D. Milbrandt of Washington University in St. Louis and his colleagues showed that the Wallerian gene mutation in these mice increases the activity of an enzyme that makes NAD [nicotinamide adenine dinucleotide], and the additional NAD appears to protect the neurons by activating Sirt1. Moreover, Milbrandt's group found that STACs [Sirtuin-activating compounds] such as resveratrol conferred a protective effect on the neurons of normal mice similar to the Wallerian mutation.

In a more recent study by Christian Nri of the French National Institute of Health and Medical Research, resveratrol and another STAC, fisetin, were shown to prevent nerve cells from dying in two different animal models (worm and mouse) of human Huntington's disease. In both cases, the protection by STACs required Sirtuin gene activity.[143]

Keeping this brief summary of the significance of Sirt1 in mind, and what we know to date about AD-related pathogenesis, the study by the Buck Institute reported that APOE4, the gene that increases the risk of late-onset AD, "significantly (i) reduces the ratio of soluble amyloid precursor protein alpha (sAPPα) to Aβ; (ii) reduces Sirtuin T1 (SirT1) expression, resulting in markedly differing ratios of neuroprotective SirT1 to neurotoxic SirT2; (iii) triggers Tau phosphorylation and APP phosphorylation; and (iv) induces programmed cell death."[144]

Thus, the APOE4 gene increases the risk of late-onset AD via processes that the plant polyphenols featured in this volume may potentially offset, including any APOE4-induced reduction of the alpha-secretase cleavage product, sAPPα, and possibly alpha-secretase/ADAM10 processing of APP, as well as inhibiting APOE4-induced triggering of tau hyperphosphorylation. This leaves the APOE4-induced reduction of Sirt1 expression as an issue that has not been addressed to this point.

Pursuant to publication of its study on APOE4 and Sirt1, the Buck Institute issued a press release in October 2013 to provide additional commentary, which reads in part: "The major genetic risk factor for Alzheimer's disease, present in about two-thirds of people who develop the disease, is ApoE4, the cholesterol-carrying protein that about a quarter of us are born with. But one of the unsolved mysteries of AD is how ApoE4 causes the risk for the incurable, neurodegenerative disease. In research published this week in *The Proceedings of the National Academy of Sciences*, researchers at the Buck Institute found a link between ApoE4 and SirT1, an 'anti-aging protein' that is targeted by resveratrol, present in red wine." The press release also reported: "The Buck group also found that the abnormalities associated with ApoE4 and AD, such as the creation of phospho-tau and amyloid-beta, could be prevented by increasing SirT1." Rammohan V. Rao, a coauthor of the study and associate research professor at the Buck Institute, reported: "One of our goals is to identify a safe, non-toxic treatment that could be given to anyone who carries the ApoE4 gene to prevent the development of AD."[145]

At least four of the plant polyphenols presented in this volume—curcumin, *Ginkgo biloba* extract 761, green tea extract (EGCG), and resveratrol—have been shown to activate Sirt1 expression, as indicated below:

CURCUMINOIDS (↑) SIRTUIN 1

CURCUMIN (↑) SIRT1
Biochemical and Biophysical Research Communications (United States) (2014)
Department of Human Anatomy and Histo-Embryology, Xi'an Jiaotong University Health Science Center, China.
[W]e found that application of curcumin activated the expression of SIRT1 and subsequently decreased the expression of Bax in the presence of Aβ25-35. The protective effect of curcumin was blocked by SIRT1 siRNA. **Taken together, our results suggest that activation of SIRT1 is involved in the neuroprotective action of curcumin.**[146]

CURCUMIN (↑) SIRT1
Brain Research Bulletin (United States) (2015)
Department of Radiology, The First Affiliated Hospital of Chongqing Medical University, China; Department of Radiology, The First Affiliated Hospital of Inner Mongolia Medical University, Huhhot, China.

The aim of this study was to investigate whether curcumin attenuates inflammation and mitochondrial dysfunction in a rat model of cerebral ischemia/reperfusion injury and whether Sirt1 is involved in these potential protective effects... Our results suggested that curcumin exerted a neuroprotective effect, as shown by reduced infarct volumes and brain edema and improved neurological scores... **Moreover, curcumin upregulated SIRT1 and Bcl-2 expression and downregulated Ac-p53 and Bax expression. These effects of curcumin were abolished by sirtinol. In conclusion, our results demonstrate that curcumin treatment attenuates ischemic stroke-induced brain injury via activation of SIRT1.**[147]

Ginkgo biloba EXTRACT 761 (↑) SIRTUIN 1

Ginkgo biloba EXTRACT 761 (↑) SIRT1
Free Radical Biology and Medicine (United States) (2006)
INRS, Institut Armand Frappier, Université du Québec, 245 Boulevard Hymus, Pointe Claire, Univ. Laval, Quebec, Canada.
We demonstrate on neuroblastoma cell line N2a that EGb 761 could prevent the activation of NF-kB, ERK1/2, and JNK pathways induced by Abeta. **Furthermore, our results show that EGb 761 can also activate SIRT1.** This activation could explain the reduction of NF-kB activity by promoting the deacetylation of Lys310 of subunit p65.[148]

GREEN TEA EXTRACTS (↑) SIRTUIN 1

GREEN TEA EXTRACT (EGCG) (↑) SIRT1
Aging Cell (England) (2013)
Department of Nutrition and Food Hygiene, College of Public Health, Harbin Medical University, China.
The aim of this study was to investigate the effects of long-term consumption of the phytochemical, epigallocatechin gallate (EGCG), on body growth, disease protection, and lifespan in healthy rats. 68 male weaning Wistar rats were randomly divided into the control and EGCG groups... **The median lifespan of controls was 92.5 weeks. EGCG increased median lifespan to 105.0 weeks and delayed death by approximately 8–12 weeks...** Damage to liver and kidney function was significantly alleviated in the EGCG group. In addition, EGCG decreased the mRNA and protein expressions of transcription factor NF-κB and increased the upstream protein expressions of silent mating type information regulation two homolog one (SIRT1) and forkhead box class O 3a (FOXO3a). **In conclusion, EGCG extends lifespan in healthy rats by reducing liver and kidney damage and improving age-associated**

inflammation and oxidative stress through the inhibition of NF-κB signaling by activating the longevity factors FoxO3a and SIRT1.[149]

GREEN TEA EXTRACT (EGCG) (↑) SIRT1
Oncotarget (United States) (2015)
Biosafety Research Institute, College of Veterinary Medicine, Chonbuk National University, Jeonju, Jeonbuk, South Korea; Department of Bioactive Material Sciences and Research Center of Bioactive Materials, Chonbuk National University, Jeonbuk, South Korea.
Prion diseases caused by aggregated misfolded prion protein (PrP) are transmissible neurodegenerative disorders that occur in both humans and animals. Epigallocatechin-3-gallate (EGCG) has preventive effects on prion disease; however, the mechanisms related to preventing prion diseases are unclear. We investigated whether EGCG, the main polyphenol in green tea, prevents neuron cell damage induced by the human prion protein... The results showed that EGCG protects the neuronal cells against human prion protein-induced damage through inhibiting Bax and cytochrome c translocation and autophagic pathways... **We demonstrated that EGCG activated the autophagic pathways by inducing Sirt1, and had protective effects against human prion protein-induced neuronal cell toxicity. These results suggest that EGCG may be a therapeutic agent for treatment of neurodegenerative disorders including prion diseases.**[150]

RESVERATROL (↑) SIRTUIN 1

RESVERATROL (↑) SIRT1
The EMBO Journal (England) (2007)
Howard Hughes Medical Institute, Massachusetts Institute of Technology, Boston, Massachusetts; Picower Insitute for Learning and Memory, Massachusetts Institute of Technology; Riken-MIT Neuroscience Research Center (MIT); Department of Brain and Cognitive Sciences, Massachusetts Institute of Technology.
In cell-based models for Alzheimer's disease/tauopathies and ALS [amyotrophic lateral sclerosis], SIRT1 and resveratrol, a SIRT1-activating molecule, both promote neuronal survival. In the inducible p25 transgenic mouse—a model of Alzheimer's disease and tauopathies—resveratrol reduced neurodegeneration in the hippocampus, prevented learning impairment, and decreased the acetylation of the known SIRT1 substrates PGC-1alpha and p53.[151]

RESVERATROL (↑) SIRT1
Recent Patents on CNS Drug Discovery (United Arab Emirates) (2008)
Unitat de Farmacologia i Farmacognòsia i Institut de Biomedicina (IBUB), Facultat de Farmàcia, Universitat de Barcelona, España.

[I]ncreasing SIRT1 has been found to protect cells against amyloid-beta-induced ROS production and DNA damage, thereby reducing apoptotic death in vitro. Moreover, **it has been demonstrated that Alzheimer's and Huntington's disease neurons are rescued by the over-expression of SIRT1, induced by either caloric restriction or administration of resveratrol, a potential activator of this [SIRT1] enzyme.**[152]

RESVERATROL (↑) SIRT1
Journal of Neurochemistry (England) (2009)

Department of Neuroscience, Mario Negri Institute for Pharmacological Research, Milan, Italy.
Human sirtuins are a family of seven conserved proteins (SIRT1-7). The most investigated is the silent mating type information regulation-2 homolog (SIRT1, NM_012238), which was associated with neuroprotection in models of polyglutamine toxicity or Alzheimer's disease (AD) and whose activation by the phytocompound resveratrol has been described... **Resveratrol prevented toxicity triggered by hydrogen peroxide or 6-hydroxydopamine (6-OHDA). This action was likely mediated by SIRT1 activation,** as the protection was lost in the presence of the SIRT1 inhibitor sirtinol and when SIRT1 expression was down-regulated by siRNA approach... **We conclude that SIRT1 activation by resveratrol can prevent in our neuroblastoma model the deleterious effects triggered by oxidative stress or alpha-syn(A30P) aggregation, while resveratrol displayed a SIRT1-independent protective action against Abeta42.**[153]

RESVERATROL (↑) SIRT1
Molecular Neurobiology (United States) (2010)

Department of Medical Pharmacology and Physiology, University of Missouri, Columbia.
In this review, we first describe oxidative mechanisms associated with stroke, AD [Alzheimer's disease], and PD [Parkinson's disease], and subsequently, we place emphasis on recent studies implicating neuroprotective effects of resveratrol, a polyphenolic compound derived from grapes and red wine. **These studies show that the beneficial effects of resveratrol are not only limited to its antioxidant and anti-inflammatory action but also include activation of sirtuin 1 (SIRT1) and vitagenes, which can prevent the deleterious effects triggered by oxidative stress.** In fact, SIRT1 activation by resveratrol is gaining importance in the development of innovative treatment strategies for stroke and other neurodegenerative disorders.[154]

RESVERATROL (↑) SIRT1
Current Opinion in Psychiatry (United States) (2012)

School of Psychiatry, University of New South Wales, Prince of Wales Hospital, Sydney, Australia.

Sirtuins are a family of enzymes highly conserved in evolution and involved in mechanisms known to promote healthy ageing and longevity. This review aims to discuss recent advances in understanding the role of sirtuins, in particular mammalian SIRT1, in promoting longevity and its potential molecular basis for neuroprotection against cognitive ageing and Alzheimer's disease pathology… **[R]esveratrol strongly stimulates SIRT1 deacetylase activity in a dose-dependent manner by increasing its binding affinity to both the acetylated substrate and NAD(+).** Recently, SIRT1 has been shown to affect amyloid production through its influence over the ADAM10 gene. Upregulation of SIRT1 can also induce the Notch pathway and inhibit mTOR signaling.[155]

RESVERATROL (↑) SIRT1
Age (The Netherlands) (2013)

Unitat de Farmacologia i Farmacognòsia, Facultat de Farmàcia, Institut de Biomedicina (IBUB), Centros de Investigación Biomédica en Red de Enfermedades Neurodegenerativas (CIBERNED), Barcelona, Spain.

Resveratrol is a polyphenol that is mainly found in grapes and red wine and has been reported to be a caloric restriction (CR) mimetic driven by Sirtuin 1 (SIRT1) activation… In this study, we examined the role of dietary resveratrol in SAMP8 mice, a model of age-related AD… **[W]e found that long-term dietary resveratrol activates AMPK pathways and pro-survival routes such as SIRT1 in vivo.**[56]

RESVERATROL (↑) SIRT1
PLoS One (United States) (2013)

Key Laboratory of the Ministry of Education for Experimental Teratology and Department of Anatomy, School of Medicine, Shandong University, China.

PC12 cells were injured by $A\beta(25–35)$, and resveratrol at different concentrations was added into the culture medium… **$A\beta(25–35)$-suppressed silent information regulator 1 (SIRT1) activity was significantly reversed by resveratrol,** resulting in the downregulation of Rho-associated kinase 1 (ROCK1). **Our results clearly revealed that resveratrol significantly protected PC12 cells and inhibited the β-amyloid-induced cell apoptosis through the upregulation of SIRT1.**[156]

RESVERATROL (↑) SIRT1
Frontiers in Pharmacology (Switzerland) (2014)

Vittorio Erspamer School of Physiology and Pharmacology, SAPIENZA University of Rome, Italy; Department of Drug Chemistry and Technologies, SAPIENZA University of Rome; Institute Pasteur, Cenci Bolognetti Foundation, SAPIENZA University of Rome.

Sirtuins (SIRTs) are a family of NAD(+)-dependent enzymes, implicated in the control of a variety of biological processes that have the potential to modulate neurodegeneration. Here we tested the hypothesis that activation of SIRT1 or inhibition of SIRT2 would prevent reactive gliosis which is considered one of the most important hallmarks of AD. **Primary rat astrocytes were activated with beta amyloid 1–42 (Aβ 1–42) and treated with resveratrol or AGK-2, a SIRT1 activator and a SIRT2-selective inhibitor, respectively. Results showed that both resveratrol and AGK-2 were able to reduce astrocyte activation as well as the production of pro-inflammatory mediators.** These data disclose novel findings about the therapeutic potential of SIRT modulators, and suggest novel strategies for AD treatment.[157]

The detailed arrangement of evidence presented in Chapters 3–5 demonstrating nearly complete coverage of the pathogenic hallmarks of AD by the plant polyphenols permits a bit of further analysis in the following chapter with regard to their effect on the underlying supporting cast of the hallmark elements. This will extend the already extensive pleiotropic range of the plant polyphenols with respect to AD. It will also show an extraordinary affinity of the plant polyphenols toward health-sustaining and disease-denying biological processes, consistent with the epidemiology and population studies reviewed in Chapter 2.

REFERENCES

1. Albert, DeKosky, Dickson, et al. The diagnosis of mild cognitive impairment due to Alzheimer's disease: recommendations from the National Institute on Aging-Alzheimer's Association Workgroups on diagnostic guidelines for Alzheimer's disease. *Alzheimer's and Dementia* May 2011;**7**(3):270–9.
2. Diseases and Conditions: Mild Cognitive Impairment (MCI): Definition, Mayo Clinic. At: http://www.mayoclinic.org/diseases-conditions/mild-cognitive-impairment/basics/definition/con-20026392.
3. Diseases and Conditions: Mild Cognitive Impairment (MCI): Causes, Mayo Clinic. At: http://www.mayoclinic.org/diseases-conditions/mild-cognitive-impairment/basics/causes/con-20026392.
4. Alzheimer's Disease: Unraveling the Mystery, National Institute on Aging, September 2011; 73.
5. William Klein: Department of Neurobiology: Northwestern University. At: http://www.neurobiology.northwestern.edu/people/core-faculty/william-klein.html.
6. Lambert, Barlow, Chromy, et al. Diffusible, nonfibrillar ligands derived from Abeta$_{1-42}$ are potent central nervous system neurotoxins. *Proceedings of the National Academy of Sciences* May 26, 1998;**95**(11):6448–53.
7. Zahs, Ashe. β-amyloid oligomers in aging and Alzheimer's disease. *Frontiers in Aging Neuroscience* July 4, 2013;**5**:28.
8. Selkoe. Alzheimer's disease is a synaptic failure. *Science* October 25, 2002;**298**(5594):789–91.
9. Rowan, Klyubin, Wang, et al. Synaptic memory mechanisms: Alzheimer's disease amyloid beta-peptide-induced dysfunction. *Biochemical Society Transactions* November 2007;**35**(Pt. 5):1219–23.
10. Selkoe. Soluble oligomers of the amyloid beta-protein impair synaptic plasticity and behavior. *Behavioural Brain Research* September 1, 2008;**192**(1):106–13.
11. Hu, Smith, Walsh, Rowan. Soluble amyloid-beta peptides potently disrupt hippocampal synaptic plasticity in the absence of cerebrovascular dysfunction in vivo. *Brain: A Journal of Neurology* September 2008;**131**(Pt. 9):2414–24.

12. Nimmrich, Ebert. Is Alzheimer's disease a result of presynaptic failure? Synaptic dysfunctions induced by oligomeric beta-amyloid. *Reviews in the Neurosciences* 2009;**20**(1):1–12.

13. Klein. Synaptotoxic amyloid-β oligomers: a molecular basis for the cause, diagnosis, and treatment of Alzheimer's disease? *Journal of Alzheimer's Disease* 2013;**33**(Suppl. 1):S49–65.

14. Pozueta, Lefort, Shelanski. Synaptic changes in Alzheimer's disease and its models. *Neuroscience* October 22, 2013;**251**:51–65.

15. Price, Varghese, Sowa, et al. Altered synaptic structure in the hippocampus in a mouse model of Alzheimer's disease with soluble amyloid-β oligomers and no plaque pathology. *Molecular Neurodegeneration* October 13, 2014;**9**:41.

16. Balducci, Forloni. In vivo application of beta amyloid oligomers: a simple tool to evaluate mechanisms of action and new therapeutic approaches. *Current Pharmaceutical Design* 2014;**20**(15):2491–505.

17. Cerasoli, Ryadnov, Austen. The elusive nature and diagnostics of misfolded Aβ oligomers. *Frontiers in Chemistry* March 19, 2015;**3**:17.

18. Viola, Klein. Amyloid β oligomers in Alzheimer's disease pathogenesis, treatment, and diagnosis. *Acta Neuropathologica* February 2015;**129**(2):183–206.

19. Frautschy, Hu, Kim, et al. Phenolic and anti-inflammatory antioxidant reversal of Abeta-induced cognitive deficits and neuropathology. *Neurobiology of Aging* November–December 2001;**22**(6): 993–1005.

20. Cole, Morihara, Lim, et al. NSAID and antioxidant prevention of Alzheimer's disease: lessons from in vitro and animal models. *Annals of the New York Academy of Sciences* December 2004;**1035**:68–84.

21. Ahmed, Enam, Gilani. Curcuminoids enhance memory in an amyloid-infused rat model of Alzheimer's disease. *Neuroscience* September 1, 2010;**169**(3):1296–306.

22. Ahmed, Gilani, Hosseinmardi, et al. Curcuminoids rescue long-term potentiation impaired by amyloid peptide in rat hippocampal slices. *Synapse* July 2011;**65**(7):572–82.

23. Hoppe, Haag, Whalley, et al. Curcumin protects organotypic hippocampal slice cultures from Aβ$_{1-42}$-induced synaptic toxicity. *Toxicology in Vitro* December 2013;**27**(8):2325–30.

24. Hoppe, Frozza, Pires, et al. The curry spice curcumin attenuates beta-amyloid-induced toxicity through beta-catenin and PI3K signaling in rat organotypic hippocampal slice culture. *Neurological Research* October 2013;**35**(8):857–66.

25. Yamada, Ono, Hamaguchi, Noguchi-Shinohara. Natural phenolic compounds as therapeutic and preventive agents for cerebral amyloidosis. *Advances in Experimental Medicine and Biology* 2015;**863**:79–94.

26. Bate, Tayebi, Williams. Ginkgolides protect against amyloid-beta 1–42-mediated synapse damage in vitro. *Molecular Neurodegeneration* January 7, 2008;**3**:1.

27. Tchantchou, Lacor, Cao, et al. Stimulation of neurogenesis and synaptogenesis by bilobalide and quercetin via common final pathway in hippocampal neurons. *Journal of Alzheimer's Disease* 2009;**18**(4):787–98.

28. Müller, Heiser, Leuner. Effects of the standardized *Ginkgo biloba* extract EGb 761® on neuroplasticity. *International Psychogeriatrics* August 2012;**24**(Suppl. 1):S21–4.

29. Liu, Hao, Qin, et al. Long-term treatment with *Ginkgo biloba* extract EGb 761 improves symptoms and pathology in a transgenic mouse model of Alzheimer's disease. *Brain, Behavior, and Immunity* May 2015;**46**:121–31.

30. Li, Zhao, Zhang, et al. Long-term green tea catechin administration prevents spatial learning and memory impairment in senescence-accelerated mouse Prone-8 mice by decreasing Abeta$_{1-42}$ oligomers and upregulating synaptic plasticity-related proteins in the hippocampus. *Neuroscience* October 20, 2009;**163**(3):741–9.

31. Menard, Bastianetto, Quirion. Neuroprotective effects of resveratrol and epigallocatechin gallate polyphenols are mediated by the activation of protein kinase C gamma. *Frontiers in Cellular Neuroscience* December 26, 2013;**7**:281.

32. Pitt, Roth, Lacor, et al. Alzheimer's-associated Abeta oligomers show altered structure, immunoreactivity and synaptotoxicity with low doses of oleocanthal. *Toxicology and Applied Pharmacology* October 15, 2009;**240**(2):189–97.

33. Luccarini, Grossi, Rigacci, et al. Oleuropein aglycone protects against pyroglutamylated-3 amyloid-β toxicity: biochemical, epigenetic and functional correlates. *Neurobiology of Aging* February 2015;**36**(2): 648–63.

34. Maurer K, Maurer U. *Alzheimer: the life of a physician and the career of a disease*, vol. ix. Columbia University Press; 2003.
35. Maurer K, Maurer U. *Alzheimer: the life of a Physician and the career of a disease*, vol. ix. Columbia University Press; 2003. p. 159.
36. Maurer K, Maurer U. *Alzheimer: the life of a physician and the career of a disease*, vol. ix. Columbia University Press; 2003. p. 163.
37. Maurer K, Maurer U. *Alzheimer: the life of a physician and the career of a disease*, vol. ix. Columbia University Press; 2003. p. 164.
38. Deleted in review.
39. Tau a driver of Alzheimer's disease, study of thousands of brains reveal. *ScienceDaily* March 24, 2015. At: http://www.sciencedaily.com/releases/2015/03/150324084339.htm.
40. Gong, Iqbal. Hyperphosphorylation of microtubule-associated protein tau: a promising therapeutic target for Alzheimer disease. *Current Medicinal Chemistry* 2008;**15**(23):2321–8.
41. Himmelstein, Ward, Lancia, et al. Tau as a therapeutic target in neurodegenerative disease. *Pharmacology & Therapeutics* October 2012;**136**(1):8–22.
42. Iqbal, Gong, Liu. Microtubule-associated protein tau as a therapeutic target in Alzheimer's disease. *Expert Opinion on Therapeutic Targets* March 2014;**18**(3):307–18.
43. Targeting Tau as a Treatment for Tauopathies: Laboratories: Neurodegenerative Diseases: Leonard Petrucelli, Mayo Clinic. At: http://www.mayo.edu/research/labs/neurodegenerative-diseases/targeting-tau-treatment-tauopathies.
44. Lasagna-Reeves, Castillo-Carranza, Jackson, Kayed. Tau oligomers as potential targets for immunotherapy for Alzheimer's disease and tauopathies. *Current Alzheimer Research* September 2011;**8**(6):659–65.
45. Ward, Himmelstein, Lancia, Binder. Tau oligomers and tau toxicity in neurodegenerative disease. *Biochemical Society Transactions* August 2012;**40**(4):667–71.
46. Cowan, Quraishe, Mudher. What is the pathological significance of tau oligomers? *Biochemical Society Transactions* August 2012;**40**(4):693–7.
47. Tian, Davidowitz, Lopez, et al. Trimeric tau is toxic to human neuronal cells at low nanomolar concentrations. *International Journal of Cell Biology* 2013;**2013**:260787.
48. Iqbal, Gong, Lie. Hyperphosphorylation-induced tau oligomers. *Frontiers in Neurology* August 15, 2013;**4**:112.
49. Park, Kim, Cho, et al. Curcumin protected PC$_{12}$ cells against beta-amyloid-induced toxicity through the inhibition of oxidative damage and tau hyperphosphorylation. *Food and Chemical Toxicology* August 2008;**46**(8):2881–7.
50. Ma, Yang, Rosario, et al. Beta-amyloid oligomers induce phosphorylation of tau and inactivation of insulin receptor substrate via c-Jun N-Terminal kinase signaling: suppression by omega-3 fatty acids and curcumin. *The Journal of Neuroscience* July 15, 2009;**29**(28):9078–89.
51. Huang, Tang, Xu, Jiang ZF. Curcumin attenuates amyloid-β-induced tau hyperphosphorylation in human neuroblastoma SH-SY5Y cells involving PTEN/Akt/Gsk-3β signaling pathway. *Journal of Receptor and Signal Transduction Research* February 2014;**34**(1):26–37.
52. Chen, Wang, Hu, et al. Effects of ginkgolide a on okadaic acid-induced tau hyperphosphorylation and the PI3K-Akt signaling pathway in N2a cells. *Planta Medica* August 2012;**78**(12):1337–41.
53. Kwon, Lee, Cho, et al. *Ginkgo biloba* extract (EGb 761) attenuates zinc-induced tau phosphorylation at Ser262 by regulating GSK3β activity in rat primary cortical neurons. *Food and Function* June 2015;**6**(6):2058–67.
54. Santa-Maria, Diaz-Ruiz, Ksiezak-Reding, et al. GSPE interferes with tau aggregation in vivo: implication for treating tauopathy. *Neurobiology of Aging* September 2012;**33**(9):2072–81.
55. Rezai-Zadeh, Arendash, Hou, et al. Green tea epigallocatechin-3-gallate (EGCG) reduces beta-amyloid mediated cognitive impairment and modulates tau pathology in Alzheimer transgenic mice. *Brain Research* June 12, 2008;**1214**:177–87.
56. Porquet, Casadesús, Bayod, et al. Dietary resveratrol prevents Alzheimer's markers and increases life span in SAMP8. *Age* October 2013;**35**(5):1851–65.
57. Du, Xie, Cheng, et al. Activation of sirtuin 1 attenuates cerebral ventricular streptozotocin-induced tau hyperphosphorylation and cognitive injuries in rat hippocampi. *Age* April 2014;**36**(2):613–23.

58. Dolai, Shi, Corbo, et al. "Clicked" sugar-curcumin conjugate: modulator of amlyoid-β and tau peptide aggregation at ultralow concentrations. *American Chemical Neuroscience* December 21, 2011;**2**(12):694–9.

59. Ho, Yemul, Wang, Pasinetti. Grape seed polyphenolic extract as a potential novel therapeutic agent in tauopathies. *Journal of Alzheimer's Disease* 2009;**16**(2):43–9.

60. Wang, Santa-Maria, Ho, et al. Grape derived polyphenols attenuate tau neuropathology in a mouse model of Alzheimer's disease. *Journal of Alzheimer's Disease* 2010;**2**(2):653–61.

61. Ksiezak-Reding, Ho, Santa-Maria, et al. Ultrastructural alterations of Alzheimer's disease paired helical filament by grape seed-derived polyphenols. *Neurobiology of Aging* July 2012;**33**(7):1427–39.

62. Wobst, Sharma, Diamond, et al. The green tea polyphenol (-)-epigallocatechin gallate prevents the aggregation of tau protein into toxic oligomers at substoichiometric ratios. *FEBS Letters* January 2, 2015;**589**(1):77–83.

63. Daccache, Lion, Sibille, et al. Oleuropein and derivatives from olives as tau aggregation inhibitors. *Neurochemistry International* May 2011;**58**(6):700–7.

64. Monti, Margarucci, Tosco, et al. New insights on the interaction mechanism between tau protein and oleocanthal, an extra-virgin olive-oil bioactive component. *Food and Function* July 2011;**2**(7):423–8.

65. Monti, Margarucci, Riccio, Casapullo. Modulation of tau protein fibrillization by oleocanthal. *Journal of Natural Products* September 28, 2012;**75**(9):1584–8.

66. Ma, Zuo, Yang, et al. Curcumin suppresses soluble tau dimers and corrects molecular chaperone, synaptic, and behavioral deficits in aged human tau transgenic mice. *The Journal of Biological Chemistry* February 8, 2013;**288**(6):4056–65.

67. Patil, Tran, Geekiyanage, et al. Curcumin-induced upregulation of the anti-tau cochaperone BAG2 in primary rat cortical neurons. *Neuroscience Letters* October 25, 2013;**554**:121–5.

68. Chesser, Ganeshan, Yang, Johnson. Epigallocatechin-3-Gallate enhances clearance of phosphorylated tau in primary neurons. *Nutritional Neuroscience* July 24, 2015. [Epub ahead of print].

69. Blurton-Jones, Laferla. Pathways by which Abeta facilitates tau pathology. *Current Alzheimer Research* December 2006;**3**(5):437–48.

70. King, Kan, Baas, et al. Tau-dependent microtubule disassembly initiated by prefibrillar beta-amyloid. *The Journal of Cell Biology* November 20, 2006;**175**(4):541–6.

71. Götz, Chen, van Dorpe, Nitsch. Formation of neurofibrillary tangles in P301l tau transgenic mice induced by Abeta 42 fibrils. *Science* August 24, 2001;**293**(5534):1491–5.

72. Lewis, Dickson, Lin, et al. Enhanced neurofibrillary degeneration in transgenic mice expressing mutant tau and APP. *Science* August 24, 2001;**293**(5534):1487–91.

73. Viola, Velasco, Klein. Why Alzheimers is a disease of memory: the attack on synapses by A beta oligomers (ADDLs). *The Journal of Nutrition, Health, and Aging* January 2008;**12**(1):51S–7S.

74. Lasagna-Reeves, Castillo-Carranza, Guerrero-Muoz, et al. Preparation and characterization of neurotoxic tau oligomers. *Biochemistry* November 30, 2010;**49**(47):10039–41.

75. Dennis Selkoe on the amyloid hypothesis of Alzheimer's disease: special topic of Alzheimer's disease interview. Part 3. *Science Watch* March 2011. http://archive.sciencewatch.com/ana/st/alz2/11marSTAlz2Selk2/.

76. Jin, Shepardson, Yang, et al. Soluble amyloid beta-protein dimers isolated from Alzheimer cortex directly induce tau hyperphosphorylation and neuritic degeneration. *Proceedings of the National Academy of Sciences* April 5, 2011;**108**(14):5819–24.

77. Chabrier, Blurton-Jones, Agazaryan, et al. Soluble Aβ promotes wild-type tau pathology in vivo. *The Journal of Neuroscience* November 28, 2012;**32**(48):17345–50.

78. Zhang, Lu, Jia, et al. A lifespan observation of a novel mouse model: in vivo evidence supports Aβ oligomer hypothesis. *PLoS One* January 21, 2014;**9**(1):e85885.

79. Bloom. Amyloid-β and tau: the trigger and bullet in Alzheimer disease pathogenesis. *JAMA Neurology* April 2014;**71**(4):505–8.

80. Bright, Hussain, Dang, et al. Human secreted tau increases amyloid-beta production. *Neurobiology of Aging* February 2015;**36**(2):693–709.

81. Akiyama, Barger, Barnum, et al. Inflammation and Alzheimer's disease. *Neurobiology of Aging* May–June 2000;**21**(3):383–421.

82. Rosenberg. Clinical aspects of inflammation in Alzheimers disease. *International Review of Psychiatry* December 2005;**17**(6):503–14.

83. Griffin, Barger. Neuroinflammatory cytokines—the common thread in Alzheimer's pathogenesis. *US Neurology* 2010;**6**(2):19–27.

84. McGeer, McGeer. Neuroinflammation in Alzheimer's disease and mild cognitive impairment: a field in its infancy. *Journal of Alzheimer's Disease* 2010;**19**(1):355–61.

85. Kitazawa, Cheng, Tsukamoto, et al. Blocking IL-1 signaling rescues cognition, attenuates tau pathology, and restores neuronal β-catenin pathway function in an Alzheimer's disease model. *The Journal of Immunology* December 15, 2011;**187**(12):6539–49.

86. Ghosh, Wu, Shaftel, et al. Sustained Interleukin-1β overexpression exacerbates tau pathology despite reduced amyloid burden in an Alzheimer's mouse model. *The Journal of Neuroscience* March 13, 2013;**33**(11):5053–64.

87. Rubio-Perez, Morillas-Ruiz. A review: inflammatory process in Alzheimer's disease, role of cytokines. *The Scientific World Journal* 2012;**2012**:756357.

88. Lee, McGeer, McGeer. Activated human microglia stimulate neuroblastoma cells to upregulate production of beta amyloid protein and tau: implications for Alzheimer's disease pathogenesis. *Neurobiology of Aging* January 2015;**36**(1):42–52.

89. Heneka, Carson, El Khoury, et al. Neuroinflammation in Alzheimer's disease. *The Lancet Neurology* April 2015;**14**(4):388–405.

90. Misonou, Morishima-Kawashima, Ihara. Oxidative stress induces intracellular accumulation of amyloid beta-protein (Abeta) in human neuroblastoma cells. *Biochemistry* June 13, 2000;**39**(23):6951–9.

91. Reich, Markesbery, Roberts, et al. Brain regional quantification of F-Ring and D-/E-Ring isoprostanes and neuroprostanes in Alzheimer's disease. *The American Journal of Pathology* January 2001;**158**(1):293–7.

92. Tamagno, Guglielmotto, Aragno, et al. Oxidative stress activates a positive feedback between the gamma- and beta-secretase cleavages of the beta-amyloid precursor protein. *Journal of Neurochemistry* February 2008;**104**(3):683–95.

93. Quiroz-Baez, Rojas, Arias. Oxidative stress promotes JNK-dependent amyloidogenic processing of normally expressed human APP by differential modification of alpha-, beta- and gamma-secretase expression. *Neurochemistry International* December 2009;**55**(7):662–70.

94. Arimon, Takeda, Post, et al. Oxidative stress and lipid peroxidation are upstream of amyloid pathology. *Neurobiology of Disease* December 2015;**84**:109–19.

95. Behl, Davis, Cole, Schubert. Vitamin E protects nerve cells from amyloid beta protein toxicity. *Biochemical and Biophysical Research Communications* July 31, 1992;**186**(2):944–50.

96. Sultana, Butterfield. Oxidatively modified, mitochondria-relevant brain proteins in subjects with Alzheimer disease and mild cognitive impairment. *Journal of Bioenergetics and Biomembranes* October 2009;**41**(5):441–6.

97. Guglielmotto, Monteleone, Giliberto, et al. Amyloid-β42 activates the expression of BACE1 through the JNK pathway. *Journal of Alzheimer's Disease* 2011;**27**(4):871–83.

98. Wang, Zhao, Zhang, et al. PPARgamma agonist curcumin reduces the amyloid-beta-stimulated inflammatory responses in primary astrocytes. *The Journal of Alzheimer's Disease* 2010;**20**(4):1189–99.

99. Tegenge, Rajbhandari, Shrestha, et al. Curcumin protects axons from degeneration in the setting of local neuroinflammation. *Experimental Neurobiology* March 2014;**253**:102–10.

100. Nahar, Slitt, Seeram. Anti-inflammatory effects of novel standardized solid lipid curcumin formulations. *Journal of Medicinal Food* July 2015;**18**(7):786–92.

101. Shi, Zheng, Li, et al. Curcumin inhibits Aβ-induced microglial inflammatory response in vitro: involvement of ERK1/2 and p38 signaling pathways. *Neuroscience Letters* May 6, 2015;**594**:105–10.

102. Wang, Thomas, Zhong, et al. Consumption of grape seed extract prevents amyloid-beta deposition and attenuates inflammation in brain of an Alzheimer's disease mouse. *Neurotoxicity Research* January 2009;**15**(1):3–14.

103. Lee, Choi, Yun, et al. Epigallocatechin-3-Gallate prevents systemic inflammation-induced memory deficiency and amyloidogenesis via its anti-neuroinflammatory properties. *The Journal of Nutritional Biochemistry* January 2013;**24**(1):298–310.

104. Candelario-Jalil, de Oliveira, Gräf, et al. Resveratrol potently reduces prostaglandin E2 production and free radical formation in lipopolysaccharide-activated primary rat microglia. *Journal of Neuroinflammation* October 10, 2007;**4**:25.

105. Zhang, Liu, Shi. Anti-inflammatory activities of resveratrol in the brain: role of resveratrol in microglial activation. *European Journal of Pharmacology* June 25, 2010;**636**(1–3):1–7.

106. Lu, Ma, Ruan, et al. Resveratrol differentially modulates inflammatory responses of microglia and astrocytes. *Journal of Neuroinflammation* August 17, 2010;**7**:46.

107. Capiralla, Vingtdeux, Zhao, et al. Resveratrol mitigates lipopolysaccharide- and Aβ-mediated microglial inflammation by inhibiting the TLR4/NF-kB/STAT signaling cascade. *Journal of Neurochemistry* February 2012;**120**(3):461–72.

108. Zhang, Wang, Wu, et al. Resveratrol protects cortical neurons against microglia-mediated neuroinflammation. *Phytotherapy Research* March 2013;**27**(3):344–9.

109. Frozza, Bernardi, Hoppe, et al. Lipid-core nanocapsules improve the effects of resveratrol against Abeta-induced neuroinflammation. *Journal of Biomedical Nanotechnology* December 2013;**9**(12): 2086–104.

110. Kim, Park, Kim. Curcuminoids from *Curcuma longa* L. (Zingiberaceae) that protect PC_{12} rat pheochromocytoma and normal human umbilical vein endothelial cells from beta(1–42) insult. *Neuroscience Letters* April 27, 2001;**303**(1):57–61.

111. Calabrese, Scapagnini, Colombrita, et al. Redox regulation of heat shock protein expression in aging and neurodegenerative disorders associated with oxidative stress: a nutritional approach. *Amino Acids* December 2003;**25**(3–4):437–44.

112. Baum, Ng. Curcumin interaction with copper and iron suggests one possible mechanism of action in Alzheimer's disease animal models. *Journal of Alzheimer's Disease* August 2004;**6**(4):367–77.

113. Atamna, Boyle. Amyloid-beta peptide binds with heme to form a peroxidase: relationship to the cytopathologies of Alzheimer's disease. *Proceedings of the National Academy of Sciences* February 28, 2006;**103**(9):3381–6.

114. Mancuso, Scapagini, Currò, et al. Mitochondrial dysfunction, free radical generation and cellular stress response in neurodegenerative disorders. *Frontiers in Bioscience* January 1, 2007;**12**:1107–23.

115. Shimmyo, Kihara, Akaike, et al. Epigallocatechin-3-Gallate and curcumin suppress amyloid beta-induced beta site APP cleaving enzyme-1-upregulation. *Neuroreport* August 27, 2008;**19**(13): 1329–33.

116. Kumar, Dogra, Prakash. Protective effect of curcumin (*Curcuma longa*), against aluminum toxicity: possible behavioral and biochemical alteration in rats. *Behavioral Brain Research* December 28, 2009;**205**(2):384–90.

117. Aggarwal, Mishra, Tyagi, et al. Effect of curcumin on brain insulin receptors and memory functions in STZ (ICV) induced dementia model of rat. *Pharmacological Research* March 2010;**61**(3):247–52.

118. Kim, Lee, Lee. Naturally occurring phytochemicals for the prevention of Alzheimer's disease. *Journal of Neurochemistry* March 2010;**112**(6):1415–30.

119. Huang, Chang, Dai, Jiang. Protective effects of curcumin on amyloid-β-induced neuronal oxidative damage. *Neurochemical Research* July 2012;**37**(7):1584–97.

120. Doggui, Sahni, Arseneault, et al. Neuronal uptake and neuroprotective effect of curcumin-loaded PLGA nanoparticles on the human SK-N-SH cell line. *Journal of Alzheimer's Disease* 2012;**30**(2): 377–92.

121. Zhou, Zhu. Reactive oxygen species-induced apoptosis in PC_{12} cells and protective effect of bilobalide. *The Journal of Pharmacology and Experimental Therapeutics* June 2000;**293**(3):982–8.

122. Bastianetto, Zheng, Quirion. The *Ginkgo biloba* extract (EGb 761) protects and rescues hippocampal cells against nitric oxide-induced toxicity: involvement of its flavonoid constituents and protein kinase C. *Journal of Neurochemistry* June 2000;**74**(6):2268–77.

123. Bridi, Crossetti, Steffen, Henriques. The antioxidant activity of standardized extract of *Ginkgo biloba* (EGb 761) in rats. *Phytotherapy Research* August 2001;**15**(5):449–51.

124. Smith, Luo. Elevation of oxidative free radicals in Alzheimer's disease models can be attenuated by *Ginkgo biloba* extract EGb 761. *Journal of Alzheimer's Disease* August 2003;**5**(4):287–300.

125. Garcia-Alloza, Dodwell, Meyer-Luehmann, et al. Plaque-derived oxidative stress mediates distorted neurite trajectories in the Alzheimer mouse model. *Journal of Neuropathology and Experimental Neurology* November 2006;**65**(11):1082–9.

126. Russo, Palumbo, Aliano, et al. Red wine micronutrients as protective agents in Alzheimer-like induced insult. *Life Sciences* April 11, 2003;**72**(21):2369–79.

127. Li, Jang, Sun, Surh. Protective effects of oligomers of grape seed polyphenols against beta-amyloid-induced oxidative cell death. *Annals of the New York Academy of Sciences* December 2004;**1030**: 317–29.

128. Choi, Jung, Lee, et al. The green tea polyphenol (-)-epigallocatechin gallate attenuates beta-amyloid-induced neurotoxicity in cultured hippocampal neurons. *Life Sciences* December 21, 2001;**70**(5):603–14.

129. Kim, Lee, Park, Jang. Neuroprotective effect of epigallocatechin-3-gallate against beta-amyloid-induced oxidative and nitrosative cell death via augmentation of antioxidant defense capacity. *Archives of Pharmacal Research* June 2009;**32**(6):869–81.

130. Choi, Kim, Cho, et al. Effects of flavonoid compounds on β-amyloid-peptide-induced neronal death in cultured mouse cortical neurons. *Chonnam Medical Journal* August 2014;**50**(2):45–51.

131. Savaskan, Olivieri, Meier, et al. Red wine ingredient resveratrol protects from beta-amyloid neurotoxicity. *Gerontology* November–December 2003;**49**(6):380–3.

132. Kumar, Naidu, Seghal, Padi. Neuroprotective effects of resveratrol against intracerebroventricular colchicine-induced cognitive impairment and oxidative stress in rats. *Pharmacology* 2007;**79**(1):17–26.

133. Raval, Lin, Dave, et al. Resveratrol and ischemic preconditioning in the brain. *Current Medicinal Chemistry* 2008;**15**(15):1545–51.

134. Kwon, Kim, Shin, Han. Melatonin potentiates the neuroprotective properties of resveratrol against beta-amyloid-induced neurodegeneration by modulating AMP-activated protein kinase pathways. *Journal of Clinical Neurology* September 2010;**6**(3):127–37.

135. Kwon, Kim, Kim, et al. Melatonin synergistically increases resveratrol-induced heme oxygenase-1 expression through the inhibition of ubiquitin-dependent proteasome pathway: a possible role in neuroprotection. *Journal of Pineal Research* March 2011;**50**(2):110–23.

136. Ma, Sun, Liu, et al. Resveratrol improves cognition and reduces oxidative stress in rats with vascular dementia. *Neural Regeneration Research* August 5, 2013;**8**(22):2050–9.

137. Rege, Geetha, Broderick, Babu. Resveratrol protects β amyloid-induced oxidative damage and memory associated proteins in H_{19-7} hippocampal neuronal cells. *Current Alzheimer Research* 2015;**12**(2):147–56.

138. Alzheimer's Disease Genetics Fact Sheet, National Institute on Aging. At: https://www.nia.nih.gov/alzheimers/publication/alzheimers-disease-genetics-fact-sheet.

139. Strittmatter, Saunders, Schmechel, et al. Apolipoprotein E: high-avidity binding to beta-amyloid and increased frequency of type 4 allele in late-onset familial Alzheimer disease. *Proceedings of the National Academy of Sciences* March 1, 1993;**90**(5):1977–81.

140. Seshadri, Drachman, Lippa. Apolipoprotein E epsilon 4 allele and the lifetime risk of Alzheimer's disease. What physicians know, and what they should know. *Archives of Neurology* November 1995;**52**(11):1074–9.

141. Tai, Ghura, Koster, et al. APOE-modulated Aβ-induced neuroinflammation in Alzheimer's disease: current landscape, novel data, and future perspective. *Journal of Neurochemistry* May 2015;**133**(4):465–88.

142. Teter, LaDu, Sullivan, et al. Apolipoprotein E isotype-dependent modulation of microRNA-146a in plasma and brain. *Neuroreport* August 3, 2016;**27**(11):791–5.

143. Sinclair, Guarente. Unlocking the secrets of longevity genes. *Scientific American* March 2006;**294**(3). 48–51, 54–57.

144. Theendakara, Patent, Peters, et al. Neuroprotective sirtuin ratio reversed by ApoE4. *Proceedings of the National Academy of Sciences* November 5, 2013;**110**(45):18303–8. The findings of the Buck Institute researchers with respect to the effects of APOE4 on Sirt1 were confirmed a few months later in a study published in 2014 in *Neuroscience*. See, Lattanzio, Carboni, Carretta, et al. Human apolipoprotein E4 modulates the expression of Pin1, Sirtuin 1, and Presenilin 1 in brain regions of targeted replacement ApoE mice. *Neuroscience* January 3, 2014;**256**:360–9.

145. Major Alzheimer's Risk Factor Linked to Red Wine Target, Buck Institute for Research on Aging; October 21, 2013. At: http://www.buckinstitute.org/buck-news/major-alzheimers-risk-factor-linked-red-wine-target.

146. Sun, Jia, Wang, et al. Activation of SIRT1 by curcumin blocks the neurotoxicity of amyloid-β25–35 in rat cortical neurons. *Biochemical and Biophysical Research Communications* May 23, 2014;**448**(1):89–94.

147. Miao, Zhao, Gao, et al. Curcumin pretreatment attenuates inflammation and mitochondrial dysfunction in experimental stroke: the possible role of Sirt1 signaling. *Brain Research Bulletin* November 27, 2015;**121**:9–15.

148. Longpré, Garneau, Christen, Ramassamy. Protection by EGb 761 against beta-amyloid-induced neurotoxicity: involvement of NF-kappaB, SIRT1, and MAPKs pathways and inhibition of amyloid fibril formation. *Free Radical Biology and Medicine* December 15, 2006;**41**(12):1781–94.

149. Niu, Na, Feng, et al. The phytochemical, EGCG, extends lifespan by reducing liver and kidney function damage and improving age-associated inflammation and oxidative stress in healthy rats. *Aging Cell* December 2013;**12**(6):1041–9.

150. Lee, Moon, Kim, et al. EGCG-mediated autophagy flux has a neuroprotection effect via a class III histone deacetylase in primary neuron cells. *Oncotarget* April 30, 2015;**6**(12):9701–17.

151. Kim, Nguyen, Dobbin, et al. SIRT1 deacetylase protects against neurodegeneration in models for Alzheimer's disease and amyotrophic lateral sclerosis. *The EMBO Journal* July 11, 2007;**26**(13):3169–79.

152. Pallàs, Verdaguer, Tajes, et al. Modulation of sirtuins: new targets for antiaging. *Recent Patents on CNS Drug Discovery* January 2008;**3**(1):61–9.

153. Albani, Polito, Batelli, et al. The SIRT1 activator resveratrol protects SK-N-be cells from oxidative stress and against toxicity caused by alpha-synuclein or amyloid-beta (1–42) peptide. *Journal of Neurochemistry* September 2009;**110**(5):1445–56.

154. Sun, Wang, Simonyi, Sun. Resveratrol as a therapeutic agent for neurodegenerative diseases. *Molecular Neurobiology* June 2010;**41**(2–3):375–83.

155. Braidy, Jayasena, Poljak, Sachdev. Sirtuins in cognitive ageing and Alzheimer's disease. *Current Opinion in Psychiatry* May 2012;**25**(3):226–30.

156. Feng, Liang, Zhu, et al. Resveratrol inhibits β-amyloid-induced neuronal apoptosis through regulation of SIRT1-ROCK1 signaling pathway. *PLoS One* 2013;**8**(3):e59888.

157. Scuderi, Stecca, Bronzuoli, et al. Sirtuin modulators control reactive gliosis in an in vitro model of Alzheimer's disease. *Frontiers in Pharmacology* May 13, 2014;**5**:89.

CHAPTER 6

Pleiotropic Theory

Contents

A Paradigm Shift to Prevent and Treat Alzheimer's Disease
ISBN 978-0-12-812259-4
http://dx.doi.org/10.1016/B978-0-12-812259-4.00006-0

We mentioned previously that a phase III clinical trial of the gamma-secretase inhibitor semagacestat for the treatment of Alzheimer's disease (AD) did not improve cognitive status and that the drug caused serious side effects.[1] A review in *Current Medical Research and Opinion* of safety-related outcomes reported that semagacestat treatment "was associated with increased reporting of Notch-related adverse events (gastrointestinal, infection, and skin cancer related)" and that "other relevant safety findings associated with semagacestat treatment included cognitive and functional worsening, skin-related TEAEs [treatment-emergent adverse events], renal and hepatic changes, increased QT interval [in the heart's electrical cycle], and weight loss." The review concluded, probably incorrectly, that "many of these safety findings can be attributed to γ-secretase inhibition,"[2] as opposed to the poor design of the drug itself. Other reviews of the semagacestat trial have asserted, also erroneously from our standpoint, that it "represents a direct test of the amyloid hypothesis,"[3] that its failure "bring[s] into question the long-hypothesized association between AD and Aβ production,"[4] and "suggests that the amyloidogenic hypothesis needs to be revised."[5]

A better assessment, at least in part, of the poor trial outcome is that a majority of gamma-secretase inhibitors, including semagacestat, "block γ-secretase nonselectively," including as involved with normal gamma-secretase processing of Notch receptors, and that "nonspecificity has disadvantages,"[6] as indicated in the adverse effects of the semagacestat-treated patients. Another possible explanation is that semagacestat—like most or all AD drugs tested thus far in clinical trials, including the humanized monoclonal antibodies—lacked pleiotropism, and thus acted too narrowly in its targeting of gamma secretase. A better therapeutic model would feature *both* broader targeting and beneficial collateral effects as opposed to narrow targeting and the kind of unintended consequences that resulted from the semagacestat trial noted above.

While this scenario seems beyond the reach of the pharmaceutical industry, it describes core capabilities of the plant polyphenols presented in this volume. Furthermore, although we subscribe in general to the amyloid hypothesis of AD, we also note the therapeutic applications of the plant polyphenols to each of the major hallmarks of AD pathophysiology as listed in the tables above. Thus, for us, consensus on a "right hypothesis" that functioned mainly as prologue for a "right drug" for a "right target" approach to AD prevention and treatment would be less auspicious than a robust pleiotropic prevention-plus-treatment framework that could safely, selectively, and therapeutically address the major pathogenic dimensions of AD, in addition to the supporting cast of pathological processes, including those profiled below.

In this chapter, we first present a case study of the collateral beneficial effects of curcumin and (-)-epigallocatechin-3-gallate (EGCG) on modulating notch processing and signaling, specifically the anticarcinogenic effects, as compared to the indiscriminate action of semagacestat. We then extend the range of the plant polyphenol pleiotropism to other key cellular and molecular processes that are implicated in AD, including glycogen synthase kinase 3 beta (GSK-3β), mitogen-activated protein kinases (MAPK), *N*-methyl-D-aspartate (NMDA) receptors, nuclear factor (NF)-kappa B, microglial activation, tumor necrosis factor (TNF), interleukin-1beta (IL-1β), interleukin-6 (IL-6), and iron and copper toxicity. This chapter will thus deepen the already deep bench of pleiotropic activities of the plant polyphenols while simultaneously underscoring the nearly perfect pitch of the plant-based compounds in favorably modifying these additional AD-related biofactors with few if any serious adverse effects.

Semagacestat is a classic mono-targeting, nonselective drug, given that its limited function was to inhibit gamma-secretase activity and that its nonselectivity led to serious Notch-related adverse effects, including skin cancer. In Chapter 3, we presented evidence that curcumin and EGCG act as gamma-secretase inhibitors. Here, we provide indications that the corrective effects of curcumin and EGCG on dysregulations in Notch signaling, unlike semagacestat, may help prevent various cancers.

CURCUMINOIDS (↓) NOTCH-1 SIGNALING

CURCUMIN (↓) NOTCH-1 SIGNALING
Cancer (United States) (2006)
Department of Pathology, Karmanos Cancer Institute, Wayne State University School of Medicine, Detroit, Michigan
Notch signaling plays a critical role in maintaining the balance between cell proliferation, differentiation, and apoptosis, and thereby may contribute to the development of pancreatic cancer. Therefore, the down-regulation of Notch signaling may be a novel approach for pancreatic cancer therapy… **Curcumin inhibited cell growth and induced apoptosis in pancreatic cancer cells. Notch-1, Hes-1, and Bcl-XL expression levels concomitantly were down-regulated by curcumin treatment…** These results suggest that the down-regulation of Notch signaling by curcumin may be a novel strategy for the treatment of patients with pancreatic cancer.[7]

CURCUMIN, EGCG, AND RESVERATROL (↓) NOTCH SIGNALING
Cellular Signalling (England) (2009)
Department of Pathology, Karmanos Cancer Institute, Wayne State University School of Medicine, Detroit, Michigan.
Cancer cells are known to have alterations in multiple cellular signaling pathways and because of the complexities in the communication between multiple signaling networks, the treatment and the cure for most human malignancies is still an open question. Perhaps this is the reason why specific inhibitors that target only one pathway have typically failed in cancer treatment. However, the in vitro and in vivo studies

have demonstrated that some natural products such as isoflavones, indole-3-carbinol (I3C), 3,3'-diindolylmethane (DIM), curcumin, (-)-epigallocatechin-3-gallate (EGCG), resveratrol, lycopene, etc., have inhibitory effects on human and animal cancers through targeting multiple cellular signaling pathways and thus these "natural agents" could be classified as multi-targeted agents. This is also consistent with the epidemiological studies showing that the consumption of fruits, soybean and vegetables is associated with reduced risk of several types of cancers. **By regulating multiple important cellular signaling pathways including NF-kappaB, Akt, MAPK, Wnt, Notch, p53, AR, ER, etc., these natural products are known to activate cell death signals and induce apoptosis in pre-cancerous or cancer cells without affecting normal cells.**[8]

CURCUMIN (↓) NOTCH-1 SIGNALING
Journal of Cellular Biochemistry (United States) (2011)
Department of Stomatology, First Affiliated Hospital of Bengbu Medical College, Anhui, China.
In the present study, we assessed the effects of curcumin on cell viability and apoptosis in oral cancer... **[W]e conclude that the down-regulation of Notch-1 by curcumin could be an effective approach, which will cause down-regulation of NF-κB, resulting in the inhibition of cell growth and invasion. These results suggest that antitumor activity of curcumin is mediated through a novel mechanism involving inactivation of Notch-1 and NF-κB signaling pathways.**[9]

CURCUMIN (↓) NOTCH-1 SIGNALING
PLoS One (United States) (2012)
Department of Molecular and Integrative Physiology, University of Kansas Medical Center, Kansas City.
Curcumin inhibits the growth of esophageal cancer cell lines; however, the mechanism of action is not well understood. It is becoming increasingly clear that aberrant activation of Notch signaling has been associated with the development of esophageal cancer. **Here, we have determined that curcumin inhibits esophageal cancer growth via a mechanism mediated through the Notch signaling pathway... Curcumin is a potent inhibitor of esophageal cancer growth that targets the Notch-1 activating γ-secretase complex proteins. These data suggest that Notch signaling inhibition is a novel mechanism of action for curcumin during therapeutic intervention in esophageal cancers.**[10]

CURCUMIN (↓) NOTCH-1 SIGNALING
The FEBS Journal (England) (2012)

Department of Orthopedics, Qilu Hospital, Shandong University, Jinan, China.

The Notch signaling pathway plays critical roles in human cancers, including osteosarcoma, suggesting that the discovery of specific agents targeting Notch would be extremely valuable for osteosarcoma… The results showed that curcumin caused marked inhibition of osteosarcoma cell growth and G2/M phase cell cycle arrest. This was associated with concomitant attenuation of Notch-1 and downregulation of its downstream genes, such as matrix metalloproteinases, resulting in the inhibition of osteosarcoma cell invasion through Matrigel… **These results suggest that antitumor activity of curcumin is mediated through a novel mechanism involving inactivation of the Notch-1 signaling pathway. Our data provide the first evidence that the downregulation of Notch-1 by curcumin may be an effective approach for the treatment of osteosarcoma.**[11]

CURCUMIN (↓) NOTCH-1 SIGNALING
Biofactors (England) (2013)
School of Life Sciences, College of Natural Sciences, Kyungpook National University, Daegu, South Korea.

Several reports have demonstrated that curcumin inhibits animal and human cancers, suggesting that it may serve as a chemopreventive agent. **Numerous in vitro and in vivo experimental models have also revealed that curcumin regulates several molecules in cell signal transduction pathway, including NF-κB, Akt, MAPK, p53, Nrf2, Notch-1, JAK/STAT, β-catenin, and AMPK.** Modulation of cell signaling pathways through the pleiotropic effects of curcumin likely activate cell death signals and induce apoptosis in cancer cells, thereby inhibiting the progression of disease.[12]

CURCUMIN (↓) NOTCH-1 SIGNALING
Molecular Nutrition and Food Research (Germany) (2013)
Department of Cancer Studies and Molecular Medicine, University of Leicester, UK.

Many cancers contain cell subpopulations that display characteristics of stem cells. These cells are characterised by their ability to self-renew, form differentiated progeny and develop resistance to chemotherapeutic strategies. Cancer stem cells may utilise many of the same signalling pathways as normal stem cells including Wnt, Notch and Hedgehog… **Emerging evidence suggests that curcumin can exert its anti-carcinogenic activity via targeting cancer stem cells through the disruption of stem cell signalling pathways. In this review, we summarise the ability of curcumin to interfere with signalling pathways Wnt, Hedgehog, Notch, Signal Transducers and Activator (STAT) and interleukin-8, and report curcumin-induced changes in function and properties of cancer stem cells.**[13]

CURCUMIN (↓) NOTCH-1 SIGNALING
Molecular Biology Reports (The Netherlands) (2014)
Department of Pharmacy, The Second People's Hospital of Hefei, Anhui, People's Republic of China.
Extensive studies revealed that curcumin has chemopreventive properties, which are mainly due to its ability to arrest cell cycle and to induce apoptosis in cancer cells either alone or in combination with chemotherapeutic agents or radiation... **By regulating multiple important cellular signalling pathways including NF-κB, TRAIL, PI3K/Akt, JAK/STAT, Notch-1, JNK, etc., curcumin [is] known to activate cell death signals and induce apoptosis in pre-cancerous or cancer cells without affecting normal cells, thereby inhibiting tumor progression.**[14]

GREEN TEA EXTRACTS (↓) NOTCH-1 SIGNALING

GREEN TEA EXTRACT (EGCG) (↓) NOTCH-1 SIGNALING
Biochemical Pharmacology (England) (2011)
Department of Pharmacology, Toxicology and Therapeutics, and Medicine, The University of Kansas Cancer Center, The University of Kansas Medical Center, Kansas City.
Apparently, EGCG functions as a powerful antioxidant, preventing oxidative damage in healthy cells, but also as an antiangiogenic and antitumor agent and as a modulator of tumor cell response to chemotherapy. Much of the cancer chemopreventive properties of green tea are mediated by EGCG that induces apoptosis and promotes cell growth arrest by altering the expression of cell cycle regulatory proteins, activating killer caspases, and suppressing oncogenic transcription factors and pluripotency maintain factors [*sic*]. **In vitro studies have demonstrated that EGCG blocks carcinogenesis by affecting a wide array of signal transduction pathways including JAK/STAT, MAPK, PI3K/AKT, Wnt and Notch.** EGCG stimulates telomere fragmentation through inhibiting telomerase activity. Various clinical studies have revealed that treatment by EGCG inhibits tumor incidence and multiplicity in different organ sites such as liver, stomach, skin, lung, mammary gland and colon.[15]

GREEN TEA EXTRACT (EGCG) (↓) NOTCH-1 SIGNALING
Nutrition and Cancer (United States) (2011)
Department of Hematology and Oncology, First Hospital, Jilin University, China.
We investigated the anticancer activity of green tea extract (GTE) and EGCG on 3 human squamous carcinoma cell lines (CAL-27, SCC-25, and KB) in vitro. We also examined the effects of GTE and EGCG on cell signaling networks using our newly developed Pathway Array technology, which is an innovative proteomic assay to globally screen changes in protein expression and phosphorylation... The Pathway

Array assessment of 107 proteins indicated that different signaling pathways were activated in different cell lines, suggesting heterogeneity at the signaling network level. After treatment with GTE or EGCG, a total of 21 proteins and phosphorylations altered significantly in all 3 cell lines based on analysis of variance (ANOVA) ($P < 0.05$). **The major signaling pathways affected by GTE [green tea extract] and EGCG were EGFR and Notch pathways, which, in turn, affected cell cycle-related networks. These results suggested that GTE and EGCG target multiple pathways or global networks in cancer cells, which resulted in collective inhibition of cancer cell growth.**[16]

GREEN TEA EXTRACT (EGCG) (↓) NOTCH SIGNALING
European Journal of Cancer (England) (2013)
Department of Otorhinolaryngology-Head and Neck Surgery, Kangbuk Samsung Hospital, Sungkyunkwan University School of Medicine, Seoul, South Korea.
Most solid cancers including head and neck squamous carcinoma (HNSC) are believed to be initiated from and maintained by cancer stem cells (CSCs) that are responsible for treatment resistance, resulting in tumour relapse. Epigallocatechin-3-gallate (EGCG), the most abundant polyphenol in green tea, can potently inhibit cancer growth and induce apoptosis in various cancers, including HNSC... We demonstrated that EGCG inhibits the self-renewal capacity of HNSC CSCs by suppressing their sphere forming capacity, and attenuates the expression of stem cell markers, such as Oct4, Sox2, Nanog and CD44... **As one of [the] mechanisms of suppression of HNSC CSC traits, EGCG decreased the transcriptional level of Notch, resulting in the inhibition of Notch signalling.** Collectively, our data suggest that EGCG in combination with cisplatin can be used for the management of HNSC CSCs.[17]

These reports, in addition to some of those catalogued in the chapters above, indicate that curcumin and EGCG are pleiotropic, selectively acting on multiple cellular signaling pathways—including the JAK/STAT cascade that conveys extracellular signals into a transcriptional response, AP-1 and NF-kappa B transcription factors regulating inflammation, PI3K→Akt regulating insulin signaling, Notch-1, Wnt/β-catenin, the tumor suppressor protein p53, antioxidant response elements such as Nrf2, MAP and c-jun N-terminal kinases, and Epidermal Growth Factor Receptors dysregulated in some cancers—in ways that are beneficial in both AD and cancer. Curcumin and EGCG are thereby regarded among many expert AD and cancer researchers as promising, nontoxic agents in the prevention and treatment of AD and numerous cancers in ways that are again consistent with the epidemiological evidence and population studies summarized in Chapter 2.

Pursuant to the above reports that the curcuminoids and EGCG may act as gamma-secretase inhibitors while having anticancer effects, in part by selectively modifying

Notch-1 signaling, we look below at the functions in AD pathogenesis of GSK-3β, MAPK, and NMDA receptors, then at the manner in which these kinases and NMDA receptors are beneficially influenced by the plant polyphenols. The idea, as stated above, is to document further the extensive therapeutic pleiotropism of the plant polyphenols, including as applied to AD pathogenesis.

GLYCOGEN SYNTHASE KINASE-3 AND ALZHEIMER'S DISEASE

GSK-3β AND ALZHEIMER'S DISEASE
Neuroscience Letters (Ireland) (2001)
Department of Neurology, Gunma University School of Medicine, Japan.
To clarify how Abeta deposits induce secondary tauopathy, the presence of phosphorylated tau, glycogen synthase kinase 3alpha (GSK3alpha), GSK3beta, cyclin-dependent kinase 5 (CDK5), mitogen-activated protein kinase (MAPK) and fyn were examined in the Tg2576 brain showing substantial brain Abeta amyloidosis and behavioral abnormalities. Phosphorylated tau at Ser199, Thr231/Ser235, Ser396 and Ser413 accumulated in the dystrophic neurites of senile plaques. **The major kinase for tau phosphorylation was GSK3beta**. Smaller contributions of GSK3alpha, CDK5 and MAPK were suggested. **Thus, brain Abeta amyloidosis has a potential role in the induction of tauopathy leading to the mental disturbances of Alzheimer's disease.**[18]

GSK-3β AND ALZHEIMER'S DISEASE
Cellular and Molecular Life Sciences (Switzerland) (2006)
Division of Digestive Diseases, Emory University, Atlanta, Georgia.
Several studies have implicated molecular and cellular signaling cascades involving the serine-threonine kinase, glycogen synthase kinase beta (GSK-3beta), in the pathogenesis of AD. GSK-3beta may play an important role in the formation of neurofibrillary tangles and senile plaques, the two classical pathological hallmarks of AD. In this review, we discuss the interaction between GSK-3beta and several key molecules involved in AD, including the presenilins, amyloid precursor protein, tau, and beta-amyloid. We identify the signal transduction pathways involved in the pathogenesis of AD, including Wnt, Notch, and the PI3 kinase/Akt pathway. These may be potential therapeutic targets in AD.[19]

GSK-3β AND ALZHEIMER'S DISEASE
Journal of Alzheimer's Disease (The Netherlands) (2006)
Laboratory for Alzheimer's Disease, Riken Brain Science Institute, Saitama, Japan.
Glycogen synthase kinase-3 (GSK-3) is a pivotal molecule in the development of Alzheimer's disease (AD). GSK-3beta is involved in the formation of paired helical filament (PHF)-tau, which is an integral component of

the neurofibrillary tangle (NFT) deposits that disrupt neuronal function, and a marker of neurodegeneration in AD... In the present review, we discuss our initial in vitro results and additional investigations showing that Abeta activates GSK-3beta through impairment of phosphatidylinositol-3 (PI3)/Akt signaling; that Abeta-activated GSK-3beta induces hyperphosphorylation of tau, NFT formation, neuronal death, and synaptic loss (all found in the AD brain); that GSK-3beta can induce memory deficits in vivo; and that inhibition of GSK-3alpha (an isoform of GSK-3beta) reduces Abeta production. These combined results strongly suggest that GSK-3 activation is a critical step in brain aging and the cascade of detrimental events in AD, preceding both the NFT and neuronal death pathways. Therefore, therapeutics targeted to inhibiting GSK-3 may be beneficial in the treatment of this devastating disease.[20]

GSK-3 AND ALZHEIMER'S DISEASE
Expert Review of Neurotherapeutics (England) (2007)
Centro de Biología Molecular Severo Ochoa, CSIC/UAM, Universidad Autónoma de Madrid, Spain.

Glycogen synthase kinase (GSK)-3 has been proposed as the link between the two histopathological hallmarks of Alzheimer's disease, the extracellular senile plaques made of beta-amyloid and the intracellular neurofibrillary tangles made of hyperphosphorylated tau. Thus, GSK-3 is one of the main tau kinases and it modifies several sites of tau protein present in neurofibrillary tangles. Furthermore, GSK-3 is able to modulate the generation of beta-amyloid as well as to respond to this peptide. The use of several transgenic models overexpressing GSK-3 has been associated with neuronal death, tau hyperphosphorylation and a decline in cognitive performance.[21]

GSK-3 AND ALZHEIMER'S DISEASE
Journal of Alzheimer's Disease (The Netherlands) (2008)
Instituto de Química Médica-CSIC, Madrid, Spain.

Increased activity and/or overexpression of this enzyme [GSK-3] in AD is associated with increased tau hyperphosphorylation and alterations in amyloid-beta processing that are thought to precede the formation of neurofibrillary tangles and senile plaques, respectively. Furthermore, over-activity of GSK-3 is also involved in neuronal loss. These data clearly identify GSK-3 inhibitors as one of the most promising new approaches for the future treatment of AD and a reduction of the aberrant over-activity of this enzyme might decrease several aspects of the neuronal pathology in AD.[22]

GSK-3 AND ALZHEIMER'S DISEASE
IUBMB Life (England) (2009)

Centre for the Cellular Basis of Behaviour, Institute of Psychiatry, King's College London, England.

Abnormalities in molecular signalling have been implicated in neurodegeneration. **It is emerging that glycogen synthase kinase-3 (GSK-3) is a key signalling molecule that induces neurodegeneration and deficits in memory formation related to Alzheimer's disease (AD).** Early stages of AD are associated with deficits in memory formation before neuronal cell death is detectable. Recent studies in rodents have suggested that these impairments in memory formation might result from increased GSK-3 signalling, because enhanced GSK-3 activity impairs hippocampal memory formation. GSK-3 activity blocks synaptic long-term potentiation and induces long-term depression. Furthermore, increased GSK-3 signalling is likely to be a key contributor to the formation of the pathological hallmarks in AD, neurofibrillary tangles (NFTs) and amyloid plaques. Recent studies with mouse models have indicated that GSK-3, but not cyclin-dependent kinase 5, is critical for hyperphosphorylation of the cytoskeletal protein tau, which is the prerequisite for NFT formation in AD... **Taken together, the current evidences suggest that increased GSK-3 signalling may be responsible for the deficits in memory formation in early stages of AD and neurodegeneration in later stages of the disease.**[23]

GSK-3β AND ALZHEIMER'S DISEASE

Current Alzheimer Research (United Arab Emirates) (2012)

Department of Pharmaceutical Sciences, Texas A&M Health Science Center, Kingsville.

Evidence from basic molecular biology has noted a critical role of GSK-3 in Alzheimer's disease (AD) pathogenesis such as beta-amyloid (Aβ) production and accumulation, the formation of neurofibrillary tangle (NFT), and neuronal degeneration. Aβ generation and deposition represents a key feature and is generated from APP by the sequential actions of two proteolytic enzymes: β-secretase and γ-secretase. GSK-3 could play a critical role in Aβ production via enhancing β-secretase activity. GSK-3 not only modulates APP processing in the process of Aβ generation, but regulates Aβ production by interfering with APP cleavage at the γ-secretase complex step since the APP and PS1 (a component of γ-secretase complex) are substrates of GSK-3 as well. GSK-3 may downregulate α-secretase through inhibiting PKC and ADAMs activity which are the substrates of GSK-3 contributing to Aβ production. Meanwhile, Aβ accumulation can induce GSK-3 activation through Aβ-mediated neuroinflammation and oxidative stress. Considering that active GSK-3 and some common GSK-3-shared factors induce the hyperphosphorylation of tau and neurofibrillary lesions, GSK-3 is a possible linking between amyloid plaques and NFT pathology. Additionally, GSK-3 could disrupt acetylcholine activity, and accelerate axon degeneration and failures in axonal transport, and lead to cognitive impairment in AD. **Preclinical and clinical studies have supported that GSK-3β inhibitors could be useful in the treatment of AD. Consequently, an effective measure to inhibit GSK-3 activity may be a very attractive drug target in AD.**[24]

CURCUMINOIDS (↓) GLYCOGEN SYNTHASE KINASE-3

CURCUMIN (↓) GSK-3β

Journal of Enzyme Inhibition and Medicinal Chemistry (England) (2009)

Faculty of Pharmacy, University of Jordan, Amman.

Curcumin was investigated as an inhibitor of glycogen synthase kinase-3beta (GSK-3beta) in an attempt to explain some of its interesting multiple pharmacological effects, such as its anti-diabetic, anti-inflammatory, anti-cancer, anti-malarial and anti-Alzheimer's properties. The investigation included simulated docking experiments to fit curcumin within the binding pocket of GSK-3beta followed by experimental in vitro and in vivo validations. Curcumin was found to optimally fit within the binding pocket of GSK-3beta via several attractive interactions with key amino acids. **Experimentally, curcumin was found to potently inhibit GSK-3beta (IC50 = 66.3 nM)... Our findings strongly suggest that the diverse pharmacological activities of curcumin are at least partially mediated by inhibition of GSK-3beta.**[25]

CURCUMIN (↓) GSK-3β

European Journal of Pharmaceutical Sciences (The Netherlands) (2011)

Institute of Neuroscience, Chongqing Medical University, China.

Wnt/β-catenin signaling pathway plays an important role in the genesis and development of Alzheimer's disease. The study aims to investigate the effect of curcumin on the expression of GSK-3β, β-catenin and CyclinD1 in vitro, which are tightly correlated with Wnt/β-catenin signaling pathway, and also to explore the mechanisms, which will provide a novel therapeutic intervention for treatment of Alzheimer's disease... RT-PCR and Western blot results showed that the expression of GSK-3β mRNA and protein significantly decreased in the transfected cells treated with curcumin, and that the changes were in a dose and time-dependent manner (P<0.05); however, the protein expression of GSK-3β-Ser9 was increased (P<0.05)... **GSK-3β is a potential target for treatment of AD. Curcumin could activate the Wnt/β-catenin signaling pathway through inhibiting the expression of GSK-3β and inducing the expression of β-catenin and CyclinD1, which will provide a new theory for treatment of neurodegenerative diseases by curcumin.**[26]

CURCUMIN (↓) GSK-3β
Pharmacological Reports (Poland) (2011)
Department of Pathology, Chongqing Medical University, China.

The present study aimed to investigate the effects of curcumin on the generation of Aβ in cultured neuroblastoma cells and on the in vitro expression of PS1 and GSK-3β… **Curcumin treatment was found to markedly reduce the production of Aβ(40/42). Treatment with curcumin also decreased both PS1 and GSK-3β mRNA and protein levels in a dose- and time-dependent manner. Furthermore, curcumin increased the inhibitory phosphorylation of GSK-3β protein at Ser9. Therefore, we propose that curcumin decreases Aβ production by inhibiting GSK-3β-mediated PS1 activation.**[27]

CURCUMIN (↓) GSK-3β
Journal of Alzheimer's Disease (The Netherlands) (2012)
College of Life Sciences and Technology, Beijing University of Chemical Technology, Beijing, China.

In this study, we investigated the protective effects of curcumin against mitochondrial dysfunction induced by Aβ… **This study demonstrates curcumin-mediated neuroprotection against Aβ-induced mitochondrial metabolic deficiency and abnormal alteration of oxidative stress. Inhibition of GSK-3β is involved in the protection of curcumin against Aβ-induced mitochondrial dysfunction.**[28]

CURCUMIN (↓) GSK-3β
Journal of Receptor and Signal Transduction Research (England) (2014)
Beijing Key Laboratory of Bioactive Substances and Functional Foods, Beijing Union University, People's Republic of China.

In this study, cellular signaling of tau phosphorylation induced by Aβ and the inhibiting effects of curcumin on this signaling were investigated on human neuroblastoma SH-SY5Y cells. **The results indicated that curcumin inhibits Aβ-induced tau phosphorylation at Thr231 and Ser396, over-expression of HDAC6, and decrease in phosphorylation of glycogen synthase kinase-3β (GSK-3β) at Ser9… Curcumin depresses Aβ-induced up-regulation of PTEN induced by Aβ. These results imply that curcumin inhibits Aβ-induced tau hyperphosphorylation involving PTEN/Akt/GSK-3β pathway.**[29]

CURCUMIN (↓) GSK-3β
Journal of Medicinal Chemistry (United States) (2016)
Department of Pharmacy and Biotechnology, Alma Mater Studiorum, University of Bologna, Italy; Department for Life Quality Studies, Alma Mater Studiorum, University of Bologna;

Istituto Italiano di Tecnologia, Genova, Italy; Centro de Investigaciones Biologicas, Madrid, Spain.

The multitarget approach has gained increasing acceptance as a useful tool to address complex and multifactorial maladies such as Alzheimer's disease (AD). **The concurrent inhibition of the validated AD targets β-secretase (BACE-1) and glycogen synthase kinase-3β (GSK-3β) by attacking both β-amyloid and tau protein cascades has been identified as a promising AD therapeutic strategy. In our study, curcumin was identified as a lead compound for the simultaneous inhibition of both targets.**[30]

GREEN TEA EXTRACTS (↓) GLYCOGEN SYNTHASE KINASE-3

Green Tea Extract (EGCG) (↓) Gsk-3
Molecular Brain Research (The Netherlands) (2003)
Department of General Toxicology, National Institute of Toxicological Research, KFDA, Seoul, South Korea.

The effects of epigallocatechin gallate (EGCG) on the phosphoinositide 3-kinase (PI3K)/Akt and glycogen synthase kinase-3 (GSK-3) pathway during oxidative-stress-induced injury were studied using H2O2-treated PC12 cells, which were differentiated by nerve growth factor (NGF)… Upon examination of the PI3K/Akt and GSK-3 upstream pathway, Western blots of EGCG pretreated cells showed decreased immunoreactivity (IR) of Akt and GSK-3 and increased IR of p85a PI3K, phosphorylated Akt and phosphorylated GSK-3… **These results show that EGCG affects the PI3K/Akt, GSK-3 pathway as well as downstream signaling, including the cytochrome c and caspase-3 pathways. Therefore, it is suggested that EGCG-mediated activation of PI3K/Akt and inhibition of GSK-3 could be a new potential therapeutic strategy for neurodegenerative diseases associated with oxidative injury.**[31]

Green Tea Extract (EGCG) (↓) Gsk-3β
Neurobiology of Aging (United States) (2009)
Institute of Biochemistry and Molecular Biology, College of Medicine, National Taiwan University, Taipei.

Here, we used a human neuronal cell line MC65 conditional expression of an amyloid precursor protein fragment (APP-C99) to investigate the protection mechanism of epigallocatechin gallate (EGCG), the main constituent of green tea. We demonstrated that treatment with EGCG reduced the A beta levels by enhancing endogenous APP nonamyloidogenic proteolytic processing. **Furthermore, EGCG also decreased nuclear translocation of c-Abl and blocked APP-C99-dependent GSK3 beta activation, and these inhibitory effects** occurred through the interruption of c-Abl/Fe65 interaction. **Our results indicated that the neuroprotective action of EGCG may take place through some mechanisms other than the promotion of APP nonamyloidogenic proteolysis, as was reported previously.**[32]

GREEN TEA EXTRACT (EGCG) (↓) GSK-3β
Molecular and Cellular Biochemistry (The Netherlands) (2013)
Department of Neurology, First Affiliated Hospital of China Medical University, Shenyang, China.
Alzheimer's disease (AD) fundamentally represents a metabolic disease associated with brain insulin resistance. TNF-α/c-Jun N-terminal kinase (JNK) signaling plays a central role in serine phosphorylation of insulin receptor substrate-1 (IRS-1). (-)-Epigallocatechin-3-gallate (EGCG), a potent antioxidant, has been verified to attenuate peripheral insulin resistance by reducing IRS-1 signaling blockage. This study aimed to investigate the effects and possible mechanisms of EGCG on central IRS-1 signaling in vivo… Our results showed that EGCG ameliorated the impaired learning and memory in APP/PS1 mice. Notably, we found a significant reduction of IRS-1pS636 level accompanied with decreased Aβ42 levels in the hippocampus of 13-month-old female APP/PS1 mice after treatment with EGCG (2 or 6 mg/kg/day) for 4 weeks. **Furthermore, EGCG treatment inhibited TNF-α/JNK signaling and increased the phosphorylation of Akt and glycogen synthase kinase-3β in the hippocampus of APP/PS1 mice. In conclusion, our study provides evidence that long-term consumption of EGCG may alleviate AD-related cognitive deficits by effectively attenuating central insulin resistance.**[33]

RESVERATROL (↓) GLYCOGEN SYNTHASE KINASE-3

RESVERATROL (↓) GSK-3β
International Journal of Food Sciences and Nutrition (England) (2014)
Division of Nutritional Sciences, Cornell University, Ithaca, New York.
This exploratory work investigates if dietary resveratrol, previously shown to have broad anti-aging effects and improve AD pathology in vivo, leads to neuroprotective changes in specific protein targets in the mouse brain. Both wild-type and APP/PS1 mice, a transgenic AD mouse model, received control AIN-93G diet or AIN-93G supplemented with resveratrol. Pathology parameters and AD risk were assessed via measurements on plaque burden, levels of phosphorylated glycogen synthase kinase 3-β (GSK3-β), tau, transthyretin and drebrin. Dietary resveratrol treatment did not decrease plaque burden in APP/PS1 mice. **However, resveratrol-fed mice demonstrated increases in GSK3-β phosphorylation, a 3.8-fold increase in**

protein levels of transthyretin, and a 2.2-fold increase in drebrin. This study broadens our understanding of specific mechanisms and targets whereby resveratrol provides neuroprotection.[34]

MITOGEN-ACTIVATED PROTEIN KINASE AND ALZHEIMER'S DISEASE

MAPK AND ALZHEIMER'S DISEASE
The EMBO Journal (England) (1992)
Max Planck Unit for Structural Molecular Biology, Hamburg, Germany.
The microtubule-associated protein tau is a major component of the paired helical filaments (PHFs) observed in Alzheimer's disease brains. The pathological tau is distinguished from normal tau by its state of phosphorylation, higher apparent M(r) and reaction with certain antibodies. However, the protein kinase(s) have not been characterized so far. **Here we describe a protein kinase from brain which specifically induces the Alzheimer-like state in tau protein. The 42 kDa protein belongs to the family of mitogen activated protein kinases (MAPKs)** and is activated by tyrosine phosphorylation. It is capable of phosphorylating Ser-Pro and Thr-Pro motifs in tau protein (approximately 14-16 P1 per tau molecule). By contrast, other proline directed Ser/Thr kinases such as p34(cdc2) combined with cyclin A or B have only minor effects on tau phosphorylation. **We propose that MAP kinase is abnormally active in Alzheimer brain tissue, or that the corresponding phosphatases are abnormally passive, due to a breakdown of the normal regulatory mechanisms.**[35]

MAPK AND ALZHEIMER'S DISEASE
Annals of the New York Academy of Sciences (United States) (1993)
Max Planck Unit for Structural Molecular Biology, Hamburg, Germany.
This paper summarizes our recent studies on microtubule-associated protein tau and its pathological state resembling that of the paired helical filaments of Alzheimer's disease. **The Alzheimer-like state of tau protein can be identified and analyzed in terms of certain phosphorylation sites and phosphorylation-dependent antibody epitopes. It can be induced by protein kinases which tend to phosphorylate serine or threonine residues followed by a proline; this includes mitogen-activated protein kinase (MAPK) and glycogen-synthase kinase 3 (GSK-3).** Both of these are tightly associated with microtubules as well as with paired helical filaments.[36]

MAPK AND ALZHEIMER'S DISEASE
Journal of Neuropathology and Experimental Neurology (England) (1999)
Alzheimer's Disease Research Unit, Massachusetts General Hospital East, Charlestown.

Indirect evidence suggests that activation of the mitogen activated protein kinase (MAPK) cascade contributes to the hyperphosphorylation of tau found in paired helical filaments in Alzheimer disease (AD). We report colocalization of the activated form of MAPK with Ser 199/202 and Ser 396/404 phosphotau immunoreactive neurofibrillary tangles and neuritic plaques in the Alzheimer brain. Fluorescence resonance energy transfer studies (FRET) demonstrate a tight intermolecular association of activated MAPK with these phosphotau epitopes. **These data support the hypothesis that activation of MAPK contributes directly to phosphorylation of tau in AD. Moreover, the stable nature of this association in postmortem human brain may suggest a stable interaction in which activated MAPK becomes tightly linked to neurofibrillary tangles.**[37]

MAPK AND ALZHEIMER'S DISEASE
Neurochemistry International (England) (2001)

The Donald W. Reynolds Department of Geriatrics, University of Arkansas for Medical Sciences, Little Rock.

Activated (phosphorylated) mitogen-activated protein kinase p38 (MAPK-p38) and interleukin-1 (IL-1) have both been implicated in the hyperphosphorylation of tau, a major component of the neurofibrillary tangles in Alzheimer's disease. This, together with findings showing that IL-1 activates MAPK-p38 in vitro and is markedly overexpressed in Alzheimer brain suggest a role for IL-1-induced MAPK-p38 activation in the genesis of neurofibrillary pathology in Alzheimer's disease. We found frequent colocalization of hyperphosphorylated tau protein (AT8 antibody) and activated MAPK-p38 in neurons and in dystrophic neurites in Alzheimer brain, and frequent association of these structures with activated microglia overexpressing IL-1. Tissue levels of IL-1 mRNA as well as of both phosphorylated and non-phosphorylated isoforms of tau were elevated in these brains. Significant correlations were found between the numbers of AT8- and MAPK-p38-immunoreactive neurons, and between the numbers of activated microglia overexpressing IL-1 and the numbers of both AT8- and MAPK-p38-immunoreactive neurons... **These results suggest that microglial activation and IL-1 overexpression are part of a feedback cascade in which MAPK-p38 overexpression and activation leads to tau hyperphosphorylation and neurofibrillary pathology in Alzheimer's disease.**[38]

MAPK AND ALZHEIMER'S DISEASE
Neuro-Signals (Switzerland) (2002)

Institute of Pathology, Case Western Reserve University, Cleveland, Ohio.

Given the critical role of mitogen-activated protein kinase (MAPK) pathways in regulating cellular processes that are affected in Alzheimer's disease (AD), the importance of MAPKs in disease pathogenesis is being increasingly recognized. All MAPK pathways, i.e., the extracellular signal-regulated kinase (ERK), c-Jun N-terminal kinase (JNK) and p38 pathways, are activated in vulnerable neurons in patients with AD suggesting that MAPK pathways are involved in the pathophysiology and pathogenesis of AD. **Here we review recent findings implicating the MAPK pathways in AD and discuss the relationship between these pathways and the prominent pathological processes, i.e., tau phosphorylation and amyloid-beta deposition, as well as the functional association to amyloid beta protein precursor. We suggest that regulation of these pathways may be a central facet to any potential treatment for the disease.**[39]

MAPK AND ALZHEIMER'S DISEASE
Sheng Li Xue Bao (China) (2008)
Department of Pathophysiology, Tongji Medical College, Huazhong University of Science and Technology, Wuhan, China.

One of the pathological features of Alzheimer's disease (AD) is neurofibrillary tangles (NFTs), which consist of paired helical filaments (PHFs) formed by hyperphosphorylated microtubule-associated protein tau. To study the role of mitogen-activated protein kinase (MAPK) in tau hyperphosphorylation and the underlying mechanism, wild type mouse neuroblastoma cells (N2a) were dealt with different concentrations (0.1 microg/mL, 0.2 microg/mL and 0.4 microg/mL) of anisomycin (an activator of MAPK) for 6h… **In conclusion, over-activation of MAPK up to a certain degree induces tau hyperphosphorylation at Ser-198/199/202 and Ser-396/404 sites, and this is probably related to the effect of activated GSK-3 by MAPK.**[40]

MAPK AND ALZHEIMER'S DISEASE
Neuropharmacology (England) (2010)
Department of Pharmaceutical Chemistry, Faculty of Pharmacy, The University of Sydney, Australia.

Accumulating evidence indicates that p38 mitogen-activated protein kinase (MAPK) could play more than one role in Alzheimer's disease (AD) pathophysiology and that patients suffering from AD dementia could benefit from p38 MAPK inhibitors. The p38 MAPK signalling has been widely accepted as a cascade contributing to neuroinflammation. However, deepening insight into the underlying biology of Alzheimer's disease reveals that p38 MAPK operates in other events related to AD, such as excitotoxicity, synaptic plasticity and tau phosphorylation. Although quantification of behavioural improvements upon p38 MAPK inhibition and in vivo evaluation of p38 MAPK significance to various aspects of AD pathology is still missing, **the p38 MAPK is emerging as a new Alzheimer's disease treatment strategy.**[41]

CURCUMINOIDS (↓) MITOGEN-ACTIVATED PROTEIN KINASE

DEMETHOXYCURCUMIN (↓) MAPK
International Immunopharmacology (The Netherlands) (2009)

Department of Pharmacology, Shenyang Pharmaceutical University, Shenyang, People's Republic of China.

In the present study, the effect and possible mechanism of demethoxycurcumin on the production of pro-inflammatory mediators in LPS-activated N9 microglial cells were further investigated. The results showed that demethoxycurcumin significantly suppressed the NO production induced by LPS in N9 microglial cells through inhibiting the protein and mRNA expression of inducible NO synthase (iNOS). Demethoxycurcumin also decreased LPS-induced TNF–alpha and IL–1beta expression at both transcriptional and protein level in a concentration-dependent manner. Further studies revealed that demethoxycurcumin blocked IkappaBalpha phosphorylation and degradation, inhibited the phosphorylation of mitogen-activated protein kinases (MAPKs)... **In summary, these data suggest that demethoxycurcumin exerts its in vitro anti-inflammatory effect in LPS-activated N9 microglial cells by blocking nuclear factor-kappaB (NF-kappaB) and MAPKs activation, which may be partly due to its potent down-regulation of the NADPH-derived iROS production.**[42]

CURCUMIN (↓) p38–MAPK
Neuroscience Letters (Ireland) (2015)

Guangdong Key Laboratory for Diagnosis and Treatment of Major Neurological Diseases, Department of Neurology, The First Affiliated Hospital, Sun Yat-sen University, Guangzhou, People's Republic of China; Department of Neurology, Zengcheng People's Hospital, People's Republic of China.

In the present study, we found that curcumin improved microglial viability against Aβ42 in a time- and dose-dependent manner and remarkably suppressed Aβ42-induced CD68 expression. Moreover, curcumin concentration-dependently abolished Aβ42-induced interleukin-1β (IL-1β), interleukin-6 (IL-6) and tumor necrosis factor-α (TNF-α) production in mRNA and protein levels in microglia. **Besides, curcumin exerted an inhibitory effect on phosphorylation of ERK1/2 and p38 in Aβ42-activated microglia. Further experiments indicated that blockage of ERK1/2 and p38 pathways reduced inflammatory cytokines production from microglia. These results show that curcumin suppresses ERK1/2 and p38 signaling, thus attenuating inflammatory responses of brain microglia.**[43]

CURCUMIN (↓) MAPK
Cellular and Molecular Neurobiology (United States) (2016)

Key Lab of Cerebral Microcirculation in Universities of Shandong, Taishan Medical University, China; School of Basic Medicine, Taishan Medical University; Departments of Rehabilitation, Taian Central Hospital, China.

The present study evaluated the protective effect of curcumin against Aβ-induced cytotoxicity and apoptosis in PC12 cells and investigated the underlying mechanism. The results showed that curcumin markedly reduced Aβ-induced cytotoxicity by inhibition of mitochondria-mediated apoptosis through regulation of Bcl-2 family. The PARP cleavage, caspases activation, and ROS-mediated DNA damage induced by Aβ were all significantly blocked by curcumin. **Moreover, regulation of p38 MAPK and AKT pathways both contributed to this protective potency.**[44]

Curcumin (↓) MAPK
Molecular Neurobiology (United States) (2016)
School of Basic Medicine, Taishan Medical University, Taian, Shandong, China; Key Lab of Cerebral Microcirculation in Universities of Shandong, Taishan Medical University, Taian; Shandong, China; Department of Neurosurgery, Huxi Hospital, Jining Medical University, Shanxian, Shandong, China; Department of Neurology and Center of Cerebrovascular Disease Research, University of Pittsburgh, Pennsylvania; Department of Neurology, Affiliated Hospital of Taishan Medical University, China.

Herein, we evaluated the protective effect of curcumin on PC12 cells against H2O2-induced neurotoxicity and investigated its underlying mechanism. The results indicated that curcumin pre-treatment significantly suppressed H2O2-induced cytotoxicity, inhibited the loss of mitochondrial membrane potential (Δψm) through regulation of Bcl-2 family expression, and ultimately reversed H2O2-induced apoptotic cell death in PC12 cells. Attenuation of caspase activation, poly(ADP-ribose) polymerase (PARP) cleavage, DNA damage, and accumulation of reactive oxygen species (ROS) all confirmed its protective effects. **Moreover, curcumin markedly alleviated the dysregulation of the MAPK and AKT pathways induced by H2O2.**[45]

MICROGLIAL ACTIVATION AND ALZHEIMER'S DISEASE

Microglial Activation and Alzheimer's Disease
International Journal of Alzheimer's Disease (United States) (2011)
Clinical Neurochemistry Laboratory, Department of Neurochemistry and Psychiatry, Institute of Neuroscience and Physiology, The Sahlgrenska Academy at University of Gothenburg, Mölndal, Sweden.

Intensive research over the last decades has provided increasing evidence for neuro-inflammation as an integral part in the pathogenesis of neurodegenerative diseases such as Alzheimer's disease (AD). **Inflammatory responses in the central** nervous system (CNS) are initiated by activated microglia, representing the first line of the innate immune defence of the brain. Therefore, biochemical markers of microglial activation may help us understand the underlying mechanisms of neuroinflammation in AD as well as the double-sided qualities of microglia, namely, neuroprotection and neurotoxicity.[46]

MICROGLIAL ACTIVATION AND ALZHEIMER'S DISEASE
Frontiers in Pharmacology (Switzerland) (2012)

Centre for Translational Medicine and Therapeutics, William Harvey Research Institute, Barts and The London, Queen Mary's School of Medicine and Dentistry London, England.

The microglia, the resident "macrophages" of the brain's innate immune system, are most responsive, and increasing evidence suggests that they enter a hyper-reactive state in neurodegenerative conditions and aging. As sustained over-production of microglial pro-inflammatory mediators is neurotoxic, this raises great concern that systemic inflammation (that also escalates with aging) exacerbates or possibly triggers neurological diseases (Alzheimer's, prion, motor-neuron disease)... **On one hand, microglia may delay the progression of AD by contributing to the clearance of Aβ, since they phagocyte Aβ and release enzymes responsible for Aβ degradation. Microglia also secrete growth factors and anti-inflammatory cytokines, which are neuroprotective. In addition, microglia removal of damaged cells is a very important step in the restoration of the normal brain environment... On the other hand, as we age microglia become steadily less efficient at these processes, tending to become over-activated in response to stimulation and instigating too potent a reaction, which may cause neuronal damage in its own right.**[47]

MICROGLIAL ACTIVATION AND ALZHEIMER'S DISEASE
Neurobiology of Aging (United States) (2013)

European Institute for Molecular Imaging, Westfalian Wilhelms University Münster, Germany.
In Alzheimer's disease (AD), persistent microglial activation as [a] sign of chronic neuroinflammation contributes to disease progression.[48]

MICROGLIAL ACTIVATION AND ALZHEIMER'S DISEASE
Alzheimer's and Dementia: The Journal of the Alzheimer's Association (United States) (2015)

Neurology Imaging Unit, Department of Medicine, Imperial College London, Hammersmith Hospital, England; Inserm-EPHE-University of Caen/Basse-Normandie, France; Department of Neurology, National Parkinson Foundation Centre of Excellence, King's College Hospital, and King's Health Partners, London, England.

Alzheimer's disease (AD) and Parkinson's disease (PD) are the two common neurode-generative diseases characterized by progressive neuronal dysfunction in the presence of pathological microglial activation… **AD, MCI, and PDD [Parkinson's disease dementia] subjects demonstrated significant correlation between increased microglial activation and reduced glucose metabolism (rCMRGlc). AD and MCI subjects also showed significant positive correlation between amyloid and microglial activation. Levels of cortical microglial activation were negatively correlated with Mini-Mental State Examination in both AD and PDD.** The significant inverse correlations between cortical levels of microglial activation and glucose metabolism in AD and PDD suggest cortical neuroinflammation may drive neuronal dysfunction in these dementias.[49]

MICROGLIAL ACTIVATION AND ALZHEIMER'S DISEASE
Journal of Neuroinflammation (England) (2015)
Department of Pharmacology, Shanghai Jiao Tong University School of Medicine, Shanghai, China; Shanghai Institute of Immunology, Shanghai Jiao Tong University School of Medicine, Shanghai.

Aggregated forms of amyloid-β (Aβ) peptides are important triggers for microglial activation, which is an important pathological component in the brains of Alzheimer's patients. Cu(II) [copper oxide] ions are reported to be coordinated to monomeric Aβ, drive Aβ aggregation, and potentiate Aβ neurotoxicity. Here we investigated whether Cu(II) binding modulates the effect of Aβ on microglial activation and the subsequent neurotoxicity… **Our observations suggest that Cu(II) enhances the effect of Aβ on microglial activation and the subsequent neurotoxicity. The Cu(II)-Aβ-triggered microglial activation involves NF-κB activation and mitochondrial ROS production.**[50]

MICROGLIAL ACTIVATION AND ALZHEIMER'S DISEASE
Journal of Alzheimer's Disease (The Netherlands) (2016)
Neurology Imaging Unit, Imperial College London, England; Department of Nuclear Medicine, Aarhus University, Denmark.

The influence of neuroinflammation on neuronal function and hippocampal atrophy in Alzheimer's disease and Parkinson's disease dementia is still unclear. Here we investigated whether microglial activation measured by [11C]PK11195 PET is associated with neuronal function measured by cerebral glucose metabolic rate (rCMRGlc) using FDG-PET and hippocampal volume measurements… **These findings indicate that microglial activation inversely correlated with hippocampal volume and hippocampal rCMRGlc in neurodegenerative diseases with dementia, providing further evidence for the central role of microglial activation in neurodegenerative diseases.**[51]

MICROGLIAL ACTIVATION AND ALZHEIMER'S DISEASE

Brain (England) (2016)

Unit of Memory and Language, Université Paris Descartes, Sorbonne, Paris.

While emerging evidence suggests that neuroinflammation plays a crucial role in Alzheimer's disease, the impact of the microglia response in Alzheimer's disease remains a matter of debate. We aimed to study microglial activation in early Alzheimer's disease and its impact on clinical progression using a second-generation 18-kDa translocator protein positron emission tomography radiotracer together with amyloid imaging using Pittsburgh compound B positron emission tomography… **Microglial activation appears at the prodromal and possibly at the preclinical stage of Alzheimer's disease, and seems to play a protective role in the clinical progression of the disease at these early stages.** The extent of microglial activation appears to differ between patients, and could explain the overlap in translocator protein binding values between patients with Alzheimer's disease and amyloidosis controls.[52]

CURCUMINOIDS (↓) MICROGLIAL ACTIVATION

CURCUMIN (↓) MICROGLIAL ACTIVATION

Annals of the New York Academy of Sciences (United States) (2004)

Greater Los Angeles Healthcare System, Veterans Administration Medical Center, North Hills, California.

Curcumin proved to be immunomodulatory, simultaneously inhibiting cytokine and microglial activation indices related to neurotoxicity, but increasing an index of phagocytosis. Curcumin directly targeted Abeta and was also effective in other models, warranting further preclinical and clinical exploration.[53]

CURCUMIN (↓) MICROGLIAL ACTIVATION

Die Pharmazie (Germany) (2007)

Pharmacology Department, National Institute of Toxicological Research, Seoul, South Korea.

In this study, curcumin's neuroprotective effect was carefully examined using a coculture system, based on reports that curcumin-containing plants are neuroprotective. Coculturing neuronal cells and activated microglial cells enhanced dopamine-induced neuronal cell death from 30% up to 50%. **However, curcumin did not protect dopamine-directed neuronal cell death and sodium nitroprosside (SNP)-induced NO generation, but only blocked activated microglial cell-mediated neuronal cell damage under inflammatory conditions. Indeed, curcumin blocked the production of pro-inflammatory and cytotoxic mediators such as NO, TNF-alpha, IL-1alpha, and IL-6 produced from Abeta(25-35)/IFN-gamma- and LPS-stimulated microglia, in a dose-dependent manner.**

Therefore, our results suggest that curcumin-mediated neuroprotective effects may be mostly due to its anti-inflammatory effects.[54]

CURCUMIN (↓) MICROGLIAL ACTIVATION
Acta Pharmacologica Sinica (China/United States) (2007)
Department of Microbiology, College of Natural Sciences, Pusan National University, Busan, South Korea.

Curcumin significantly inhibited the release of NO, PGE2, and pro-inflammatory cytokines in a dose-dependent manner. Curcumin also attenuated the expressions of inducible NO synthase and cyclooxygenase-2 mRNA and protein levels. Moreover, curcumin suppressed NF-kappaB activation via the translocation of p65 into the nucleus. Our data also indicate that curcumin exerts anti-inflammatory properties by suppressing the transcription of proinflammatory cytokine genes through the NF-kappaB signaling pathway. **Anti-inflammatory properties of curcumin may be useful for treating the inflammatory and deleterious effects of microglial activation in response to LPS stimulation.**[55]

CURCUMIN (↓) MICROGLIAL INFLAMMATORY RESPONSES
Journal of Neuroinflammation (England) (2011)
Institute of Human Genetics, University of Regensburg, Germany.

To study the immuno-modulatory effects of curcumin on a transcriptomic level, DNA-microarray analyses were performed with resting and LPS-challenged microglial cells after short-term treatment with curcumin... Curcumin treatment markedly changed the microglial transcriptome with 49 differentially expressed transcripts in a combined analysis of resting and activated microglial cells. Curcumin effectively triggered anti-inflammatory signals as shown by induced expression of Interleukin 4 and Peroxisome proliferator activated receptor α. Several novel curcumin-induced genes, including Netrin G1, Delta-like 1, Platelet endothelial cell adhesion molecule 1, and Plasma cell endoplasmic reticulum protein 1, have been previously associated with adhesion and cell migration. **Consequently, curcumin treatment significantly inhibited basal and activation-induced migration of BV-2 microglia**. Curcumin also potently blocked gene expression related to pro-inflammatory activation of resting cells including Toll-like receptor 2 and Prostaglandin-endoperoxide synthase 2. Moreover, transcription of NO synthase 2 and Signal transducer and activator of transcription 1 was reduced in LPS-triggered microglia. These transcriptional changes in curcumin-treated LPS-primed microglia also lead to decreased neurotoxicity with reduced apoptosis of 661W photoreceptor cultures. **Collectively, our results suggest that curcumin is a potent modulator of the microglial transcriptome. Curcumin attenuates microglial migration and triggers a phenotype with anti-inflammatory and neuroprotective properties.**[56]

CURCUMIN (↓) MICROGLIAL ACTIVATION

International Immunopharmacology (The Netherlands) (2016)

Department of Biosciences, Biotechnologies and Biopharmaceutics, University of Bari, Italy; Department of Clinical and Experimental Medicine, University of Foggia, Italy.

We hypothesized that curcumin supplementation could reduce the inflammatory responses of activated microglial cells modulating PI3K/Akt pathway. Different curcumin concentrations were administered as BV-2 microglia pre-treatment 1h prior to LPS stimulation… Curcumin significantly attenuated, in a dose-dependent manner, LPS-induced release of NO and pro-inflammatory cytokines, as well as iNOS expression. Interestingly, curcumin was able to reduce, again in a dose-dependent manner, PI3K/Akt phosphorylation as well as NF-κB activation in LPS-activated microglial cells. **Overall these results suggest that curcumin plays an important role in the attenuation of LPS-induced inflammatory responses in microglial cells and that the mechanisms involve down-regulation of the PI3K/Akt signalling.**[57]

CURCUMIN (↓) MICROGLIAL ACTIVATION

Experimental and Therapeutic Medicine (Greece) (2016)

Ningxia Key Laboratory of Cerebrocranial Diseases, Ningxia Medical University, People's Republic of China; Department of Neurology, The General Hospital of Ningxia Medical University; Department of ICU, The First People's Hospital of Yinchuan, Ningxia.

In the present study, curcumin demonstrated marked suppression of the LPS-induced expression of MyD88, NF-κB, caspase-3, inducible nitric oxide synthase, tumor necrosis factor-α, interleukin (IL)-1β and IL-6 in the microglia. **These results indicate that curcumin may exert its neuroprotective and anti-inflammatory effects by inhibiting microglial activation through the HSP60/TLR-4/MyD88/NF-κB signaling pathway. Therefore, curcumin may be useful for the treatment of neurodegenerative diseases that are associated with microglial activation.**[58]

GREEN TEA EXTRACTS (↓) MICROGLIAL ACTIVATION

GREEN TEA EXTRACT (EGCG) (↓) MICROGLIAL ACTIVATION

Journal of Neuroscience Research (United States) (2004)

Health Science Center, Shanghai Institute for Biological Science, Chinese Academy of Science, People's Republic of China.

We provide evidence that (–)-epigallocatechin gallate (EGCG), a major monomer of green tea polyphenols, potently inhibits lipopolysaccharide (LPS)-activated microglial secretion of nitric oxide (NO) and tumor necrosis factor-alpha (TNF-alpha) through the down-regulation of

inducible NO synthase and TNF-alpha expression. In addition, EGCG exerted significant protection against microglial activation-induced neuronal injury both in the human dopaminergic cell line SH-SY5Y and in primary rat mesencephalic cultures. Our study demonstrates that EGCG is a potent inhibitor of microglial activation and thus is a useful candidate for a therapeutic approach to alleviating microglia-mediated dopaminergic neuronal injury in PD.[59]

GREEN TEA EXTRACT (EGCG) (↓) MICROGLIAL ACTIVATION
European Journal of Pharmacology (The Netherlands) (2016)
Division of Allergy, Immunology and Rheumatology, Chung Shan Medical University Hospital, Taichung, Taiwan; Institute of Integrative Medicine, China Medical University, Taichung, Taiwan; Institute of Medicine, Chung Shan Medical University, Taichung, Taiwan; Department of Applied Science, National Hsinchu University of Education, Hsinchu, Taiwan; Department of Biomedical Sciences, Chung Shan Medical University, Taichung, Taiwan; Division of Basic Medical Science, Hungkuang University, Taichung, Taiwan; Department of Medical Research, Chung Shan Medical University Hospital, Taichung, Taiwan.
In this study, we investigated the inhibitory effects of EGCG on amyloid β (Aβ)-induced microglial activation and neurotoxicity. **Our results indicated that EGCG significantly suppressed the expression of tumor necrosis factor α (TNFα), interleukin-1β, interleukin-6, and inducible nitric oxide synthase (iNOS) in Aβ-stimulated EOC 13.31 microglia.** EGCG also restored the levels of intracellular antioxidants nuclear erythroid-2 related factor 2 (Nrf2) and heme oxygenase-1 (HO-1), thus inhibiting reactive oxygen species-induced nuclear factor-κB (NF-κB) activation after Aβ treatment. Furthermore, **EGCG effectively protected neuro-2a neuronal cells from Aβ-mediated, microglia-induced cytotoxicity by inhibiting mitogen-activated protein kinase-dependent, Aβ-induced release of TNFα. Taken together, our findings suggested that EGCG suppressed Aβ-induced neuroinflammatory response of microglia and protected against indirect neurotoxicity.**[60]

RESVERATROL (↓) MICROGLIAL ACTIVATION

RESVERATROL (↓) MICROGLIAL ACTIVATION
European Journal of Pharmacology (The Netherlands) (2010)
Shanghai University of Traditional Chinese Medicine, Shanghai, China.
Microglia serve the role of immune surveillance under normal conditions, but after brain damage or exposure to inflammation, microglia are activated and secrete pro-inflammatory and neurotoxic mediators. Sustained production of these factors contributes to neuronal damage. Therefore, inhibition of microglia-mediated neuroinflammation may become a promising therapeutic target for neurological

disorders.... **This mini-review summarized the anti-inflammatory activities of resveratrol in the brain from both in vivo and in vitro studies, and highlighted the inhibition of activated microglia as a potential mechanism of neuroprotection.... Taken together, microglia are an important target for anti-inflammatory activities of resveratrol in the brain.**[61]

RESVERATROL (↓) MICROGLIAL ACTIVATION
Journal of Neurochemistry (England) (2012)
The Litwin-Zucker Research Center for the Study of Alzheimer's Disease, The Feinstein Institute for Medical Research, Manhasset, New York.
In this study, we show that resveratrol, a natural polyphenol associated with anti-inflammatory effects and currently in clinical trials for AD, prevented the activation of murine RAW 264.7 macrophages and microglial BV-2 cells treated with the TLR4 ligand, lipopolysaccharide (LPS). Resveratrol preferentially inhibited nuclear factor κ-light-chain-enhancer of activated B cells (NF-κB) activation upon LPS stimulation by interfering with IKK and IκB phosphorylation, an effect that potently reduced the transcriptional stimulation of several NF-κB target genes, including tumor necrosis factor-α and interleukin-6. Consequently, downstream phosphorylation of signal transducer and activator of transcription (STAT)1 and STAT3 upon LPS stimulation was also inhibited by resveratrol. We found that resveratrol acted upstream in the activation cascade by interfering with TLR4 oligomerization upon receptor stimulation. Resveratrol treatment also prevented the pro-inflammatory effect of fibrillar Aβ on macrophages by potently inhibiting the effect of Aβ on IκB phosphorylation, activation of STAT1 and STAT3, and on tumor necrosis factor-α and interleukin-6 secretion. **Importantly, orally administered resveratrol in a mouse model of cerebral amyloid deposition lowered microglial activation associated with cortical amyloid plaque formation. Together this work provides strong evidence that resveratrol has in vitro and in vivo anti-inflammatory effects against Aβ-triggered microglial activation.**[62]

RESVERATROL (↓) MICROGLIAL ACTIVATION
Phytotherapy Research (England) (2013)
Department of Pharmacology and Key Lab of Basic Pharmacology of Guizhou, Zunyi Medical College, People's Republic of China.
Here, by using rat primary cortical neuron-glia cultures, results showed that resveratrol attenuated lipopolysaccharide (LPS)-induced cortical neurotoxicity. Further studies revealed that microglia were responsible for resveratrol-mediated neuroprotection. **Resveratrol significantly inhibited LPS-induced microglial activation and subsequent production of multiple pro-inflammatory and cytotoxic factors such as tumor necrosis factor-α, nitric oxide, and interleukin-1β. Collectively, resveratrol produced neuroprotection against microglia-induced neurotoxicity.**[63]

RESVERATROL (↓) MICROGLIAL ACTIVATION

Journal of Alzheimer's Disease (The Netherlands) (2014)

Departments of Biochemistry and Molecular Biology, University of New Mexico School of Medicine, Albuquerque; The Center for Magnetic Resonance Research and Department of Radiology, University of Minnesota Medical School, Minneapolis; Department of Chemistry and Chemical Biology, University of New Mexico, Albuquerque; Quatros LLC, Albuquerque, New Mexico; Departments of Pathology, University of New Mexico School of Medicine, Albuquerque.

Since the pro-inflammatory transcription factor NF-κB is one of the major regulators of Aβ-induced inflammation, we treated transgenic amyloid-β protein protein/presenilin-1 (AβPP/PS1) mice for one year with a low dose (0.01% by weight in the diet) of either of two trans-stilbene NF-κB inhibitors, resveratrol or a synthetic analog LD55… The MRI measurements were confirmed by optical microscopy of thioflavin-stained brain tissue sections and indicated that supplementation with either of the two trans-stilbenes lowered Aβ plaque density in the cortex, caudoputamen, and hippocampus by 1.4 to 2-fold. The optical measurements also included the hippocampus and indicated that resveratrol and LD55 reduced average Aβ plaque density by 2.3-fold and 3.1-fold, respectively. **Ex vivo measurements of the regional distribution of microglial activation by Iba-1 immunofluorescence of brain tissue sections showed that resveratrol and LD55 reduced average microglial activation by 4.2- fold and 3.5-fold, respectively.**[64]

RESVERATROL (↓) MICROGLIAL ACTIVATION

Molecular Medicine Reports (Greece) (2015)

Department of Medical Chemistry, School of Basic Medical Science, Ningxia Medical University, Yinchuan, Ningxia, People's Republic of China; Department of Pharmacology, School of Pharmacy, Ningxia Medical University; Department of Basic Traditional Chinese Medicine, School of Traditional Chinese Medicine, Ningxia Medical University; National Resource Center of Chinese Materia Medica, Academy of Chinese Medical Sciences, Beijing, People's Republic of China.

The present study aimed to investigate the effects of resveratrol on the activation of oligomeric amyloid β (oAβ)-induced BV-2 microglia, and to determine the role of NADPH oxidase in these effects… **The results of the present study demonstrated that resveratrol inhibited the proliferation of oAβ-induced microglia and the production of pro-inflammatory factors, including ROS, NO, TNF-α and IL-1β.** Subsequent mechanistic investigations demonstrated that resveratrol inhibited the oAβ-induced mRNA and protein expression levels of p47phox and gp91phox. These results suggested that NADPH oxidase may be a potential target for AD treatment, and resveratrol may be a valuable natural product possessing therapeutic potential against AD.[65]

RESVERATROL (↓) MICROGLIAL ACTIVATION
Current Pharmaceutical Design (The Netherlands) (2015)

Department of Biosciences, Biotechnologies and Biopharmaceutics, University of Bari, Italy.
It is possible that M1/M2 polarization of microglia may play an important role in controlling the balance between promoting and resolving neuroinflammation in the CNS. Immunomodulatory strategies capable of redirecting the microglial response toward the neuroprotective M2 phenotype could offer attractive options for neurodegenerative diseases with inflammatory components. **The neuroprotective actions of resveratrol seem to be attributable to its anti-inflammatory properties, due not only to its direct scavenger effects versus toxic molecules but also to a capacity to upregulate natural anti-inflammatory defences, thus counteracting excessive responses of classically activated M1 microglia. The goal of this review is to summarize recent insights into the therapeutic potential of resveratrol as a natural modulator of microglia-mediated neurotoxicity.**[66]

RESVERATROL (↓) MICROGLIAL ACTIVATION
International Immunopharmacology (The Netherlands) (2015)

Department of Biosciences, Biotechnologies and Biopharmaceutics, University of Bari, Italy; Department of Clinical and Experimental Medicine, University of Foggia, Italy; Department of Biological and Environmental Sciences and Technologies, Section of Human Anatomy, University of Salento, Lecce, Italy.

In this paper, we demonstrate that in LPS–stimulated microglia resveratrol pretreatment reduced in a dose-dependent manner pro-inflammatory cytokines IL–1β, TNF–α and IL–6 mRNA expression and increased the release of anti-inflammatory interleukin (IL)–10. Moreover, resveratrol pretreatment up-regulated the phosphorylated forms of JAK1 and STAT3, as well as suppressor of cytokine signaling (SOCS)3 protein expression in LPS activated cells, demonstrating that the JAK-STAT signaling pathway is involved in the anti-inflammatory effect exerted by resveratrol. By supplementing the cultures with an IL-10 neutralizing antibody (IL-10NA) we obtained the opposite effect. **Taken together, these data allow us to conclude that the LPS-induced pro-inflammatory response in microglial cells can be markedly reduced by resveratrol through IL-10 dependent up-regulation of SOCS3, requiring the JAK-STAT signaling pathway.**[67]

RESVERATROL (↓) MICROGLIAL ACTIVATION
Cellular and Molecular Neurobiology (United States) (2015)

Department of Neurosurgery, The Second Hospital of Hebei Medical University, Shijiazhuang, China.

Our findings support that resveratrol inhibits microglial over-activation and alleviates neuronal injuries induced by microglial activation. Our study suggests the use of resveratrol as an alternative intervention approach that could prevent further neuronal insults.[68]

N-METHYL-D-ASPARTATE AND ALZHEIMER'S DISEASE

NMDA AND ALZHEIMER'S DISEASE
Proceedings of the National Academy of Sciences (United States) (2006)
Dipartimento di Neuroscienze, Università di Tor Vergata, Rome, Italy.
The altered function and/or structure of tau protein is postulated to cause cell death in tauopathies and Alzheimer's disease. However, the mechanisms by which tau induces neuronal death remain unclear. Here we show that overexpression of human tau and of some of its N-terminal fragments in primary neuronal cultures leads to an N-methyl-D-aspartate receptor (NMDAR)-mediated and caspase-independent cell death... **Our findings unravel a cellular mechanism linking tau toxicity to NMDAR activation and might be relevant to Alzheimer's disease and tauopathies where NMDAR-mediated toxicity is postulated to play a pivotal role.**[69]

NMDA AND ALZHEIMER'S DISEASE
British Journal of Pharmacology (England) (2012)
Merz Pharmaceuticals GmbH, Eckenheimer Landstraße, Frankfurt am Main, Germany.
β-amyloid (Aβ) is widely accepted to be one of the major pathomechanisms underlying Alzheimer's disease (AD), although there is presently lively debate regarding the relative roles of particular species/forms of this peptide. Most recent evidence indicates that soluble oligomers rather than plaques are the major cause of synaptic dysfunction and ultimately neurodegeneration. **Soluble oligomeric Aβ has been shown to interact with several proteins, for example glutamatergic receptors of the NMDA type and proteins responsible for maintaining glutamate homeostasis such as uptake and release. As NMDA receptors are critically involved in neuronal plasticity including learning and memory, we felt that it would be valuable to provide an up to date review of the evidence connecting Aβ to these receptors and related neuronal plasticity.**[70]

NMDA AND ALZHEIMER'S DISEASE
Journal of Alzheimer's Disease (The Netherlands) (2014)
Department of Neurology, Faculty of Medicine, Albert Szent-Györgyi Clinical Center, University of Szeged, Hungary; Department of Physiology, Anatomy and Neuroscience,

Faculty of Science and Informatics, University of Szeged, Hungary; MTA-SZTE Neuroscience Research Group, Szeged, Hungary.

The impairment of glutamatergic neurotransmission plays an important role in the development of Alzheimer's disease (AD). The pathological process, which involves the production of amyloid-β peptides and hyperphosphorylated tau proteins, spreads over well-delineated neuroanatomical circuits. **The gradual deterioration of proper synaptic functioning (via GluN2A-containing *N*-methyl-D-aspartate receptors, NMDARs) and the development of excitotoxicity (via GluN2B-containing NMDARs) in these structures both accompany the disease pathogenesis.**[71]

NMDA AND ALZHEIMER'S DISEASE
Cell and Tissue Research (Germany) (2014)
Grenoble Institut des Neurosciences (GIN), Grenoble, France.

It is becoming increasingly clear that aberrant neuronal activity can be the cause and the result of amyloid beta production. **Synaptic activation facilitates non-amyloidogenic processing of amyloid precursor protein (APP) and cell survival, primarily through synaptic NMDA receptors (NMDARs) and perhaps specifically those containing GluN2A-subunits. In contrast, extrasynaptic and GluN2B-containing NMDARs promote beta-secretase cleavage of APP into amyloid-beta (Aβ). The opposing nature of these NMDAR populations is reflected in their control over cell survival and death pathways.** Subtle changes in glutamate homeostasis may shift the balance between these pathways and could play a role in Alzheimer's disease (AD). Indeed, Aβ production, regional loss of brain connectivity and neurodegeneration correlate with neuronal activity in AD patients. **From another perspective, Aβ oligomers (Aβo) alter neuronal signaling through several mechanisms involving NMDARs and intracellular calcium mishandling. While Aβo affect multiple receptors, GluN2B-NMDARs have emerged as primary mediators of altered synaptic plasticity and neurotoxicity… Recently, Aβo were shown to trigger astrocytic release of glutamate to the extrasynaptic space where it activates NMDARs to promote further Aβ production and synaptic depression. Combined with the reciprocal regulation between neuronal activity and Aβ production, extrasynaptic glutamate release adds to a maladaptive model and ultimately results in synaptotoxicity and neurodegeneration of AD. Extrasynaptic NMDAR antagonists remain as a promising therapeutic avenue by interfering with this cascade.**[72]

NMDA AND ALZHEIMER'S DISEASE
Neurological Sciences (Italy) (2016)
Hunan Provincial Tumor Hospital and the Affiliated Tumor Hospital of Xiangya Medical School, Central South University, Changsha, Hunan, China; Cancer Research Institute,

School of Basic Medical Science, Central South University, Changsha, Hunan, China; Key Laboratory of Carcinogenesis and Cancer Invasion, Ministry of Education, Changsha, Hunan, China.

N-methyl-D-aspartate receptors (NMDARs) play a pivotal role in the synaptic transmission and synaptic plasticity thought to underlie learning and memory. **NMDARs activation has been recently implicated in Alzheimer's disease (AD) related to synaptic dysfunction. Synaptic NMDARs are neuroprotective, whereas overactivation of NMDARs located outside of the synapse cause loss of mitochondrial membrane potential and cell death. NMDARs dysfunction in the glutamatergic tripartite synapse, comprising presynaptic and postsynaptic neurons and glial cells, is directly involved in AD.** This review discusses both beta-amyloid (Aβ) and tau perturb[ed] synaptic functioning of the tripartite synapse, including alterations in glutamate release, astrocytic uptake, and receptor signaling. Particular emphasis is given to the role of NMDARs as a possible convergence point for Aβ and tau toxicity and possible reversible stages of the AD through preventive and/or disease-modifying therapeutic strategies.[73]

NMDA AND ALZHEIMER'S DISEASE
Behavioural Brain Research (The Netherlands) (2016)

Department of Neuroscience, McKnight Brain Institute, University of Florida, Gainesville.

Brain regions that are vulnerable to aging exhibit the earliest pathology of AD. Senescent synaptic function is observed as a shift in Ca2+-dependent synaptic plasticity and similar mechanisms are thought to contribute to the early cognitive deficits associated with AD. **In the case of aging, intracellular redox state mediates a shift in Ca2+ regulation including N-methyl-D-aspartate (NMDA) receptor hypofunction and increased Ca2+ release from intracellular stores to alter synaptic plasticity. AD can interact with these aging processes such that molecules linked to AD, β-amyloid (Aβ) and mutated presenilin 1 (PS1), can also degrade NMDA receptor function, promote Ca2+ release from intracellular stores, and may increase oxidative stress.** Thus, age is one of the most important predictors of AD and brain aging likely contributes to the onset of AD.[74]

NMDA AND AMPA AND ALZHEIMER'S DISEASE
Neural Plasticity (United States) (2016)

Clem Jones Centre for Ageing Dementia Research, Queensland Brain Institute, The University of Queensland, Brisbane, Australia.

Evidence from neuropathological, genetic, animal model, and biochemical studies has indicated that the accumulation of amyloid-beta (Aβ) is associated with, and probably induces, profound neuronal changes in brain regions critical for memory and cognition in the development of Alzheimer's disease (AD). There is considerable evidence that synapses are particularly vulnerable to AD, establishing synaptic dysfunction as one of the earliest events in pathogenesis, prior to neuronal loss. **It is clear that excessive Aβ levels can disrupt excitatory synaptic transmission and plasticity, mainly due to dysregulation of the AMPA and NMDA glutamate receptors in the brain. Importantly, AMPA receptors are the principal glutamate receptors that mediate fast excitatory neurotransmission. This is essential for synaptic plasticity, a cellular correlate of learning and memory, which are the cognitive functions that are most disrupted in AD.** Here we review recent advances in the field and provide insights into the molecular mechanisms that underlie Aβ-induced dysfunction of AMPA receptor trafficking. This review focuses primarily on NMDA receptor- and metabotropic glutamate receptor-mediated signaling. In particular, we highlight several mechanisms that underlie synaptic long-term depression as common signaling pathways that are hijacked by the neurotoxic effects of Aβ.[75]

CURCUMINOIDS (↓) N-METHYL-ᴅ-ASPARTATE–MEDIATED EXCITOTOXICITY

CURCUMIN (↓) **NMDA-MEDIATED EXCITOTOXICITY**
Experimental Brain Research (Germany) (2005)
Department of Cell Biology and Neuroscience, Istituto Superiore di Sanità, Roma, Italy.
Curcumin, an extract from the plant *Curcuma longa* with well-known antioxidant and anti-inflammatory activities, was tested as protective agent against excitotoxicity in rat retinal cultures. **A 24 h-treatment with curcumin reduced N-methyl-ᴅ-aspartate (NMDA)-mediated excitotoxic cell damage, estimated as decrease of cell viability and increase in apoptosis. The protection was associated with decrease of NMDA receptor-mediated Ca(2+) rise and reduction in the level of phosphorylated NR1 subunit of the NMDA receptor.** These results enlighten a new pharmacological action of the plant extract, possibly mediated by a modulation of NMDA receptor activity.[76]

CURCUMIN (↓) **GLUTAMATE-MEDIATED CALCIUM INFLUX AND EXCITOTOXICITY**
FEBS Letters (England) (2006)
Department of Neuroscience for Drug Discovery, Graduate School of Pharmaceutical Sciences, Kyoto University, Japan.
Glutamate excitotoxicity is mediated by intracellular Ca(2+) overload, caspase-3 activation, and ROS generation. Here, we show that curcumin, tannic acid (TA) and (+)-catechin hydrate (CA) all inhibited glutamate-induced excitotoxicity. Curcumin inhibited PKC activity and subsequent phosphorylation of NR1 of the NMDA receptor. As a result, glutamate-mediated Ca(2+) influx was reduced. TA attenuated glutamate-mediated Ca(2+) influx only when simultaneously administered, directly interfering with Ca(2+).

Both curcumin and TA inhibited glutamate-induced caspase-3 activation. Although Ca(2+) influx was not attenuated by CA, caspase-3 was reduced by direct inhibition of the enzyme. All polyphenols reduced glutamate-induced generation of ROS.[77]

CURCUMIN (↓) NMDA-INDUCED EXCITOTOXICITY
Investigative Ophthalmology and Visual Science (United States) (2011)
Department of Cell Biology and Neuroscience, Istituto Superiore di Sanità, Rome, Italy.
Curcumin, a phenolic compound extracted from the rhizome of *Curcuma longa*, was found to attenuate NMDA-induced excitotoxicity in primary retinal cultures. This study was conducted to further characterize curcumin neuroprotective ability and analyze its effects on NMDA receptor (NMDAr)... **Curcumin dose- and time-dependently protected both retinal and hippocampal neurons against NMDA-induced cell death, confirming its anti-excitotoxic property**. In primary retinal cultures, in line with the observed reduction of NMDA-induced [Ca(2+)] (i) rise, whole-cell patch-clamp experiments showed that a higher percentage of retinal neurons responded to NMDA with low amplitude current after curcumin treatment. In parallel, curcumin induced an increase in NMDAr subunit type 2A (NR2A) level, with kinetics closely correlated to time-course of neuroprotection and decrease in [Ca(2+)](i). The relation between neuroprotection and NR2A level increase was also in line with the observation that curcumin neuroprotection required protein synthesis. Electrophysiology confirmed an increased activity of NR2A-containing NMDAr at the plasma membrane level... **These results confirm the neuroprotective activity of curcumin against NMDA toxicity, possibly related to an increased level of NR2A, and encourage further studies for a possible therapeutic use of curcumin based on neuromodulation of NMDArs**.[78]

CURCUMIN (↓) NMDA-MEDIATED INTRACELLULAR CALCIUM ELEVATION
Journal of Receptor and Signal Transduction Research (England) (2015)
Beijing Key Laboratory of Bioactive Substances and Functional Foods, College of Arts and Science, Beijing Union University, China.
This study investigates the inhibitory effects of curcumin on Aβ-induced cell damage and death involving NMDA receptor-mediated intracellular Ca(2+) elevation in human neuroblastoma SH-SY5Y cells. Cells were impaired significantly in Aβ-damaged group compared with the control group, and cell viability was decreased while the released LDH from the cytosol was increased. Curcumin promotes cell growth and decreases cell impairment induced by Aβ. **Curcumin attenuates Aβ-induced elevation of the ratio of cellular glutamate/γ-aminobutyric acid (GABA) with a concentration-dependent manner. Curcumin inhibits Aβ-induced increase of cellular Ca(2+) and depresses Aβ-induced phosphorylations of both NMDA receptor and cyclic AMP response element-binding protein (CREB) and activating transcription factor 1 (ATF-1). These results indicated that curcumin inhibits**

Aβ-induced neuronal damage and cell death involving the prevention from intracellular Ca(2+) elevation mediated by the NMDA receptor.[79]

GREEN TEA EXTRACTS (↓) *N*-METHYL-D-ASPARTATE–MEDIATED EXCITOTOXICITY

GREEN TEA EXTRACT (EGCG) (↓) QUIN-INDUCED EXCITOTOXICITY
International Brain Research Organization (The Netherlands) (2009)
Department of Physiology, Chonnam National University Medical School, Gwangju, Republic of Korea.

Excessive stimulation of the NMDA receptor induces neuronal cell death and is implicated in the development of several neurodegenerative diseases. While EGCG suppresses apoptosis induced by NMDA receptor-mediated excitotoxicity, the mechanisms underlying this process have yet to be completely determined. This study was designed to investigate whether (-)-epigallocatechin-3-gallate (EGCG) plays a neuroprotective role by inhibiting nitric oxide (NO) production and activating cellular signaling mechanisms, including MAP kinase, PI3K, and GSK-3beta and acting on the antiapoptotic and the proapoptotic genes in N18D3 neural cells. **The cells were pretreated with EGCG for 2h and then exposed to quinolinic acid (QUIN), an NMDA receptor agonist, 30 mM for 24h. MTT assay and DAPI staining were used to identify cell viability and apoptosis, respectively, and demonstrated that EGCG significantly increased cell viability and protected the cells from apoptotic death. In addition, EGCG had a capacity to reduce QUIN-induced excitotoxic cell death not only by blocking increase of intracellular calcium levels but also by inhibiting NO production.** Gene expression analysis revealed that EGCG prevented the QUIN-induced expression of the proapoptotic gene, caspase-9, and increased that of the antiapoptotic genes, Bcl-XL, Bcl-2, and Bcl-w. Further examination about potential cell signaling candidate involved in this neuroprotective effect showed that immunoreactivity of PI3K was significantly increased in the cells treated with EGCG. **These results suggest that the neuroprotective mechanism of EGCG against QUIN-induced excitotoxic cell death includes regulation of PI3K and modulation of cell survival and death genes through decreasing of intracellular calcium levels and controlling of NO production.**[80]

GREEN TEA EXTRACT (EGCG) (↓) NMDA-MEDIATED CALCIUM INFLUX
ASN Neuro (United States) (2011)
Department of Biochemistry, University of Missouri, Columbia.

Our recent study demonstrated the ability for oligomeric Aβ to stimulate the production of ROS (reactive oxygen species) in neurons through an NMDA (*N*-methyl-D-aspartate)-dependent pathway. However, whether prolonged exposure of neurons to aggregated Aβ is associated with impairment of NMDA receptor function has not been extensively investigated. In the present study, we show that prolonged exposure of primary cortical neurons to Aβ oligomers caused mitochondrial dysfunction, an attenuation of NMDA receptor-mediated Ca2+ influx and inhibition of NMDA-induced AA (arachidonic acid) release. Mitochondrial dysfunction and the decrease in NMDA receptor activity due to oligomeric Aβ are associated with an increase in ROS production… **Furthermore, Aβ-induced mitochondrial dysfunction, impairment of NMDA Ca2+ influx and ROS production were prevented by pre-treatment of neurons with EGCG, a major polyphenolic component of green tea. Taken together, these results support a role for NADPH oxidase-mediated ROS production in the cytotoxic effects of Aβ, and demonstrate the therapeutic potential of EGCG and other dietary polyphenols in delaying onset or retarding the progression of AD.**[81]

Table 6.1 (below) summarizes the findings presented in the mechanistic studies of the curcuminoids, EGCG, and resveratrol on Notch-1, GSK-3β, MAPK, microglial activation, and NMDA, in addition to the results on the plant polyphenols and SIRT1 that were presented in the previous chapter.

Table 6.2 (below) combines the results from Table 1H in the previous chapter with Table 6.1 (above) to provide a fuller picture, as presented up to this point in this

Table 6.1 Alzheimer's disease (AD)–related plant polyphenol pleiotropism (curcuminoids/EGCG/resveratrol)

Curcuminoids, EGCG, Resveratrol	Curcuminoids	EGCG	Resveratrol	Combined
(↑) SIRT1	(2)	(2)	(8)	(12)
(↓) Notch	(8)	(4)	(1)	(13)
(↓) GSK-3β	(6)	(3)	(1)	(10)
(↓) MAPK	(4)			(4)
(↓) Microglial Activation	(6)	(2)	(8)	(16)
(↓) NMDA receptors	(4)	(2)		(6)

GSK-3β, glycogen synthase kinase 3 beta; *MAPK*, mitogen-activated protein kinases; *NMDA*, N-methyl-D-aspartate.

Table 6.2 (Table 1H plus Table 6.1) Alzheimer's disease (AD)–related plant polyphenol pleiotropism (curcuminoids/EGCG/resveratrol)

Plant Polyphenols and AD	Curcuminoids	EGCG	Resveratrol	Combined
(↑) Alpha-secretase/sAPPα		(8)		(8)
(↓) Beta-secretase	(5)	(2)		(7)
(↓) Gamma-secretase	(2)	(1)		(3)
(↓) Abeta Aggregation	(7)	(6)		(13)
(↓) Abeta toxicity	(9)	(3)	(10)	(22)
(↑) Abeta clearance	(13)	(1)	(3)	(17)
(↑) Synaptic function	(7)	(2)	(1)	(10)
(↓) Tau hyperphosphorylation	(3)	(1)	(2)	(6)
(↓) Tau Aggregation	(1)	(1)		(2)
(↑) Tau clearance	(2)	(1)		(3)

(↓) Aberrant neuroinflammation	(4)	(1)	(6)	(11)
(↓) AD-related oxidative stress	(11)	(5)	(8)	(24)
(↑) SIRT1	(2)	(2)	(8)	(12)
(↓) Notch-1 signaling	(8)	(4)	(1)	(13)
(↓) GSK-3β	(6)	(3)	(1)	(10)
(↓) MAPK	(4)			(4)
(↓) Microglial Activation	(6)	(2)	(8)	(16)
(↓) NMDA excitotoxicity	(4)	(2)	(6)	(6)

Abeta, amyloid beta peptide; *APP*, amyloid precursor protein; *GSK-3β*, glycogen synthase kinase 3 beta; *MAPK*, mitogen-activated protein kinases; *NMDA*, N-methyl-D-aspartate.

volume, of the pleiotropic capabilities of the three most-studied plant polyphenols—the curcuminoids, EGCG, and resveratrol. Given the therapeutic coverage of 18 established biomarkers of AD pathophysiology represented below, Table 6.2 reflects a comprehensive approach from a pharmacological standpoint to the prevention and treatment of AD.

We now explore an additional extension of the pleiotropic range of these three plant polyphenols (curcuminoids, EGCG, and resveratrol) to address five additional AD-related biofactors—NF kappa B, IL-1β, IL-6, TNF, and iron toxicity. The results, as shown in Table 6.4 toward the end of this chapter, to our knowledge is an unprecedented display of the therapeutically expansive coverage of 23 established biomarkers of AD pathophysiology by a simple formula of three nontoxic, well-tested plant-based polyphenolic compounds.

CURCUMINOIDS (↓) NUCLEAR FACTOR-KAPPA B

Curcumin (↓) NF-kappa B
Journal of Neuroscience Research (United States) (1998)
F. Hoffmann-LaRoche AG, Pharma Division Preclinical Research, Basel, Switzerland.
Amyloid beta peptide (Abeta), a proteolytic fragment of the amyloid precursor protein (APP), is a major component of the plaques found in the brain of Alzheimer's disease (AD) patients. These plaques are thought to cause the observed loss of cholinergic neurons in the basal forebrain of AD patients. In these neurons, particularly those of the nucleus basalis of Meynert, an up-regulation of 75kD-neurotrophin receptor (p75NTR), a nonselective neurotrophin receptor belonging to the death receptor family, has been reported. p75NTR expression has been described to correlate with beta-amyloid sensitivity in vivo and in vitro, suggesting a possible role for p75NTR as a receptor for Abeta. Here we used a human neuroblastoma cell line to investigate the involvement of p75NTR in Abeta-induced cell death. Abeta peptides were found to bind to p75NTR resulting in activation of NFKB in a time- and dose-dependent manner. **Blocking the interaction of Abeta with p75NTR using NGF or inhibition of NFKB activation by curcumin or NFKB SN50 attenuated or abolished Abeta-induced apoptotic cell death.** The present results suggest that p75NTR might be a death receptor for Abeta, thus being a possible drug target for treatment of AD.[82]

Curcumin (↓) NF kappa B
Journal of Clinical Immunology (The Netherlands) (2007)
Cytokine Research Laboratory, Department of Experimental Therapeutics, The University of Texas M.D. Anderson Cancer Center, Houston.
Traditionally known for its an antiinflammatory effects, curcumin has been shown in the last two decades to be a potent immunomodulatory agent that can modulate the

activation of T cells, B cells, macrophages, neutrophils, natural killer cells, and dendritic cells. **Curcumin can also downregulate the expression of various pro-inflammatory cytokines including TNF, IL-1, IL-2, IL-6, IL-8, IL-12, and chemokines, most likely through inactivation of the transcription factor NF-kappaB.**[83]

CURCUMIN (↓) **NF** KAPPA **B**
Cellular and Molecular Life Sciences (Switzerland) (2008)
Department of Cancer Biology, Wake Forest University School of Medicine, Winston–Salem, North Carolina.
The pleiotropic activities of curcumin derive from its complex chemistry as well as its ability to influence multiple signaling pathways, including survival pathways such as those regulated by NF-kappaB, Akt, and growth factors; cytoprotective pathways dependent on Nrf2; and metastatic and angiogenic pathways.[84]

CURCUMIN (↓) **NF** KAPPA **B**
The AAPS Journal (United States) (2014)
Department of Neurology, The First Affiliated Hospital of Chongqing Medical University, People's Republic of China.
Neprilysin (NEP, EP24.11), a zinc-dependent metallopeptidase expressed relatively low in the brain, is emerging as a potent inhibitor of AKT/Protein Kinase B. In addition, hyper-methylated promoter of NEP has been reported to be associated with decreases in NEP expression. In the present study, using bisulfite-sequencing PCR (BSP) assay, we showed that the CpG sites in NEP gene were hyper-methylated both in wild-type mouse neuroblastoma N2a cells (N2a/wt) and N2a cells stably expressing human Swedish mutant amyloid precursor protein (APP) (N2a/APPswe) associated with familial early onset AD. **Curcumin treatment induced restoration of NEP gene via CpG demethylation. This curcumin-mediated upregulation of NEP expression was also concomitant with the inhibition of AKT, subsequent suppression of nuclear transcription factor-κB (NF-κB) and its downstream pro-inflammatory targets including COX-2, iNOS in N2a/APPswe cells.** This study represents the first evidence on a link between CpG demethylation effect on NEP and anti-inflammation ability of curcumin that may provide a novel mechanistic insight into the anti-inflammatory actions of curcumin as well as new basis for using curcumin as a therapeutic intervention for AD.[85]

CURCUMIN (↓) **NF** KAPPA **B**
Journal of Medicinal Food (United States) (2015)
Bioactive Botanical Research Laboratory, Department of Biomedical and Pharmaceutical Sciences, College of Pharmacy, University of Rhode Island, Kingston.

Taken together, these results show that in RAW 264.7 murine macrophages, SLCPs [solid lipid curcumin particles] have improved solubility over unformulated curcumin, and significantly decrease the LPS-induced pro-inflammatory mediators NO, PGE2, and IL-6 by inhibiting the activation of NF-κB.[86]

GREEN TEA EXTRACTS (↓) NUCLEAR FACTOR KAPPA B

GREEN TEA EXTRACT (EGCG) (↓) NF KAPPA B
The Journal of Nutritional Biochemistry (United States) (2007)

Department of Pharmacology, College of Oriental Medicine, Institute of Oriental Medicine, Kyung Hee University, Seoul, Republic of Korea.

In this study, we investigated the effects of EGCG in attenuating the inflammatory response induced by interleukin (IL)-1beta+beta-amyloid (25-35) fragment (Abeta) in human astrocytoma U373MG cells. EGCG significantly inhibited the IL-1beta+Abeta (25-35)-induced IL-6, IL-8, vascular endothelial growth factor (VEGF) and prostaglandin (PG)E(2) production at 24 h (P<0.01). The maximal inhibition rate of IL-6, IL-8, VEGF and PGE(2) production by EGCG was approximately 54.40%, 56.01%, 69.06% and 47.03%, respectively. **EGCG also attenuated the expression of cyclo-oxygenase-2 and activation of nuclear factor-kappaB induced by IL-1beta+Abeta (25-35).** We demonstrated that EGCG suppresses IL-1beta+Abeta (25-35)-induced phosphorylation of the mitogen-activated protein kinase p38 and the c-Jun N-terminal kinase. In addition, EGCG induced the expression of mitogen-activated protein kinase phosphatase-1. These results provide new insight into the pharmacological actions of EGCG and its potential therapeutic application to various neurodegenerative diseases such as AD.[87]

GREEN TEA EXTRACT (EGCG) (↓) NF KAPPA B
The Journal of Nutrition (United States) (2009)

College of Pharmacy, Chungbuk National University 12, Republic of Korea.

In this study, we investigated the possible effects of (-)-epigallocatechin-3-gallate (EGCG) on memory dysfunction caused by Abeta through the change of Abeta-induced secretase activities. Mice were pretreated with EGCG (1.5 or 3 mg/kg body weight in drinking water) for 3 wk before intracerebroventricular administration of 0.5 microg Abeta(1-42). EGCG dose-dependently reduced the Abeta(1-42)-induced memory dysfunction, which was evaluated using passive avoidance and water maze tests. **Abeta(1-42) induced a decrease in brain alpha-secretase and increases in both brain beta- and gamma-secretase activities, which were reduced by EGCG. In the cortex and the hippocampus, expression of the metabolic products of the beta- and gamma-secretases from APP, C99, and Abeta also were dose-dependently suppressed by EGCG. Paralleled with the** suppression of beta- and gamma-secretases by EGCG, we found that EGCG inhibited the activation of extracellular signal-regulated kinase and nuclear transcription factor-kappaB in the Abeta(1-42)-injected mouse brains.[88]

RESVERATROL (↓) NUCLEAR FACTOR KAPPA B

RESVERATROL (↓) NEURONAL DAMAGE FROM NF KAPPA B ACTIVATION
Neuroreport (England) (1998)

Department of Pharmacology, University of Missouri, Columbia.

The present study examined the possible signal transduction cascade leading to cell death by oxLDL and oxVLDL in PC12 cells. Using the electrophoretic mobility shift assay, we found that both oxLDL and oxVLDL activated the binding of NF-kappaB to the consensus sequence in the promoter region of the target genes, followed by apopotic cell death. **Resveratrol protects the cells from both the activation of NF-kappa-B/ DNA binding activity and apoptotic cell death**. Results indicated that oxidized lipoproteins may serve as an oxidative mediator and may activate apoptosis through a nuclear signalling pathway contributing to the pathology in Alzheimer's disease.[89]

RESVERATROL (↓) NF KAPPA B
Free Radical Biology and Medicine (2003)

Research Institute of Pharmaceutical Sciences, College of Pharmacy, Seoul National University, South Korea.

In this study, we have investigated the effects of resveratrol on beta-amyloid-induced oxidative cell death in cultured rat pheochromocytoma (PC12) cells. PC12 cells treated with beta-amyloid exhibited increased accumulation of intracellular ROI and underwent apoptotic death as determined by characteristic morphological alterations and positive in situ terminal end-labeling (TUNEL staining). Beta-amyloid treatment also led to the decreased mitochondrial membrane potential, the cleavage of poly(ADP-ribose)polymerase, an increase in the Bax/Bcl-X(L) ratio, and activation of c-Jun N-terminal kinase. **Resveratrol attenuated beta-amyloid-induced cytotoxicity, apoptotic features, and intracellular ROI accumulation. Beta-amyloid transiently induced activation of NF-kappaB in PC12 cells, which was suppressed by resveratrol pretreatment.**[90]

RESVERATROL (↓) NF KAPPA B
The Journal of Biological Chemistry (United States) (2005)

Gladstone Institute of Neurological Disease, University of California, San Francisco.

Accumulating evidence suggests that neurodegeneration induced by pathogenic proteins depends on contributions from surrounding glia. Here we demonstrate that NF-kappaB signaling in microglia is critically involved in neuronal death induced by amyloid-beta (Abeta) peptides, which are widely presumed to cause Alzheimer disease. Constitutive inhibition of NF-kappaB signaling in microglia by expression of the nondegradable IkappaBalpha super-repressor blocked neurotoxicity, indicating a pivotal role for microglial NF-kappaB signaling in mediating Abeta toxicity. Stimulation of microglia with Abeta increased acetylation of RelA/p65 at lysine 310, which regulates the NF-kappaB pathway. **Overexpression of SIRT1 deacetylase and the addition of the SIRT1 agonist resveratrol markedly reduced NF-kappaB signaling stimulated by Abeta and had strong neuroprotective effects**. Our results support a glial loop hypothesis by demonstrating a critical role for microglial NF-kappaB signaling in Abeta-dependent neurodegeneration. They also implicate SIRT1 in this pathway and highlight the therapeutic potential of resveratrol and other sirtuin-activating compounds in Alzheimer disease.[91]

RESVERATROL (↓) NF KAPPA B
International Journal of Molecular Medicine (Greece) (2006)

Department of Biochemistry, Dongeui University College of Oriental Medicine, Busan, South Korea.

In the present study, we investigated the protective effect of resveratrol on beta-amyloid-induced cytotoxicity in cultured rat astroglioma C6 cells. Preincubation of C6 cells with resveratrol concentration-dependently protected the cells from the growth inhibition induced by beta-amyloid treatment. beta-amyloid treatment led to increased nitric oxide (NO) synthesis and inducible nitric oxide synthase (iNOS) expression; however, cells pretreated with resveratrol showed a dose-dependent inhibition of NO production and iNOS expression following beta-amyloid treatment. Resveratrol also attenuated beta-amyloid-induced prostaglandin E2 (PGE2) release, which was associated with the inhibition of cyclooxygenase (COX)-2 expression. **Furthermore, beta-amyloid treatment induced nuclear translocation of NF-kappaB, which was suppressed by resveratrol pretreatment. Collectively, the present results indicate that modulation of nuclear factor-kappaB (NF-kappaB) activity is involved in the neuroprotective action of resveratrol against beta-amyloid-induced toxicity.**[92]

RESVERATROL (↓) NF KAPPA B
The Journal of Biological Chemistry (United States) (2008)

Louisiana State University Neuroscience Center, Louisiana State University Health Science Center, New Orleans.

Human brains retain discrete populations of micro RNA (miRNA) species that support homeostatic brain gene expression functions; however, specific miRNA abundance is significantly altered in neurological disorders such as Alzheimer disease (AD) when compared with age-matched controls. Here we provide evidence in AD brains of a specific up-regulation of an NF-kappaB-sensitive miRNA-146a highly complementary to the 3'-untranslated region of complement factor H (CFH), an important repressor of the inflammatory response of the brain. Up-regulation of miRNA-146a coupled to down-regulation of CFH was observed in AD brain and in interleukin-1beta, Abeta42, and/or oxidatively stressed human neural (HN) cells in primary culture. **Transfection of HN cells using an NF-kappaB-containing pre-miRNA-146a promoter-luciferase reporter construct in stressed HN cells showed significant up-regulation of luciferase activity that paralleled decreases in CFH gene expression. Treatment of stressed HN cells with the NF-kappaB inhibitor pyrollidine dithiocarbamate or the resveratrol analog CAY10512 abrogated this response.**[93]

Resveratrol (↓) NF kappa B
Journal of Inorganic Biochemistry (United States) (2011)
Neuroscience Center, Department of Genetics, Louisiana State University Health Sciences Center, New Orleans.

Micro RNAs (miRNAs) constitute a unique class of small, non-coding ribonucleic acids (RNAs) that regulate gene expression at the post-transcriptional level. The presence of two inducible miRNAs, miRNA-125b and miRNA-146a, involved in respectively, astroglial cell proliferation and in the innate immune and inflammatory response, is significantly up-regulated in human neurological disorders including Alzheimer's disease (AD). In this study we analyzed abundances miRNA-125b and miRNA-146a in magnesium-, iron-, gallium, and aluminum-sulfate-stressed human-astroglial (HAG) cells, a structural and immune-responsive brain cell type. The combination of iron-plus aluminum-sulfate was found to be significantly synergistic in up-regulating reactive oxygen species (ROS) abundance, NF-κB-DNA binding and miRNA-125b and miRNA-146a expression. **Treatment of metal-sulfate stressed HAG cells with the antioxidant phenyl butyl nitrone (PBN) or the NF-κB inhibitors curcumin, the metal chelator-anti-oxidant pyrollidine dithiocarbamate (PDTC), or the resveratrol analog CAY10512, abrogated both NF-κB signaling and induction of these miRNAs. Our observations further illustrate the potential of physiologically relevant amounts of aluminum and iron sulfates to synergistically up-regulate specific miRNAs known to contribute to AD-relevant pathogenetic mechanisms, and suggest that antioxidants or NF-κB inhibitors may be useful to quench metal-sulfate triggered genotoxicity.**[94]

Resveratrol (↓) NF kappa B
Neuroscience (United States) (2015)
Department of Nutrition and Food Hygiene, School of Public Health, Shanxi Medical University, People's Republic of China; Center for Disease Control and Prevention in Shanxi Province, People's Republic of China.

The aim of the present study is to further elucidate whether resveratrol prevents AD rats from inflammation induced by Aβ1-42 and protects the integrity of the blood-brain barrier. Rats were divided into six groups: (1) ovariectomized (OVX)+D-galactose (D-gal) 100 mg/kg group (OVX+D-gal); (2-4) OVX, D-gal and Res 20, 40 and 80 mg/kg treated groups; and (5) OVX, D-gal and estradiol valerate 0.8 mg/kg treated group (ET); (6) Sham control group. **12 weeks later, resveratrol 40 and 80 mg/kg treatment exhibited a significant decrease of Aβ1-42 compared with the OVX+D-gal rats of hippocampus, which was accompanied by decreased expression of advanced glycation endproducts (RAGE), matrix metalloprotein-9 (MMP-9), nuclear factor kappaB (NF-κB) and the increase of Claudin-5.** These results suggest that resveratrol is useful not only in protecting OVX+D-gal rats from neuroinflammation mediated by Aβ1-42 by decreasing the expression of NF-κB but also the integrity of blood-brain barrier by increasing Claudin-5 and decreasing RAGE, MMP-9.[95]

CURCUMINOIDS (↓) INTERLEUKIN-1BETA

Curcumin (↓) IL-1β
The Journal of Neuroscience (United States) (2001)
Departments of Medicine and Neurology, University of California, Los Angeles.
To evaluate whether it could affect Alzheimer-like pathology in the APPSw mice, we tested a low (160 ppm) and a high dose of dietary curcumin (5000 ppm) on inflammation, oxidative damage, and plaque pathology. **Low and high doses of curcumin significantly lowered oxidized proteins and interleukin-1beta, a proinflammatory cytokine elevated in the brains of these mice.**[96]

Curcumin and Tetrahydrocurcumin (↓) IL-1β
The Journal of Pharmacology and Experimental Therapeutics (United States) (2008)
Department of Medicine, University of California, Los Angeles.
We examined the antioxidant, anti-inflammatory, or anti-amyloidogenic effects of dietary curcumin and tetrahydrocurcumin, either administered chronically to aged Tg2576 APPsw mice or acutely to lipopolysaccharide (LPS)-injected wild-type mice. Despite dramatically higher drug plasma levels after tetrahydrocurcumin compared with curcumin gavage, resulting brain levels of parent compounds were similar, correlating with reduction in LPS-stimulated inducible nitric-oxide synthase, nitrotyrosine, F2 isoprostanes, and carbonyls. **In both the acute (LPS) and chronic inflammation (Tg2576), tetrahydrocurcumin and curcumin similarly reduced interleukin-1beta. Despite these similarities, only curcumin was effective in reducing amyloid plaque burden, insoluble beta-amyloid peptide (Abeta), and carbonyls.** Tetrahydrocurcumin had no impact on plaques or

insoluble Abeta, but both reduced Tris-buffered saline-soluble Abeta and phospho-c-Jun NH(2)-terminal kinase (JNK).[97]

DEMETHOXYCURCUMIN (↓) IL‑1β
Brain Research (The Netherlands) (2011)
Natural Products Research Unit, Department of Biological and Biomedical Sciences, The Aga Khan University Medical College, Karachi-Pakistan.

We aimed to explore the effect of a curcuminoid mixture and its individual components on inflammatory and apoptotic genes expression in AD using an Aβ+ibotenic acid-infused rat model. **After 5 days of treatment with demethoxycurcumin, hippocampal IL‑1β levels were decreased to 118.54 ± 47.48 and 136.67 ± 31.96% respectively at 30 and 10 mg/kg, compared with the amyloid treated group (373.99 ± 15.28%).** After 5 days of treatment, the curcuminoid mixture and demethoxycurcumin effectively decreased GFAP levels in the hippocampus. When studied for their effect on apoptotic genes expression, the curcuminoid mixture and bisdemethoxycurcumin effectively decreased caspase-3 level in the hippocampus after 20 days of treatment, where bisdemethoxycurcumin showed a maximal rescuing effect (92.35 ± 3.07%) at 3 mg/kg. The curcuminoid mixture at 30 mg/kg decreased hippocampal FasL level to 70.56 ± 3.36% after 5 days of treatment and 19.01 ± 2.03% after 20 days. In the case of Fas receptor levels, demethoxycurcumin decreased levels after 5 days of treatment with all three doses showing a maximal effect (189.76 ± 15.01%) at 10 mg/kg. Each compound was effective after 20 days in reducing Fas receptor levels in the hippocampus. This study revealed the important effect of curcuminoids on genes expression, showing that each component of the curcuminoid mixture distinctly affects gene expression, thus highlighting the therapeutic potential of curcuminoids in AD.[98]

CURCUMIN (↓) IL‑1β
Neuroscience Letters (Ireland) (2015)
Guangdong Key Laboratory for Diagnosis and Treatment of Major Neurological Diseases, Department of Neurology, The First Affiliated Hospital, Sun Yat-sen University, Guangzhou, People's Republic of China; Department of Neurology, Zengcheng People's Hospital, People's Republic of China.

Moreover, curcumin concentration-dependently abolished Aβ42-induced interleukin-1β (IL‑1β), interleukin-6 (IL‑6) and tumor necrosis factor-α (TNF-α) production in mRNA and protein levels in microglia.[43]

GREEN TEA EXTRACTS (↓) INTERLEUKIN-1BETA

GREEN TEA EXTRACT (EGCG) (↓) IL‑1β
European Journal of Pharmacology (The Netherlands) (2016)

Division of Allergy, Immunology and Rheumatology, Chung Shan Medical University Hospital, Taichung, Taiwan; Institute of Integrative Medicine, China Medical University, Taichung, Taiwan; Institute of Medicine, Chung Shan Medical University, Taichung, Taiwan; Department of Applied Science, National Hsinchu University of Education, Hsinchu, Taiwan; Department of Biomedical Sciences, Chung Shan Medical University, Taichung, Taiwan; Division of Basic Medical Science, Hungkuang University, Taichung, Taiwan; Department of Medical Research, Chung Shan Medical University Hospital, Taichung, Taiwan.

In this study, we investigated the inhibitory effects of EGCG on amyloid β (Aβ)-induced microglial activation and neurotoxicity. **Our results indicated that EGCG significantly suppressed the expression of tumor necrosis factor α (TNFα), interleukin-1β, interleukin-6, and inducible nitric oxide synthase (iNOS) in Aβ-stimulated EOC 13.31 microglia.**[60]

RESVERATROL (↓) INTERLEUKIN-1BETA

RESVERATROL (↓) GLUTAMATE-INDUCED IL-1β
Journal of Neurochemistry (England) (2010)

Department of Microbiology, School of Medicine, Ewha Medical Research Institute, Ewha Women's University, Seoul, South Korea.

Published evidence has linked glutamate with the pathogenesis of Alzheimer's disease (AD) and the up-regulation of a variety of chemokines, including monocyte chemotactic protein-1 (MCP-1)/chemokine ligand 2, with AD-associated pathological changes. In this study, we assessed the potential molecular basis for the role of glutamate in hippocampal inflammation by determining its effects on MCP-1 induction. We also attempted to identify the mechanism by which resveratrol (trans-3,5,4'-trihydroxystilbene), a polyphenolic phytostilbene, modulates the expression of MCP-1 in the glutamate-stimulated hippocampus. An ex vivo study using rat hippocampal slices demonstrated a time- and dose-dependent increase in MCP-1 release from glutamate-exposed hippocampus. This increase was accompanied by enhanced MCP-1 gene expression via the activation of the MEK/extracellular signal-regulated kinase (ERK) pathway and interleukin-1beta (IL-1beta) expression. The inhibition of the MEK/ERK pathway with SL327, which is capable of crossing the blood-brain barrier, nearly abolished the observed glutamate-induced effects. Furthermore, anti-IL-1beta antibodies suppressed the glutamate-induced expression of MCP-1 mRNA and protein, whereas an isotype-matched antibody exerted only minimal effects. **It is worthy of note that resveratrol, to a similar degree as SL327, down-regulated glutamate-induced IL-1beta expression and reduced the expression of MCP-1 mRNA and protein release via the inactivation of ERK1/2.** These results indicate that the activation of the MEK/ERK pathway and the consequent IL-1beta expression are essential for glutamate-stimulated MCP-1 production in the hippocampus.[99]

RESVERATROL (↓) IL-1β
Molecular Medicine Reports (Greece) (2015)

Department of Medical Chemistry, School of Basic Medical Science, Ningxia Medical University, Yinchuan, Ningxia, People's Republic of China; Department of Pharmacology, School of Pharmacy, Ningxia Medical University; Department of Basic Traditional Chinese Medicine, School of Traditional Chinese Medicine, Ningxia Medical University; National Resource Center of Chinese Materia Medica, Academy of Chinese Medical Sciences, Beijing.

The present study aimed to investigate the effects of resveratrol on the activation of oligomeric amyloid β (oAβ)-induced BV-2 microglia, and to determine the role of NADPH oxidase in these effects. Microglial proliferation was measured by high-content screening cell counting and using a bromodeoxyuridine incorporation assay. In addition, the levels of reactive oxygen species (ROS), nitric oxide (NO), tumor necrosis factor (TNF)-α and interleukin (IL)-1β were assessed. **The results of the present study demonstrated that resveratrol inhibited the proliferation of oAβ-induced microglia and the production of pro-inflammatory factors, including ROS, NO, TNF-α and IL-1β.** Subsequent mechanistic investigations demonstrated that resveratrol inhibited the oAβ-induced mRNA and protein expression levels of p47phox and gp91phox. These results suggested that NADPH oxidase may be a potential target for AD treatment, and resveratrol may be a valuable natural product possessing therapeutic potential against AD.[65]

CURCUMINOIDS (↓) INTERLEUKIN-6

CURCUMIN (↓) IL-6
Current Drug Targets (The Netherlands) (2011)

Department of Biochemistry and Molecular Biology, Louisiana State University Health Sciences Center, Shreveport.

In recent years, extensive in vitro and in vivo studies suggested curcumin has anticancer, antiviral, antiarthritic, anti-amyloid, antioxidant, and anti-inflammatory properties. The underlying mechanisms of these effects are diverse and appear to involve the regulation of various molecular targets, including transcription factors (such as nuclear factor-kB), growth factors (such as vascular endothelial cell growth factor), inflammatory cytokines (such as tumor necrosis factor, interleukin 1 and interleukin 6), protein kinases (such as mammalian target of rapamycin, mitogen-activated protein kinases, and Akt) and other enzymes (such as cyclooxygenase 2 and 5 lipoxygenase). Thus, due to its efficacy and regulation of multiple targets, as well as its safety for human use, curcumin has received considerable interest as a potential therapeutic agent for the prevention and/or treatment of various malignant diseases, arthritis, allergies, Alzheimer's disease, and other inflammatory illnesses.[100]

CURCUMIN (↓) IL-6

Journal of Medicinal Food (United States) (2015)

Bioactive Botanical Research Laboratory, Department of Biomedical and Pharmaceutical Sciences, College of Pharmacy, University of Rhode Island, Kingston.

Taken together, these results show that in RAW 264.7 murine macrophages, SLCPs [solid lipid particle formulation of curcumin] have improved solubility over unformulated curcumin, and significantly decrease the LPS-induced pro-inflammatory mediators NO, PGE2, and IL-6 by inhibiting the activation of NF-κB.[86]

CURCUMIN (↓) IL-6

Neuroscience Letters (Ireland) (2015)

Guangdong Key Laboratory for Diagnosis and Treatment of Major Neurological Diseases, Department of Neurology, The First Affiliated Hospital, Sun Yat-sen University, Guangzhou, People's Republic of China; Department of Neurology, Zengcheng People's Hospital, Zengcheng, People's Republic of China.

In the present study, we found that curcumin improved microglial viability against Aβ42 in a time- and dose-dependent manner and remarkably suppressed Aβ42-induced CD68 expression. **Moreover, curcumin concentration-dependently abolished Aβ42-induced interleukin-1β (IL-1β), interleukin-6 (IL-6) and tumor necrosis factor-α (TNF-α) production in mRNA and protein levels in microglia.**[43]

GREEN TEA EXTRACTS (↓) INTERLEUKIN-6

GREEN TEA EXTRACT (EGCG) (↓) IL-6

The Journal of Nutritional Biochemistry (United States) (2007)

Department of Pharmacology, College of Oriental Medicine, Institute of Oriental Medicine, Kyung Hee University, Seoul, Republic of Korea.

EGCG significantly inhibited the IL-1beta+Abeta (25-35)-induced IL-6, IL-8, vascular endothelial growth factor (VEGF) and prostaglandin (PG) E(2) production at 24 h (P<0.01). The maximal inhibition rate of IL-6, IL-8, VEGF and PGE(2) production by EGCG was approximately 54.40%, 56.01%, 69.06% and 47.03%, respectively.[87]

GREEN TEA EXTRACT (EGCG) (↓) IL-6

European Journal of Pharmacology (The Netherlands) (2016)

Division of Allergy, Immunology and Rheumatology, Chung Shan Medical University Hospital, Taichung, Taiwan; Institute of Integrative Medicine, China Medical University, Taichung, Taiwan; Institute of Medicine, Chung Shan Medical University, Taichung, Taiwan; Department of Applied Science, National Hsinchu University of Education, Hsinchu, Taiwan; Department of Biomedical

Sciences, Chung Shan Medical University, Taichung, Taiwan; Division of Basic Medical Science, Hungkuang University, Taichung, Taiwan; Department of Medical Research, Chung Shan Medical University Hospital, Taichung, Taiwan.

Our results indicated that EGCG significantly suppressed the expression of tumor necrosis factor α (TNFα), interleukin-1β, interleukin-6, and inducible nitric oxide synthase (iNOS) in Aβ-stimulated EOC 13.31 microglia.[60]

RESVERATROL (↓) INTERLEUKIN-6

RESVERATROL (↓) IL-6
Journal of Neurochemistry (England) (2012)
The Litwin-Zucker Research Center for the Study of Alzheimer's Disease, The Feinstein Institute for Medical Research, Manhasset, New York.
Resveratrol treatment also prevented the pro-inflammatory effect of fibrillar Aβ on macrophages by potently inhibiting the effect of Aβ on IκB [I-kappa B kinase] phosphorylation, activation of STAT1 and STAT3, and on tumor necrosis factor-α and interleukin-6 secretion.[62]

RESVERATROL (↓) IL-6
BioMed Central Neuroscience (England) (2013)
Department of Biochemistry, Faculty of Science, Alexandria University, Egypt.
Chronic administration of aluminum is proposed as an environmental factor that may affect several enzymes and other biomolecules related to neurotoxicity and Alzheimer's disease (AD). APE1, a multifunctional protein, functions in DNA repair and plays a key role in cell survival versus cell death upon stimulation with cytotoxic agent, making it an attractive emerging therapeutic target. The promising protective effect of resveratrol, which is known to exert potent anti-inflammatory effects on neurotoxicity induced by aluminum chloride, may be derived from its own antioxidant properties. In the present work we investigated the modulation of APE1 expression during aluminum chloride-induced neuroinflammation (25 mg/Kg body weight by oral gavages) in experimental rats. We tested the hypothesis that a reactive oxygen species (ROS)-scavenger, resveratrol at 0.5 mg/kg bodyweight, which is known to exert potent anti-inflammatory effects, would attenuate central inflammation and modulate APE1 expression in aluminum chloride-fed rats... Our results indicate that resveratrol may attenuate aluminum chloride-induced direct neuroinflammation in rats, and its mechanisms are, at least partly, due to maintaining high APE1 level. Resveratrol co-administration with aluminum chloride exerted more protective effect than pre-administration or treatment of induced rats. A significant elevation of APE1 at both mRNA and protein levels was observed in addition to a marked reduction in β-secretase and amyloid-β. **We found that aluminum chloride**

stimulated the expression of TNF-α, IL-6, and iNOS in rat brain in which NF-κB was involved. Resveratrol inhibited aluminum chloride-induced expression and release of TNF-α, IL-6, and iNOS in rat brain.[101]

CURCUMINOIDS (↓) TUMOR NECROSIS FACTOR-α

CURCUMIN (↓) TNF-α
Journal of Neurochemistry (England) (2004)
Department of Biochemistry and Molecular Biology, University of Southern California, Keck School of Medicine, Los Angeles.
Curcumin abrogated Abeta1-40-induced expression of cytokines (TNF-alpha and IL-1beta) and chemokines (MIP-1beta, MCP-1 and IL-8) in both peripheral blood monocytes and THP-1 cells.[102]

CURCUMIN (↓) TNF-α
Neuroscience Letters (Ireland) (2015)
Guangdong Key Laboratory for Diagnosis and Treatment of Major Neurological Diseases, Department of Neurology, The First Affiliated Hospital, Sun Yat-sen University, Guangzhou, People's Republic of China; Department of Neurology, Zengcheng People's Hospital, Zengcheng, People's Republic of China.
Moreover, curcumin concentration-dependently abolished Aβ42-induced interleukin-1β (IL-1β), interleukin-6 (IL-6) and tumor necrosis factor-α (TNF-α) production in mRNA and protein levels in microglia.[43]

GREEN TEA EXTRACTS (↓) TUMOR NECROSIS FACTOR-α

GREEN TEA EXTRACT (EGCG) (↓) TNF-α
Journal of Neuroscience Research (United States) (2004)
Health Science Center, Shanghai Institute for Biological Science, Chinese Academy of Science, Shanghai Second Medical University, Peoples Republic of China.
Microglial activation is believed to play a pivotal role in the selective neuronal injury associated with several neurodegenerative disorders, including Parkinson's disease (PD) and Alzheimer's disease. We provide evidence that (-)-epigallocatechin gallate (EGCG), a major monomer of green tea polyphenols, potently inhibits lipopolysaccharide (LPS)-activated microglial secretion of nitric oxide (NO) and tumor necrosis factor-alpha (TNF-alpha) through the down-regulation of inducible NO synthase and TNF-alpha expression.[59]

GREEN TEA EXTRACT (EGCG) (↓) TNF-α
Molecular and Cellular Biochemistry (The Netherlands) (2013)

Department of Neurology, First Affiliated Hospital of China Medical University, Shenyang, Liaoning Province, China.

EGCG treatment inhibited TNF-α/JNK signaling and increased the phosphorylation of Akt and glycogen synthase kinase-3β in the hippocampus of APP/PS1 mice.[33]

RESVERATROL (↓) TUMOR NECROSIS FACTOR-α

Resveratrol (↓) TNF-α
Biochemical and Biophysical Research Communications (United States) (2012)

Department of Biology, Ouachita Baptist University, Arkadelphia, Arkansas.

Inflammatory molecules have been implicated in the pathogenesis of neurodegenerative diseases such as Parkinson's disease, Alzheimer's disease, and multiple sclerosis. Resveratrol is an anti-fungal compound found in the skins of red grapes and other fruits and nuts. **We examined the ability of resveratrol to inhibit lipopolysaccharide (LPS)-induced production of inflammatory molecules from primary mouse astrocytes. Resveratrol inhibited LPS-induced production of nitric oxide (NO); the cytokines tumor necrosis factor-alpha (TNF-α), interleukin 1-beta (IL-1β), and IL-6; and the chemokine monocyte chemotactic protein-1 (MCP-1), which play critical roles in innate immunity, by astrocytes.** Resveratrol also suppressed astrocyte production of IL-12p40 and IL-23, which are known to alter the phenotype of T cells involved in adaptive immunity. Finally, resveratrol inhibited astrocyte production of C-reactive protein (CRP), which plays a role in a variety of chronic inflammatory disorders. Collectively, these studies suggest that resveratrol may be an effective therapeutic agent in neurodegenerative diseases initiated or maintained by inflammatory processes.[103]

Resveratrol (↓) TNF-α
BioMed Central Neuroscience (England) (2013)

Department of Biochemistry, Faculty of Science, Alexandria University, Egypt.

We found that aluminum chloride stimulated the expression of TNF-α, IL-6, and iNOS in rat brain in which NF-κB was involved. **Resveratrol inhibited aluminum chloride-induced expression and release of TNF-α, IL-6, and iNOS in rat brain.**[101]

Resveratrol (↓) TNF-α
Molecular Medicine Reports (Greece) (2015)

Department of Medical Chemistry, School of Basic Medical Science, Ningxia Medical University, Yinchuan, Ningxia, People's Republic of China; Department of Pharmacology,

School of Pharmacy, Ningxia Medical University; Department of Basic Traditional Chinese Medicine, School of Traditional Chinese Medicine, Ningxia Medical University; National Resource Center of Chinese Materia Medica, Academy of Chinese Medical Sciences, Beijing. **The results of the present study demonstrated that resveratrol inhibited the proliferation of oAβ-induced microglia and the production of pro-inflammatory factors, including ROS, NO, TNF-α and IL-1β.**[65]

Resveratrol (↓) TNF-α
Wei Sheng Yan Jiu (China) (2015)
OBJECTIVE: To explore the effects of resveratrol on astrocyte and TNF-α in hippocampus of Alzheimer's disease (AD) model rats... RESULTS: The levels of glial fibrillary acidic protein (GFAP) and TNF-α in the model group were significantly higher than those of the sham control group ($P < 0.01$). No marked difference in the production of GFAP was observed between the resveratrol 20 mg/kg group and the model group ($P > 0.05$). However, the resveratrol 40, 80 mg/kg and estradiol valerate treated groups showed a decrease in the expression of GFAP compared with the model group ($P < 0.01$). Moreover, with the increasing of resveratrol concentration, the expression of GFAP decreased gradually. **The levels of TNF-α decreased markedly in resveratrol 20, 40, 80 mg/kg and estradiol valerate group compared with the model group ($P < 0.01$). CONCLUSION: These results suggest that the activation of astrocytes and the secretion of TNF-α can be inhibited by resveratrol in AD rats.**[104]

CURCUMINOIDS (↓) IRON AND COPPER

Curcumin (↓) Iron and Copper
Journal of Alzheimer's Disease (The Netherlands) (2004)
Department of Medicine and Therapeutics, Chinese University of Hong Kong, Shatin.
Curcumin is a polyphenolic diketone from turmeric. Because of its anti-oxidant and anti-inflammatory effects, it was tested in animal models of Alzheimer's disease, reducing levels of amyloid and oxidized proteins and preventing cognitive deficits. An alternative mechanism of these effects is metal chelation, which may reduce amyloid aggregation or oxidative neurotoxicity. **Metals can induce Abeta aggregation and toxicity, and are concentrated in AD brain. Chelators desferrioxamine and clioquinol have exhibited anti-AD effects. Using spectrophotometry, we quantified curcumin affinity for copper, zinc, and iron ions. Zn2+ showed little binding, but each Cu2+ or Fe2+ ion appeared to bind at least two curcumin molecules.** The interaction of curcumin with copper reached half-maximum at approximately 3-12 microM copper and exhibited positive cooperativity, with Kd1 approximately 10-60 microM and Kd2

approximately 1.3 microM (for binding of the first and second curcumin molecules, respectively). Curcumin-iron interaction reached half-maximum at approximately 2.5-5 microM iron and exhibited negative cooperativity, with Kd1 approximately 0.5-1.6 microM and Kd2 approximately 50-100 microM. Curcumin and its metabolites can attain these levels in vivo, suggesting physiological relevance. **Since curcumin more readily binds the redox-active metals iron and copper than redox-inactive zinc, curcumin might exert a net protective effect against Abeta toxicity or might suppress inflammatory damage by preventing metal induction of NF-kappaB.**[105]

CURCUMIN (↓) IRON
Journal of Agricultural and Food Chemistry (United States) (2008)
Department of Biochemistry, Microbiology and Biotechnology and Faculty of Pharmacy, Rhodes University, Grahamstown, South Africa.
Research demonstrates that antioxidants and metal chelators may be of beneficial use in the treatment of neurodegenerative diseases, such as Alzheimer's disease (AD). **This study investigated the antioxidant and metal-binding properties of curcumin, capsaicin, and S-allylcysteine, which are major components found in commonly used dietary spice ingredients turmeric, chilli, and garlic, respectively. The DPPH assay demonstrates that these compounds readily scavenge free radicals. These compounds significantly curtail iron- (Fe2+) and quinolinic acid (QA)-induced lipid peroxidation and potently scavenge the superoxide anion generated by 1mM cyanide in rat brain homogenate. The ferrozine assay was used to measure the extent of Fe2+ chelation, and electrochemistry was employed to measure the Fe3+ binding activity of curcumin, capsaicin, and S-allylcysteine. Both assays demonstrate that these compounds bind Fe2+ and Fe3+ and prevent the redox cycling of iron, suggesting that this may be an additional method through which these agents reduce Fe2+-induced lipid peroxidation.** This study demonstrates the antioxidant and metal-binding properties of these spice ingredients, and it is hereby postulate[d] that these compounds have important implications in the prevention or treatment of neurodegenerative diseases such as AD.[106]

CURCUMIN (↓) IRON AND COPPER
Cellular and Molecular Life Sciences (Switzerland) (2008)
Department of Cancer Biology, Wake Forest University School of Medicine, Winston–Salem, North Carolina.
Curcumin is a free radical scavenger and hydrogen donor, and exhibits both pro- and antioxidant activity. It also binds metals, particularly iron

and copper, and can function as an iron chelator. Curcumin is remarkably non-toxic and exhibits limited bioavailability. Curcumin exhibits great promise as a therapeutic agent, and is currently in human clinical trials for a variety of conditions, including multiple myeloma, pancreatic cancer, myelodysplastic syndromes, colon cancer, psoriasis and Alzheimer's disease.[84]

CURCUMIN (↓) IRON AND COPPER
Journal of Alzheimer's Disease (The Netherlands) (2011)
Department of Biochemistry and Molecular Biology, University of Texas Medical Branch, Galveston.

These metals [iron and copper] inhibit the base excision activity of NEIL family DNA glycosylases by oxidizing them, changing their structure, and inhibiting their binding to downstream repair proteins. **Metal chelators and reducing agents partially reverse the inhibition, while curcumin with both chelating and reducing activities reverses the inhibition nearly completely. In this review, we have discussed the possible etiological linkage of BER/SSBR defects to neurodegenerative diseases and the therapeutic potential of metal chelators in restoring DNA repair capacity.**[107]

CURCUMIN (↓) IRON AND COPPER
Dalton Transactions (England) (2012)
Department of Chemistry, Clemson University, South Carolina.

Concentrations of labile iron and copper are elevated in patients with neurological disorders, causing interest in metal-neurotransmitter interactions. Catecholamine (dopamine, epinephrine, and norepinephrine) and amino acid (glycine, glutamate, and 4-aminobutyrate) neurotransmitters are antioxidants also known to bind metal ions... In contrast to amino acid neurotransmitters, catecholamine neurotransmitters, L-DOPA, and curcumin prevent significant iron-mediated DNA damage (IC(50) values of 3.2 to 18 μM) and are electrochemically active. However, glycine and glutamate are more effective at preventing copper-mediated DNA damage (IC(50) values of 35 and 12.9 μM, respectively) than L-DOPA, the only catecholamine to prevent this damage (IC(50) = 73 μM). This metal-mediated DNA damage prevention is directly related to the metal-binding behaviour of these compounds. When bound to iron or copper, the catecholamines, amino acids, and curcumin significantly shift iron oxidation potentials and stabilize $Fe(3+)$ over $Fe(2+)$ and $Cu(2+)$ over $Cu(+)$, a factor that may prevent metal redox cycling in vivo. **These results highlight the disparate antioxidant activities of neurotransmitters, drugs, and supplements [curcumin] and highlight the importance of considering metal binding when identifying antioxidants to treat and prevent neurodegenerative disorders.**[108]

Curcumin (↓) Iron
Expert Review of Neurotherapeutics (England) (2015)
Greater Los Angeles Healthcare System, Veteran's Administration, Geriatric Research Education and Clinical Center, Los Angeles, California.

Curcumin is a pleiotropic molecule, which not only directly binds to and limits aggregation of the β-sheet conformations of amyloid characteristic of many neurodegenerative diseases, but also restores homeostasis of the inflammatory system, boosts the heat shock system to enhance clearance of toxic aggregates, scavenges free radicals, chelates iron and induces anti-oxidant response elements.[109]

GREEN TEA EXTRACTS (↓) IRON AND COPPER

Green Tea Extract (EGCG) (↓) Iron and Copper
Neurosignals (Switzerland) (2005)
Eve Topf and USA National Parkinson Foundation Centers of Excellence for Neurodegenerative Diseases Research, Technion-Faculty of Medicine, Haifa, Israel.

Many lines of evidence suggest that oxidative stress resulting in reactive oxygen species (ROS) generation and inflammation play a pivotal role in the age-associated cognitive decline and neuronal loss in neurodegenerative diseases including Alzheimer's (AD), Parkinson's (PD) and Huntington's diseases. One cardinal chemical pathology observed in these disorders is the accumulation of iron at sites where the neurons die. The buildup of an iron gradient in conjunction with ROS (superoxide, hydroxyl radical and nitric oxide) are thought to constitute a major trigger in neuronal toxicity and demise in all these diseases. Thus, promising future treatment of neurodegenerative diseases and aging depends on availability of effective brain permeable, iron-chelatable/radical scavenger neuroprotective drugs that would prevent the progression of neurodegeneration. **Tea flavonoids (catechins) have been reported to possess potent iron-chelating, radical-scavenging and anti-inflammatory activities and to protect neuronal death in a wide array of cellular and animal models of neurological diseases... This review will focus on the multifunctional properties of green tea and its major component (-)-epigallocatechin-3-gallate (EGCG) and their ability to induce neuroprotection and neurorescue in vitro and in vivo, in particular, their transitional metal (iron and copper) chelating property and inhibition of oxidative stress.**[110]

Green Tea Extract (EGCG) (↓) Iron
Journal of Neurochemistry (England) (2006)
Eve Topf and US National Parkinson Foundation Centers for Neurodegenerative Diseases and Department of Pharmacology, Faculty of Medicine, Technion, Haifa, Israel.

Brain iron dysregulation and its association with amyloid precursor protein (APP) plaque formation are implicated in Alzheimer's disease (AD) pathology and so iron chelation could be considered a rational therapeutic strategy for AD. Here we analyzed the effect of the main polyphenol constituent of green tea, (-)-epigallocatechin-3-gallate (EGCG), which possesses metal-chelating and radical-scavenging properties, on the regulation of the iron metabolism-related proteins APP and transferrin receptor (TfR). **EGCG exhibited potent iron-chelating activity comparable to that of the prototype iron chelator desferrioxamine, and dose dependently (1-10 microm) increased TfR protein and mRNA levels in human SH-SY5Y neuroblastoma cells...** Furthermore, EGCG reduced toxic beta-amyloid peptide generation in Chinese hamster ovary cells overexpressing the APP 'Swedish' mutation. Thus, the natural non-toxic brain-permeable EGCG may provide a potential therapeutic approach for AD and other iron-associated disorders.[111]

GREEN TEA EXTRACT (EGCG) (↓) IRON
Current Alzheimer Research (United Arab Emirates) (2007)

Eve Topf and USA National Parkinson Foundation Centers of Excellence, Technion-Rappaport Family Faculty of Medicine and Department of Pharmacology, Haifa, Israel.

Accumulation of iron at sites where neurons degenerate in Parkinson's disease (PD) and Alzheimer's disease (AD) is thought to have a major role in the oxidative stress induced process of neurodegeneration. **The main polyphenol constituent of green tea, (-)-epigallocatechin-3-gallate (EGCG), which possesses iron-metal chelating, radical scavenging, and neuroprotective properties, offers potential therapeutic benefits for these diseases.** The diverse molecular mechanisms and cell signaling pathways participating in the neuroprotective/neuro-rescue and APP regulation/processing actions of EGCG make this multifunctional compound a potential neuroprotective drug for the treatment of neurodegenerative diseases, such as PD, AD, Huntington's disease, and amyotrophic lateral sclerosis.[112]

GREEN TEA EXTRACT (EGCG) (↓) IRON
Archives of Biochemistry and Biophysics (United States) (2008)

Eve Topf and USA National Parkinson Foundation Centers of Excellence for Neurodegenerative Diseases Research and Department of Pharmacology, Rappaport Family Research Institute, Technion-Faculty of Medicine, Haifa, Israel.

In this study, we have focused on specific identification of proteins involved in the neurorescue activity of the green tea polyphenol, (-)-epigallocatechin-3-gallate (EGCG), in a progressive model of neuronal death induced by long-term serum deprivation of human neuroblastoma SH-SY5Y cells. **The study was designed in an attempt to define biomarkers for the mechanism of action of EGCG, associated with its iron chelating properties and its ability to regulate metabolic energy balance and affect cell morphology.** By using mass spectrometry analysis combined with gene expression technique, we have succeeded to identify such genes and proteins (e.g. ATP synthase mitochondrial F1 complex beta, protein kinase C epsilon, nerve vascular growth factor inducible precursor and hypoxia inducible factor–1 alpha). These results strengthen the notion that the diverse molecular signaling pathways participating in the neurorescue activity of EGCG render this multifunctional compound as potential agent to reduce risk of various neurodegenerative diseases.[113]

The results of these most recently presented studies involving the curcuminoids, EGCG, and resveratrol are shown below in Table 6.3

When we add Tables 6.2 and 6.3, we arrive at Table 6.4, which represents the totality of studies presented in this volume that involve the three most-studied plant polyphenols—the curcuminoids, EGCG, and resveratrol—and AD-related pathogenesis.

As indicated in Table 6.4, there is substantial evidence in the scientific literature from which this particular cocktail of plant polyphenols as a potential frontline formula for the prevention and treatment of AD can be viewed.

Furthermore, pursuant to the analytic framework employed in this volume, and the published studies and reviews that have been provided throughout, we present below the essential elements of a "pleiotropic theory" of plant polyphenols as they relate to the prevention and treatment of AD. Our thesis asserts the therapeutic and safety-related superiority of certain multitargeting plant polyphenols, certain characteristics of which are itemized below, over the more limited-targeting and often safety-plagued pharmaceuticals.

Table 6.3 Alzheimer's disease (AD)–related plant polyphenol pleiotropism (curcuminoids/EGCG/resveratrol)

Curcuminoids, EGCG, Resveratrol	Curcuminoids	EGCG	Resveratrol	Combined
NF-kappa B ($\downarrow$)	(5)	(2)	(7)	(14)
IL-1β ($\downarrow$)	(4)	(1)	(2)	(7)
IL-6 ($\downarrow$)	(3)	(2)	(2)	(7)
TNF ($\downarrow$)	(2)	(2)	(4)	(8)
Iron and/or copper ($\downarrow$)	(6)	(4)		(10)

IL-1β, interleukin-1beta; *IL-6*, interleukin-6; *NF-kappa B*, nuclear factor-kappa B; *TNF*, tumor necrosis factor alpha.

Table 6.4 (Table 6.2 plus Table 6.3) Alzheimer's disease (AD)–related plant polyphenol pleiotropism (curcuminoids/EGCG/resveratrol)

Plant Polyphenols and AD	Curcuminoids	EGCG	Resveratrol	Combined
(↑) Alpha-secretase/sAPPα		(8)		(8)
(↓) Beta-secretase	(5)	(2)		(7)
(↓) Gamma-secretase	(2)	(1)		(3)
(↓) Abeta Aggregation	(7)	(6)		(13)
(↓) Abeta toxicity	(9)	(3)	(10)	(22)
(↑) Abeta clearance	(13)	(1)	(3)	(17)
(↑) Synaptic function	(7)	(2)	(1)	(10)
(↓) Tau hyperphosphorylation	(3)	(1)	(2)	(6)
(↓) Tau Aggregation	(1)	(1)		(2)
(↑) Tau clearance	(2)	(1)		(3)
(↓) Aberrant neuroinflammation	(4)	(1)	(6)	(11)
(↓) AD-related oxidative stress	(11)	(5)	(8)	(24)

	(Col 1)	(Col 2)	(Col 3)	(Col 4)
(↑) SIRT1	(2)	(2)	(8)	(12)
(↓) Notch–1 signaling	(8)	(4)	(1)	(13)
(↓) GSK-3β	(6)	(3)	(1)	(10)
(↓) MAPK	(4)			(4)
(↓) Microglial Activation	(6)	(2)	(8)	(16)
(↓) NMDA excitotoxicity	(4)	(2)		(6)
(↓) NF-kappa B	(5)	(2)	(7)	(14)
(↓) IL–1β	(4)	(1)	(2)	(7)
(↓) IL-6	(3)	(2)	(2)	(7)
(↓) TNF	(2)	(2)	(4)	(8)
(↓) Iron and copper	(6)	(4)		(10)

Abeta, amyloid beta peptide; *APP*, amyloid precursor protein; *GSK-3β*, glycogen synthase kinase 3 beta; *IL-1β*, interleukin–1beta; *IL-6*, interleukin–6; *MAPK*, mitogen–activated protein kinases; *NMDA*, N-methyl-D-aspartate; *NF-kappa B*, nuclear factor-kappa B; *TNF*, tumor necrosis factor alpha.

- *MultiTargeting Capabilities Match Multidimensional Pathophysiology.* Each of the plant-based polyphenolic compounds featured in this volume are beneficially pleiotropic with respect to affecting AD-related pathways, as exhibited in each of the tables presented above. For example, of the 23 AD-related pathogenic bio-markers listed in Table 6.4, the curcuminoids have been shown to therapeutically influence 22, EGCG 21, and resveratrol 14; combined, the three polyphenols interact therapeutically with all 23 biomarkers, in most cases redundantly.
- *Established Pharmacology.* As the abstracts featured in Chapters 3–5 and in this chapter illustrate, the pharmacology of the polyphenolic compounds is well doc-umented, and demonstrates high affinity for an even wider range of AD-related biofactors beyond the 23 listed in Table 6.4 with little to no toxicity to normal-functioning cellular and molecular processes. This provides a level of assurance that the plant polyphenols are capable of producing the desired therapeutic out-comes with appropriate doses with few if any significant adverse effects.
- *Aligned Results of Mechanistic, Epidemiological, and Population Studies.* Chapters 2–5 and this chapter demonstrate that the plant polyphenols and their food-based sources are well studied and have shown consistently positive results within a range of human experience and experimental venues with low toxicity.
- *Trans-Hypotheses.* The nearly global pleiotropism of the plant polyphenols, as it relates to AD pathophysiology, facilitates a range of therapeutic activity across both amyloid- and tau-based pathologies, and can accommodate various theo-ries pertaining to the etiology and progression of AD, as Chapters 3–5 and this chapter illustrate.
- *Trans-Stage.* The plant polyphenols possess the capability to address the patho-physiological elements involved in the zero, preclinical, prodromal, and clinical dementia stages of AD, as Chapters 3–5 and this chapter demonstrate.
- *Natural Compounds from Ancient Foods.* As we illustrate in Chapter 2, humans have been consuming the polyphenolic compounds found in turmeric, grapes, green tea, olive oil, and red wine for many hundreds of years with numerous health benefits, and thus appear to be well adapted biologically to the curcuminoids found in turmeric, the EGCG and other catechins found in green tea, the proan-thocyanidins found in grape seed extract, the oleocanthal and oleuropein com-pounds found in olive oil, and the resveratrol found in grapes and red wine.
- *Little to No Toxicity.* The plant polyphenols have little to no significant toxicity in normal dosing ranges, including as eaten in food and consumed as standardized extracts from reputable manufacturers and sellers of nutritional supplements.

Overall, these pharmacological properties of the plant polyphenols point toward an emerging new paradigm for the prevention and treatment of AD, and merit a higher priority among scientific investigators, clinicians, government regulators, and both national and private healthcare insurance systems. This would include more clinical studies of the plant polyphenols for the prevention and treatment of AD involving two- and three-polyphenol formulas such as those exhibited above, with a focus on bioavailability, dose, and dose adherence.

In addition to the in vitro and in vivo studies, a number of animal studies have been conducted, including the use of AD mouse models, to test the efficacy of the plant polyphenols on memory, cognition, and behavior. Several of these studies are presented below.

CURCUMINOIDS (↑) MEMORY, COGNITION, BEHAVIOR

CURCUMIN (↑) MEMORY, COGNITION, BEHAVIOR
European Neuropsychopharmacology (The Netherlands) (2009)
Department of Emergency Medicine, Brain Research Laboratory, Emory University, Atlanta, Georgia.
The aim of the present study was to examine the modulating impacts of curcumin against cognitive deficits and oxidative damage in intracerebroventricular-streptozotocin (ICV-STZ) infused rats… **ICV-STZ rats showed significant cognitive deficits, which were significantly improved by curcumin supplementation. The study suggests that curcumin is effective in preventing cognitive deficits, and might be beneficial for the treatment of sporadic dementia of Alzheimer's type.**[114]

CURCUMIN (↑) MEMORY, COGNITION, BEHAVIOR
Progress in Neuropsychopharmacology and Biological Psychiatry (England) (2013)
Department of Physiology, Shandong University, School of Medicine, Jinan, Shandong Province, China.
The present study attempts to investigate the effects of curcumin on memory decline of aged mice with a focus upon the possible contribution of the neuronal nitric oxide synthase (nNOS)/nitric oxide (NO) pathway in the memory amelioration effect of curcumin. **The results showed that chronic administration of curcumin (50 mg/kg, i.p., 21 days) significantly ameliorated the memory acquisition ability of aged male mice in the novel object recognition and passive avoidance tasks.**[115]

CURCUMIN (↑) MEMORY, COGNITION, BEHAVIOR
Neurobiology of Learning and Memory (United States) (2013)
Programa de Pós-Graduação em Bioquímica, Departamento de Bioquímica, Universidade Federal do Rio Grande do Sul (UFRGS), Porto Alegre, Brazil.

Animals [rats] received a single intracerebroventricular injection of Aβ(1-42) and they were administered either free curcumin or curcumin-loaded lipid-core nanocapsules intraperitoneally for 10 days. Aβ(1-42)-infused animals showed a significant impairment on learning-memory ability, which was paralleled by a significant decrease in hippocampal synaptophysin levels... **Our findings demonstrate that administration of curcumin was effective in preventing behavioral impairments, neuroinflammation, tau hyperphosphorylation as well as cell signaling disturbances triggered by Aβ in vivo. Of high interest, [curcumin-loaded lipid-core nanocapsules] in a dose 20-fold lower presented similar neuroprotective results compared to the effective dose of free curcumin. Considered overall, the data suggest that curcumin is a potential therapeutic agent for neurocognition and nanoencapsulation of curcumin in [lipid core nanocapsules] might constitute a promising therapeutic alternative in the treatment of neurodegenerative diseases such as AD.**[116]

Curcumin (↑) Memory, Cognition, Behavior
ACS Nano (American Chemical Society) (2014)
Council of Scientific & Industrial Research (CSIR), Indian Institute of Toxicology Research (IITR), Lucknow, India.

Herein, we report that curcumin-encapsulated PLGA nanoparticles... potently induce neural stem cell proliferation and neuronal differentiation in vitro and in the hippocampus and subventricular zone of adult rats, as compared to uncoated bulk curcumin. [Curcumin-encapsulated PLGA nanoparticles] induce neurogenesis by internalization into the hippocampal neural stem cells. [Curcumin-encapsulated PLGA nanoparticles] significantly increase expression of genes involved in cell proliferation (reelin, nestin, and Pax6) and neuronal differentiation (neurogenin, neuroD1, neuregulin, neuroligin, and Stat3). Curcumin nanoparticles increase neuronal differentiation by activating the Wnt/β-catenin pathway, involved in regulation of neurogenesis. These nanoparticles caused enhanced nuclear translocation of β-catenin, decreased GSK-3β levels, and increased promoter activity of the TCF/LEF and cyclin-D1. Pharmacological and siRNA-mediated genetic inhibition of the Wnt pathway blocked neurogenesis-stimulating effects of curcumin. **These nanoparticles reverse learning and memory impairments in an amyloid beta induced rat model of AD-like phenotypes, by inducing neurogenesis... These results suggest that curcumin nanoparticles induce adult neurogenesis through activation of the canonical Wnt/β-catenin pathway and may offer a therapeutic approach to treating neurodegenerative diseases such as AD, by enhancing a brain self-repair mechanism.** The researchers also reported: "The present study involved the entrapment of curcumin into PLGA polymer. We used PLGA due to its biodegradable and biocompatible properties. Curcumin is insoluble in water, which greatly restricts its applications in experimental and clinical situations. Encapsulation of curcumin in PLGA leads to solubility of curcumin in water, and these nanoparticles monodisperse in water... PLGA nanoparticles can cross the blood–brain barrier and localize mainly in the hippocampus."[117]

CURCUMIN (↑) MEMORY, COGNITION, BEHAVIOR
Genetics and Molecular Research (Brazil) (2014)

Department of Neurology, The 148th Hospital, Zibo, Shandong, China.

Inhibitory factors of axonal regeneration have been shown to cause a series of pathophysiological changes in the early period of AD. In this study, the co-receptor (Nogo receptor; NgR) of three axonal growth-inhibitory proteins was examined, and effects of curcumin on spatial learning and memory abilities and hippocampal axonal growth were investigated in amyloid β-protein (Aβ)1-40-induced AD rats. Results showed that the expression of NgR in the AD group significantly increased and the number of axonal protein-positive fibers significantly reduced. **The spatial learning and memory abilities of AD rats were significantly improved in the curcumin group... Together, these results suggested that curcumin could improve the spatial learning and memory abilities of AD rats.**[118]

CURCUMIN (↑) MEMORY, COGNITION, BEHAVIOR
PLoS One (United States) (2015)

Key-Disciplines Laboratory Clinical-Medicine of Henan, Zhengzhou, Henan, China; Department of Neurology, First Affiliated Hospital of Zhengzhou University; Department of Intensive Care Unit, First Affiliated Hospital, Zhengzhou University.

In the present study, we examined the working memory and spatial reference memory in rats that received a ventricular injection of amyloid-β1-42 (Aβ1-42), representing a rodent model of Alzheimer's disease (AD). **The rats treated with Aβ1-42 exhibited obvious cognitive deficits in behavioral tasks. Chronic (seven consecutive days, once per day) but not acute (once a day) curcumin treatments (50, 100, and 200 mg/kg) improved the cognitive functions in a dose-dependent manner... These findings suggest that chronic curcumin [treatment] ameliorates AD-related cognitive deficits and that upregulated BDNF-ERK signaling in the hippocampus may underlie the cognitive improvement produced by curcumin.**[119]

GRAPE PRODUCTS (↑) MEMORY, COGNITION, BEHAVIOR

GRAPE PRODUCTS (CONCORD GRAPE JUICE) (↑) MEMORY, COGNITION, BEHAVIOR
Nutrition (United States) (2006)

United States Department of Agriculture, Agricultural Research Service, Human Nutrition Research Center on Aging, Tufts University, Boston, Massachusetts.

We investigated the beneficial effects of two concentrations of Concord grape juice (10% and 50%) compared with a calorically matched placebo for their effectiveness in reversing age-related deficits in behavioral and neuronal functions in aged Fischer 344 rats. **Rats that drank the 10% grape juice from age 19 to 21 months had improvements in oxotremorine enhancement of K+-evoked release of dopamine from striatal slices and in cognitive performance on the Morris water maze, and the 50% grape juice produced improvements in motor function. These findings suggest that, in addition to their known beneficial effects on cancer and heart disease, polyphenolics in foods may be beneficial in reversing the course of neuronal and behavioral aging, possibly through a multiplicity of direct and indirect effects that can affect a variety of neuronal parameters.**[120]

GREEN TEA EXTRACT (↑) MEMORY, COGNITION, BEHAVIOR

GREEN TEA EXTRACT (EGCG) (↑) MEMORY, COGNITION, BEHAVIOR
Iranian Biomedical Journal (Iran) (2007)
Department of Anatomy, School of Medicine, Iran University of Medical Sciences, Tehran, Iran. Since epigallocatechin-3-gallate (EGCG) is a potent antioxidant agent, which its role against oxidative stress and inflammation has been shown in prior studies, we tried to determine whether EGCG administration protects against beta-amyloid-induced memory and coordination impairment in rats... **We concluded that EGCG can be effective in restoring beta-amyloid-induced behavioral derangements in rats regarding coordination and memory abilities.**[121]

GREEN TEA EXTRACT (EGCG) (↑) MEMORY, COGNITION, BEHAVIOR
Behavioral Brain Research (The Netherlands) (2013)
Departamento de Bioquímica, Instituto de Ciências Básicas da Saúde, Universidade Federal do Rio Grande do Sul, Porto Alegre, Brazil.
Numerous studies on green tea epigallocatechin gallate (EGCG) demonstrate its beneficial effects on cognition and memory. As such, this study evaluated, for the first time, the effects of sub-chronic EGCG treatment in rats that were submitted to ICV [intracerebroventricular] infusion of streptozotocin (3mg/kg). **Our results show that EGCG was not able to modify glucose uptake and glutathione content, although cognitive deficit, S100B content and secretion, acetylcholinesterase activity, glutathione peroxidase activity, nitric oxide metabolites, and reactive oxygen species content were completely reversed by EGCG administration, confirming the neuroprotective potential of this compound.**[122]

GREEN TEA EXTRACT (EGCG) (↑) MEMORY, COGNITION, BEHAVIOR
Molecular and Cellular Biochemistry (The Netherlands) (2013)
Department of Neurology, First Affiliated Hospital of China Medical University, Shenyang, Liaoning Province, China.
In conclusion, our study provides evidence that long-term consumption of EGCG may alleviate AD-related cognitive deficits by effectively attenuating central insulin resistance.[33]

GREEN TEA EXTRACT (EGCG) (↑) MEMORY, COGNITION, BEHAVIOR
Molecular Neurobiology (United States) (2014)
Department of Pharmacology, School of Pharmaceutical Sciences, China Medical University, Shenyang, Liaoning Province, China.
(-)-Epigallocatechin-3-gallate (EGCG) is used for prevention and treatment of many neurodegenerative diseases, including AD. However, whether the neuroprotective effects of EGCG treatment were via modulating the balance of TrkA/p75NTR signaling was still unknown. **In this study, we found that EGCG treatment dramatically ameliorated the cognitive impairments, reduced the overexpressions of Aβ(1-40) and amyloid precursor protein (APP), and inhibited the neuronal apoptosis in the APP/PS1 mice.**[123]

OLIVE OIL EXTRACTS (↑) MEMORY, COGNITION, BEHAVIOR

OLIVE OIL EXTRACT (OLEUROPEIN AGLYCONE) (↑) MEMORY, COGNITION, BEHAVIOR
PLoS One (United States) (2013)
Department of Neuroscience, Psychology, Drug Research and Child Health, Division of Pharmacology and Toxicology, University of Florence, Italy.
We report here that dietary supplementation of oleuropein aglycone strongly improves the cognitive performance of young/middle-aged TgCRND8 mice, a model of amyloid-ß deposition, [with] respect to age-matched littermates with un-supplemented diet... Our results support, and provide mechanistic insights into, the beneficial effects against Alzheimer-associated neurodegeneration of a polyphenol enriched in the extra virgin olive oil, a major component of the Mediterranean diet.[124]

RESVERATROL (↑) MEMORY, COGNITION, BEHAVIOR

RESVERATROL (↑) MEMORY, COGNITION, BEHAVIOR
Life Sciences (The Netherlands) (2002)
Neuropharmacology Laboratory, Department of Pharmacology, All India Institute of Medical Sciences, New Delhi, India.

We have recently shown free radical generation is associated with cognitive impairment in intracerebroventricular (ICV) streptozotocin (STZ) model of sporadic dementia of Alzheimer's type in rats… In the present study, the effect of trans-resveratrol was investigated on ICV STZ induced cognitive impairment and oxidative stress in rats… **Trans-resveratrol treatment significantly prevented ICV STZ induced cognitive impairment.**[125]

RESVERATROL (↑) MEMORY, COGNITION, BEHAVIOR
Indian Journal of Physiology and Pharmacology (India) (2005)
Neuropharmacology Laboratory, Department of Pharmacology, All India Institute of Medical Sciences, New Delhi, India.
In the present study, the effect of the antioxidants alpha lipoic acid, melatonin, and trans-resveratrol were studied against intracerebroventricular streptozotocin induced spatial memory deficit. Male Wistar rats were injected with intracerebroventricular streptozotocin bilaterally. The rats were treated chronically with alpha lipoic acid (200 mg/kg, po), melatonin (20 mg/kg, ip), and trans-resveratrol (20 mg/kg, ip) for 18 days starting from day 1 of streptozotocin injection in separate groups… **The study demonstrated the effectiveness of alpha lipoic acid, melatonin, and trans-resveratrol in preventing spatial memory deficit induced by intracerebroventricular streptozotocin and its potential in age-related neurodegenerative disorders, such as Alzheimer's disease, where oxidative stress is involved.**[126]

These studies using curcumin, EGCG, oleuropein aglycone, and resveratrol were positive to one degree or another in improving memory, cognition, and behavior in the mice or rats that were tested. Many of these studies used delivery methods of the polyphenolic compounds that were designed to enhance or ensure bioavailability. These included curcumin-loaded nanocapsules and intraperitoneal injection. Furthermore, 100% of the tested animals actually received the prescribed treatments and in the advertised doses. These common features of the successful animal experiments—enhanced bioavailability and 100% adherence to the prescribed dose—are at variance with randomized, double-blind, placebo-controlled clinical trials for humans that have involved oral doses of low-bioavailability and intention-to-treat analyses of data, which dilute bioactivity (at the beginning) then conservatively report (at the finish) whatever treatment effects are left to count.

With respect to bioavailability—intention-to-treat is addressed in the following chapter—in 2011, researchers observed in *Expert Reviews in Molecular Medicine*: "Over the past decade, research with curcumin has increased significantly. In vitro and in vivo studies have demonstrated that curcumin could target pathways involved in the pathophysiology of Alzheimer disease (AD), such as the β-amyloid cascade, tau phosphorylation, neuroinflammation or oxidative stress. These findings suggest that curcumin might be a promising compound for the development of AD therapy."[127]

In a review of clinical trials and AD that involved curcumin, researchers reported in *Scientific World Journal* in 2014 that "several reasons could be responsible for the discrepancy between in vitro and in vivo findings and human trials, such as low bioavailability and poor study design."[128]

In 2015, in "Therapeutic Applications of Curcumin Nanoformulations," researchers reported in *The AAPS Journal* (published by the American Association of Pharmaceutical Scientists): "Despite such phenomenal advances in medicinal applications, the clinical implication of native curcumin is hindered due to low solubility, physico-chemical instability, poor bioavailability, rapid metabolism, and poor pharmacokinetics. However, these issues can be overcome by utilizing an efficient delivery system. Active scientific research was initiated in 2005 to improve curcumin's pharmacokinetics, systemic bioavailability, and biological activity by encapsulating or by loading curcumin into nanoform(s) (nanoformulations). A significant number of nanoformulations exist that can be translated toward medicinal use upon successful completion of pre-clinical and human clinical trials."[129]

And in 2016, the *British Journal of Nutrition* reported: "In vitro studies have shown that Aβ metabolism is altered by curcumin, and animal studies report that curcumin may influence brain function and the development of dementia, because of its antioxidant and anti-inflammatory properties, as well as its ability to influence Aβ metabolism. However, clinical studies of curcumin have revealed limited effects to date, most likely because of curcumin's relatively low solubility and bioavailability… [N]ew curcumin formulations that increase bioavailability [are] renewing optimism concerning curcumin-based therapy."[130] For example, a lipidated curcumin formulation showed beneficial effects on cognition and mood in healthy older patients[131] and reduced Alzheimer's biomarkers in healthy middle aged patients.[132]

REFERENCES

1. Doody, Raman, Farlow, et al. A phase 3 trial of semagacestat for treatment of Alzheimer's disease. *New England Journal of Medicine* July 25, 2013;**369**(4):341–50.
2. Henley, Sundell, Sethuraman, et al. Safety profile of semagacestat, a gamma-secretase inhibitor: IDENTITY trial findings. *Current Medical Research and Opinion* October 2014;**30**(10):2021–32.
3. Alzforum: Networking for a Cure: Therapeutics: Semagacestat, At http://www.alzforum.org/therapeutics/semagacestat.
4. Mikulca, Nguyen, Gajdosik, et al. Potential novel targets for Alzheimer pharmacotherapy: II. Update on secretase inhibitors and related approaches. *Journal of Clinical Pharmacy and Therapeuitcs* February 2014;**39**(1):25–37.
5. Folch, Ettcheto, Petrov, et al. Review of the advances in treatment for Alzheimer disease: strategies for combating β-amyloid protein. *Neurología: publicación oficial de la Sociedad Española de Neurología* May 11, 2015. http://dx.doi.org/10.1016/j.nrl.2015.03.012. [Epub ahead of print].
6. De Kloe, De Strooper. Small molecules that inhibit notch signaling. *Methods in Molecular Biology* 2014;**1187**:311–22.
7. Wang, Zhang, Banerjee, et al. Notch-1 down-regulation by curcumin is associated with the inhibition of cell growth and the induction of apoptosis in pancreatic cancer cells. *Cancer* June 1, 2006;**106**(11):2503–13.

8. Sarkar, Li, Wang, Kong. Cellular signaling perturbation by natural products. *Cellular Signalling* November 2009;**21**(11):1541–7.

9. Liao, Xia, Chen, et al. Inhibitory effect of curcumin on oral carcinoma CAL-27 cells via suppression of Notch-1 and NF-κB signaling pathways. *Journal of Cellular Biochemistry* April 2011;**112**(4): 1055–65.

10. Subramaniam, Ponnurangam, Ramamoorthy, et al. Curcumin induces cell death in esophageal cancer cells through modulating notch signaling. *PLoS One* 2012;**7**(2):e30590.

11. Li, Zhang, Ma, et al. Curcumin inhibits proliferation and invasion of osteosarcoma cells through inactivation of Notch-1 signaling. *The FEBS Journal* June 2012;**279**(12):2247–59.

12. Shehzad, Lee. Molecular mechanisms of curcumin action: signal transduction. *Biofactors* January–February 2013;**39**(1):27–36.

13. Norris, Karmokar, Howells, et al. The role of cancer stem cells in the anti-carcinogenicity of curcumin. *Molecular Nutrition & Food Research* September 2013;**57**(9):1630–7.

14. Chen, Wang, Chen. Modulation of apoptosis-related cell signalling pathways by curcumin as a strategy to inhibit tumor progression. *Molecular Biology Reports* July 2014;**41**(7):4583–94. For additional studies and reports on the downregulation of Notch signaling by curcumin, see: Sordillo, Helson. Curcumin and cancer stem cells: curcumin has asymmetrical effects on cancer and normal stem cells. *Anticancer Research* February 2015;**35**(2):599–614; Koprowski, Sokolowski, Kunnimalaiyaan, et al. Curcumin-mediated regulation of Notch1/hairy and enhancer of split-1/survivin: molecular targeting in cholangiocarcinoma. *The Journal of Surgical Research* October 2015;**198**(2):434–40; Liu, Yang, Zhou, et al. Curcumin regulates hepatoma cell proliferation and apoptosis through the notch signaling pathway. *International Journal of Clinical and Experimental Medicine* March 15, 2014;**7**(3):714–8.

15. Singh, Shankar, Srivastava. Green tea catechin, epigallocatechin-3-gallate (EGCG): mechanisms, perspectives and clinical applications. *Biochemical Pharmacology* December 15, 2011;**82**(12):1807–21.

16. Liu, Zhang, Zhang, et al. The effect of green tea extract and EGCG on the signaling network in squamous cell carcinoma. *Nutrition and Cancer* 2011;**63**(3):466–75.

17. Lee, Nam, Kang, et al. Epigallocatechin-3-gallate attenuates head and neck cancer stem cell traits through suppression of notch pathway. *Oral Oncology* October 2013;**49**(15):3210–8.

18. Tomidokoro, Ishiguro, Harigaya, et al. Abeta amyloidosis induces the initial stage of tau accumulation in APP(Sw) mice. *Neuroscience Letters* February 23, 2001;**299**(3):169–72.

19. Balaraman, Limaye, Levey, Srinivasan. Glycogen synthase kinase 3Beta and Alzheimer's disease: pathophysiological and therapeutic significance. *Cellular and Molecular Life Sciences* June 2006;**63**(11): 1226–35.

20. Takashima. GSK-3 is essential in the pathogenesis of Alzheimer's disease. *Journal of Alzheimers Disease* 2006;**9**(3 Suppl.):309–17.

21. Avila, Hernández. GSK-3 inhibitors for Alzheimer's disease. *Expert Review of Neurotherapeutics* November 2007;**7**(11):1527–33.

22. Martinez, Perez. GSK-3 inhibitors: a ray of hope for the treatment of Alzheimer's disease? *Journal of Alzheimers Disease* October 2008;**15**(2):181–91.

23. Giese. GSK-3: a key player in neurodegeneration and memory. *IUBMB Life* May 2009;**61**(5):516–21.

24. Cai, Zhao, Zhao. Roles of glycogen synthase kinase 3 in Alzheimer's disease. *Current Alzheimer Research* September 2012;**9**(7):864–79. See also: Ferrer, Barrachina, Puig. Glycogen synthase kinase-3 is associated with neuronal and glial hyperphosphorylated tau deposits in Alzheimer's disease, Pick's disease, progressive supranuclear palsy and corticobasal degeneration. *Acta Neuropathologica* December 2002;**104**(6):583–91; Ferrer, Gomez-Isla, Puig, et al. Current advances on different kinases involved in tau phosphorylation, and implications in Alzheimer's disease and tauopathies. *Current Alzheimer Research* January 2005;**2**(1):3–18; Avila, Wandosell, Hernández. Role of glycogen synthase kinase-3 in Alzheimer's disease pathogenesis and glycogen synthase kinase-3 inhibitors. *Expert Review of Neurotherapeutics* May 2010;**10**(5):703–10; Llorens-Martín, Jurado, Hernández, Avila. GSK-3β, a pivotal kinase in Alzheimer disease. *Frontiers in Molecular Neuroscience* May 21, 2014;**7**:46.

25. Bustanji, Taha, Almasri, et al. Inhibition of glycogen synthase kinase by curcumin: investigation by simulated molecular docking and subsequent in vitro/in vivo evaluation. *Journal of Enzyme Inhibition and Medicinal Chemistry* June 2009;**24**(3):771–8.

26. Zhang, Yin, Shi, Li. Curcumin activates Wnt/β-catenin signaling pathway through inhibiting the activity of GSK-3β in APPswe transfected SY5Y cells. *European Journal of Pharmaceutical Sciences* April 18, 2011;**42**(5):540–6.

27. Xiong, Hongmei, Lu, Yu. Curcumin mediates presenilin-1 activity to reduce β-amyloid production in a model of Alzheimer's disease. *Pharmacological Reports* 2011;**63**(5):1101–8.

28. Huang, Xu, Jiang. Curcumin-mediated neuroprotection against amyloid-β-induced mitochondrial dysfunction involves the inhibition of GSK-3β. *Journal of Alzheimer's Disease* 2012;**32**(4):981–96.

29. Huang, Tang, Xu, Jiang. Curcumin attenuates amyloid-β-induced tau hyperphosphorylation in human neuroblastoma SH-SY5Y cells involving PTEN/Akt/GSK-3β signaling pathway. *Journal of Receptor and Signal Transduction Research* February 2014;**34**(1):26–37.

30. Di Martino, De Simone, Andrisano, et al. Versatility of the curcumin scaffold: discovery of potent and balanced dual BACE-1 and GSK-3β inhibitors. *Journal of Medicinal Chemistry* January 28, 2016;**59**(2):531–44. See also: Hoppe, Coradini, Frozza, et al. Free and nanoencapsulated curcumin suppress β-amyloid-induced cognitive impairments in rats: involvement of BDNF and Akt/GSK-3β signaling pathway. *Neurobiology of Learning and Memory* November 2013;**106**:134–44.

31. Koh, Kim, Kwon, et al. Epigallocatechin gallate protects nerve growth factor differentiated PC12 cells from oxidative-radical-stress-induced apoptosis through its effect on phosphoinositide 3-kinase/Akt and glycogen synthase Kinase-3. *Brain Research. Molecular Brain Research* October 21, 2003;**118**(1–2):72–81.

32. Lin, Chen, Chiu, et al. Epigallocatechin gallate (EGCG) suppresses beta-amyloid-induced neurotoxicity through inhibiting c-Abl/FE65 nuclear translocation and GSK3 beta activation. *Neurobiology of Aging* January 2009;**30**(1):81–92.

33. Jia, Han, Kong, et al. (-)-Epigallocatechin-3-gallate alleviates spatial memory impairment in APP/PS1 mice by restoring IRS-1 signaling defects in the hippocampus. *Molecular and Cellular Biochemistry* August 2013;**380**(1–2):211–8.

34. Varamini, Sikalidis, Bradford. Resveratrol increases cerebral glycogen synthase kinase phosphorylation as well as protein levels of drebrin and transthyretin in mice: an exploratory study. *International Journal of Food Sciences and Nutrition* February 2014;**65**(1):89–96.

35. Drewes, Lichtenberg-Kraag, Döring, et al. Mitogen activated protein (MAP) kinase transforms tau protein into an Alzheimer-like state. *The EMBO Journal* June 1992;**11**(6):2131–8.

36. Mandelkow, Biernat, Drewes, et al. Microtubule-associated protein tau, paired helical filaments, and phosphorylation. *Annals of the New York Academy of Sciences* September 24, 1993;**695**:209–16.

37. Knowles, Chin, Ruff, Hyman. Demonstration by fluorescence resonance energy transfer of a close association between activated MAP kinase and neurofibrillary tangles: implications for MAP kinase activation in Alzheimer disease. *Journal of Neuropathology and Experimental Neurology* October 1999;**58**(10):1090–8.

38. Sheng, Jones, Zhou, et al. Interleukin-1 promotion of MAPK-p38 overexpression in experimental animals and in Alzheimer's disease: potential significance for tau protein phosphorylation. *Neurochemistry International* November–December 2001;**39**(5–6):341–8.

39. Zhu, Lee, Raina, et al. The role of mitogen-activated protein kinase pathways in Alzheimer's disease. *Neuro-signals* September–October 2002;**11**(5):270–81.

40. Wang, Zhang, Liu, Li. Overactivated mitogen-activated protein kinase by anisomycin induces tau hyperphosphorylation. *Sheng Li Xue Bao [Acta Physiologica Sinica]* August 25, 2008;**60**(4):485–91.

41. Munoz, Ammit. Targeting p38 MAPK pathway for the treatment of Alzheimer's disease. *Neuropharmacology* March 2010;**58**(3):561–8.

42. Zhang, Wu, Zhao, et al. Demethoxycurcumin, a natural derivative of curcumin, attenuates LPS-induced pro-inflammatory responses through down-regulation of intracellular ROS-related MAPK/NF-kappaB signaling pathways in N9 microglia induced by lipopolysaccharide. *International Immunopharmacology* March 2010;**10**(3):331–8.

43. Shi, Zheng, Li, et al. Curcumin inhibits Aβ-induced microglial inflammatory responses in vitro: involvement of ERK1/2 and p38 signaling pathways. *Neuroscience Letters* May 6, 2015;**594**:105–10.

44. Fan, Li, Fu, et al. Reversal of beta-amyloid-induced neurotoxicity in PC12 cells by curcumin, the important role of ROS-mediated signaling and ERK pathway. *Cellular and Molecular Neurobiology* March 14, 2016;**37**. [Epub ahead of print].

45. Fu, Yang, Cao, et al. Strategy to suppress oxidative damage-induced neurotoxicity in PC12 cells by curcumin: the role of ROS-mediated DNA damage and the MAPK and AKT pathways. *Molecular Neurobiology* January 2016;**53**(1):369–78.

46. Lautner, Mattsson, Schöll, et al. Biomarkers for microglial activation in Alzheimer's disease. *International Journal of Alzheimer's Disease* 2011;**2011**. 939426.

47. Solito, Sastre. Microglia function in Alzheimer's disease. *Frontiers in Pharmacology* February 10, 2012;**3**:14.

48. Rapic, Backes, Viel, et al. Imaging microglial activation and glucose consumption in a mouse model of Alzheimer's disease. *Neurobiology of Aging* January 2013;**34**(1):351–4.

49. Fan, Aman, Ahmed, et al. Influence of microglial activation on neuronal function in Alzheimer's and Parkinson's disease dementia. *Alzheimer's and Dementia: The Journal of the Alzheimer's Association* June 2015;**11**(6). 608–621.e7.

50. Yu, Gong, Hu, et al. Cu(II) enhances the effect of Alzheimer's amyloid-β peptide on microglial activation. *Journal of Neuroinflammation* June 24, 2015;**12**:122.

51. Femminella, Ninan, Atkinson, et al. Does microglial activation influence hippocampal volume and neuronal function in Alzheimer's disease and Parkinson's disease dementia? *Journal of Alzheimer's Disease* 2016;**51**(4):1275–89.

52. Hamelin, Lagarde, Dorothée, et al. Early and protective microglial activation in Alzheimer's disease: a prospective study using 18F-DPA-714 PET imaging. *Brain: a Journal of Neurology* April 2016;**139**(Pt 4):1252–64.

53. Cole, Morihara, Lim, et al. NSAID and antioxidant prevention of Alzheimer's disease: lessons from in vitro and animal models. *Annals of the New York Academy of Sciences* December 2004; **1035**:68–84.

54. Lee, Jung, Cho, et al. Neuroprotective effect of curcumin is mainly mediated by blockade of microglial cell activation. *Die Pharmazie* December 2007;**62**(12):937–42.

55. Jin, Lee, Park, et al. Curcumin attenuates the release of pro-inflammatory cytokines in lipopolysaccharide-stimulated BV2 microglia. *Acta Pharmacologica Sinica* October 2007;**28**(10):1645–51.

56. Karlstetter, Lippe, Walczak, et al. Curcumin is a potent modulator of microglial gene expression and migration. *Journal of Neuroinflammation* September 29, 2011;**8**:125.

57. Cianciulli, Calvello, Porro, et al. PI3k/Akt signalling pathway plays a crucial role in the anti-inflammatory effects of curcumin in LPS-activated microglia. *International Immunopharmacology* July 2016;**36**:282–90.

58. Ding, Li, Li, et al. HSP60 mediates the neuroprotective effects of curcumin by suppressing microglial activation. *Experimental and Therapeutic Medicine* August 2016;**12**(2):823–8.

59. Li, Huang, Fang, Le. (-)-Epigallocatechin gallate inhibits lipopolysaccharide-induced microglial activation and protects against inflammation-mediated dopaminergic neuronal injury. *Journal of Neuroscience Research* December 1, 2004;**78**(5):723–31.

60. Cheng-Chung Wei, Huang, Chen, et al. Epigallocatechin gallate attenuates amyloid β-induced inflammation and neurotoxicity in EOC 13.31 microglia. *European Journal of Pharmacology* January 5, 2016;**770**:16–24.

61. Zhang, Liu, Shi. Anti-inflammatory activities of resveratrol in the brain: role of resveratrol in microglial activation. *European Journal of Pharmacology* June 25, 2010;**636**(1–3):1–7.

62. Capiralla, Vingtdeux, Zhao, et al. Resveratrol mitigates lipopolysaccharide- and Aβ-mediated microglial inflammation by inhibiting the TLR4/NF-κB/STAT signaling cascade. *Journal of Neurochemistry* February 2012;**120**(3):461–72.

63. Zhang, Wang, Wu, et al. Resveratrol protects cortical neurons against microglia-mediated neuroinflammation. *Phytotherapy Research* March 2013;**27**(3):344–9.

64. Solberg, Chamberlin, Vigil, et al. Optical and SPION-enhanced MR imaging shows that trans-stilbene inhibitors of NF-κB concomitantly lower Alzheimer's disease plaque formation and microglial activation in AβPP/PS-1 transgenic mouse brain. *Journal of Alzheimers Disease* 2014;**40**(1):191–212.

65. Yao, Li, Niu, et al. Resveratrol inhibits oligomeric Aβ-induced microglial activation via NADPH oxidase. *Molecular Medicine Reports* October 2015;**12**(4):6133–9.

66. Porro, Cianciulli, Calvello, Panaro. Reviewing the role of resveratrol as a natural modulator of microglial activities. *Current Pharmaceutical Design* 2015;**21**(36):5277–91.

67. Cianciulli, Dragone, Calvello, et al. IL-10 plays a pivotal role in anti-inflammatory effects of resveratrol in activated microglia cells. *International Immunopharmacology* February 2015;**24**(2):369–76.

68. Wang, Cui, Yang, et al. Resveratrol rescues the impairments of hippocampal neurons stimulated by microglial over-activation in vitro. *Cellular and Molecular Neurobiology* April 22, 2015;**35**. [Epub ahead of print].

69. Amadoro, Ciotti, Costanzi, et al. NMDA receptor mediates tau-induced neurotoxicity by calpain and ERK/MAPK activation. *Proceedings of the National Academy of Sciences* February 21, 2006;**103**(8): 2892–7.

70. Danysz, Parsons. Alzheimer's disease, β-amyloid, glutamate, NMDA receptors and memantine—searching for the connections. *British Journal of Pharmacology* September 2012;**167**(2):324–52.

71. Zádori, Veres, Szalárdy, et al. Glutamatergic dysfunctioning in Alzheimer's disease and related therapeutic targets. *Journal of Alzheimer's Disease* 2014;**42**(Suppl. 3):S177–87.

72. Rush, Buisson. Reciprocal disruption of neuronal signaling and Aβ production mediated by extrasynaptic NMDA receptors: a downward spiral. *Cell and Tissue Research* May 2014;**356**(2):279–86.

73. Zhang, Li, Feng, Wu. Dysfunction of NMDA receptors in Alzheimer's disease. *Neurological Sciences* July 2016;**37**(7):1039–47.

74. Foster, Kyritsopoulos, Kumar. Central role for NMDA receptors in redox mediated impairment of synaptic function during aging and Alzheimer's disease. *Behavioural Brain Research* May 11, 2016;**16**:30278–9. pii:S0166-4328.

75. Guntupalli, Widagdo, Anggono. Amyloid-β-induced dysregulation of AMPA receptor trafficking. *Neural Plasticity* 2016;**2016**. 3204519.

76. Matteucci, Frank, Domenici, et al. Curcumin treatment protects rat retinal neurons against excitotoxicity: effect on N-methyl-D-aspartate-induced intracellular Ca(2+) increase. *Experimental Brain Research* December 2005;**167**(4):641–8.

77. Yazawa, Kihara, Shen, et al. Distinct mechanisms underlie distinct polyphenol-induced neuroprotection. *FEBS Letters* December 11, 2006;**580**(28–29):6623–8.

78. Matteucci, Cammarota, Paradisi. Curcumin protects against NMDA-induced toxicity: a possible role for NR2A subunit. *Investigative Ophthalmology and Visual Science* February 22, 2011;**52**(2):1070–7.

79. Huang, Chang, Lu, et al. Protection of curcumin against amyloid-β-induced cell damage and death involves the prevention from NMDA receptor-mediated intracellular Ca^{2+} elevation. *Journal of Receptor and Signal Transduction Research* 2015;**35**(5):450–7. See also: Lin, Hung, Chiu, et al. Curcumin enhances neuronal survival in N-methyl-D-aspartic acid toxicity by inducing RANTES expression in astrocytes via PI-3K and MAPK signaling pathways. *Progress in Neuro-Psychopharmacology and Biological Psychiatry* June 1, 2011;**35**(4):931–8; Cheng, Guo, Xie, et al. Curcumin rescues aging-related loss of hippocampal synapse input specificity of long term potentiation in mice. *Neurochemical Research* January 2013;**38**(1):98–107.

80. Jang, Jeong, Park, et al. Neuroprotective effects of (-)-epigallocatechin-3-gallate against quinolinic acid-induced excitotoxicity via PI3K pathway and no inhibition. *Brain Research* February 8, 2010;**1313**:25–33.

81. He, Cui, Lee, et al. Prolonged exposure of cortical neurons to oligomeric amyloid-β impairs NMDA receptor function via NADPH oxidase-mediated ROS production: protective effect of green tea (-)-epigallocatechin-3-gallate. *ASN Neuro* February 8, 2011;**3**(1):e00050.

82. Kuner, Schubenel, Hertel. Beta-amyloid binds to p57NTR and activates NFkappaB in human neuroblastoma cells. *Journal of Neuroscience Research* December 15, 1998;**54**(6):798–804.

83. Jagetia, Aggarwal. 'Spicing up' of the immune system by curcumin. *Journal of Clinical Immunology* January 2007;**27**(1):19–35.

84. Hatcher, Planalp, Cho, et al. Curcumin: from ancient medicine to current clinical trials. *Cellular and Molecular Life Sciences* June 2008;**65**(11):1631–52.

85. Deng, Lu, Wang, et al. Curcumin inhibits the AKT/NF-κB signaling via CpG demethylation of the promoter and restoration of NEP in the N2a cell line. *The AAPS Journal* July 2014;**16**(4):649–57.

86. Nahar, Slitt, Seeram. Anti-inflammatory effects of novel standardized solid lipid curcumin formulations. *Journal of Medicinal Food* July 2015;**18**(7):786–92.

87. Kim, Jeong, Lee, et al. Epigallocatechin-3-gallate suppresses NF-kappaB activation and phosphorylation of p38 MAPK and JNK in human astrocytoma U373MG cells. *The Journal of Nutritional Biochemistry* September 2007;**18**(9):587–96.

88. Lee, Lee, Ban, et al. Green tea (-)-epigallocatechin-3-gallate inhibits beta-amyloid-induced cognitive dysfunction through modification of secretase activity via inhibition of ERK and NF-kappaB pathways in mice. *The Journal of Nutrition* October 2009;**139**(10):1987–93.

89. Draczynska-Lusiak, Chen, Sun, et al. Oxidized lipoproteins activate NF-kappaB binding activity and apoptosis in PC12 cells. *Neuroreport* February 16, 1998;**9**(3):527–32.
90. Jang, Surh. Protective effect of resveratrol on beta-amyloid-induced oxidative PC12 cell death. *Free Radical Biology and Medicine* April 15, 2003;**34**(8):1100–10.
91. Chen, Zhou, Mueller-Steiner, et al. SIRT1 protects against microglia-dependent amyloid-beta toxicity through inhibiting NF-kappaB signaling. *The Journal of Biological Chemistry* December 2, 2005;**280**(48):40364–74.
92. Kim, Lim, Rhee, et al. Resveratrol inhibits inducible nitric oxide synthase and cyclooxygenase-2 expression in beta-amyloid-treated C6 glioma cells. *International Journal of Molecular Medicine* June 2006;**17**(6):1069–75.
93. Lukiw, Zhao, Cui. An NF-kappaB-sensitive micro RNA-146a-mediated inflammatory circuit in Alzheimer disease and in stressed human brain cells. *The Journal of Biological Chemistry* November 14, 2008;**283**(46):31315–22.
94. Pogue, Percy, Cui, et al. Up-regulation of NF-kB-sensitive miRNA-125b and miRNA-146a in metal sulfate-stressed human astroglial (HAG) primary cell cultures. *Journal of Inorganic Biochemistry* November 2011;**105**(11):1434–7.
95. Zhao, Li, Wang, et al. Resveratrol decreases the insoluble $A\beta1$-42 level in hippocampus and protects the integrity of the blood–brain barrier in AD rats. *Neuroscience* December 3, 2015;**310**: 641–9.
96. Lim, Chu, Yang, et al. The curry spice curcumin reduces oxidative damage and amyloid pathology in an Alzheimer transgenic mouse. *The Journal of Neuroscience* November 1, 2001;**21**(21):8370–7.
97. Begum, Jones, Lim, et al. Curcumin structure-function, bioavailability, and efficacy in models of neuroinflammation and Alzheimer's disease. *The Journal of Pharmacology and Experimental Therapeutics* July 2008;**326**(1):196–208.
98. Ahmed, Gilani. A comparative study of curcuminoids to measure their effect on inflammatory and apoptotic gene expression in an $A\beta$ plus ibotenic acid-infused rat model of Alzheimer's disease. *Brain Research* July 11, 2011;**1400**:1–18.
99. Lee, Park, Kang, et al. Resveratrol reduces glutamate-mediated monocyte chemotactic protein-1 expression via inhibition of extracellular signal-regulated kinase 1/2 pathway in rat hippocampal slice cultures. *Journal of Neurochemistry* March 2010;**112**(6):1477–87.
100. Zhou, Beevers, Huang. The targets of curcumin. *Current Drug Targets* March 1, 2011;**12**(3):332–47.
101. Zaky, Mohammad, Moftah, et al. Apurinic/apyrimidinic endonuclease 1 is a key modulator of aluminum-induced neuroinflammation. *BioMed Central Neuroscience* March 11, 2013;**14**:26.
102. Giri, Rajagopal, Kalra. Curcumin, the active constituent of turmeric, inhibits amyloid peptide-induced cytochemokine gene expression and CCR5-mediated chemotaxis of THP-1 monocytes by modulating early growth response-1 transcription factor. *Journal of Neurochemistry* December 2004;**91**(5):1199–210.
103. Wright, Tull, Deel, et al. Resveratrol effects on astrocyte function: relevance to neurodegenerative diseases. *Biochemical and Biophysical Research Communications* September 14, 2012;**426**(1):112–5.
104. Cheng, Wang, Li, Zhao. Effects of resveratrol on hippocampal astrocytes and expression of TNF-α in Alzheimer's disease model rate. *Wei Sheng Yan Jiu* July 2015;**44**(4):610–4.
105. Baum, Ng. Curcumin interaction with copper and iron suggests one possible mechanism of action in Alzheimer's disease animal models. *Journal of Alzheimer's Disease* August 2004;**6**(4):367–77. discussion 443–9.
106. Dairam, Fogel, Daya, Limson. Antioxidant and iron-binding properties of curcumin, capsaicin, and S-allylcysteine reduce oxidative stress in rat brain homogenate. *Journal of Agricultural and Food Chemistry* May 14, 2008;**56**(9):3350–6.
107. Hedge, Hedge, Rao, Mitra. Oxidative genome damage and its repair in neurodegenerative diseases: function of transition metals as a double-edged sword. *Journal of Alzheimers Disease* 2011;**24**(Suppl. 2):183–98.
108. García, Angelé-Martínez, Wilkes, et al. Prevention of iron- and copper-mediated DNA damage by catecholamine and amino acid neurotransmitters, L-DOPA, and curcumin: metal binding as a general antioxidant mechanism. *Dalton Transactions* June 7, 2012;**41**(21):6458–67.

109. Hu, Maiti, Ma, et al. Clinical development of curcumin in neurodegenerative disease. *Expert Review of Neurotherapeutics* June 2015;**15**(6):629–37.

110. Mandel, Avramovich-Tirosh, Reznichenko, et al. Multifunctional activities of green tea catechins in neuroprotection. Modulation of cell survival genes, iron-dependent oxidative stress and PKC signaling pathway. *Neuro-signals* 2005;**14**(1–2):46–60.

111. Reznichenko, Amit, Zheng, et al. Reduction of iron-regulated amyloid precursor protein and beta-amyloid peptide by (-)-epigallocatechin-3-gallate in cell cultures: implications for iron chelation in Alzheimer's disease. *Journal of Neurochemistry* April 2006;**97**(2):527–36.

112. Avramovich-Tirosh, Reznichenko, Mit, et al. Neurorescue activity, APP regulation and amyloid-beta peptide reduction by novel multi-functional brain permeable iron-chelating antioxidants, M-30 and green tea polyphenol. *Current Alzheimer Research* September 2007;**4**(4):403–11.

113. Weinreb, Amit, Youdim. The application of proteomics for studying the neurorescue activity of the polyphenol (-)-epigallocatechin-3-gallate. *Archives of Biochemistry and Biophysics* August 15, 2008;**476**(2):152–60.

114. Ishrat, Hoda, Khan, et al. Amelioration of cognitive deficits and neurodegeneration by curcumin in rat model of sporadic dementia of Alzheimer's type (SDAT). *European Neuropsychopharmacology* September 2009;**19**(9):636–47.

115. Yu, Zhang, Luo, et al. Curcumin ameliorates memory deficits via neuronal nitric oxide synthase in aged mice. *Progress in Neuropsychopharmacology and Biological Psychiatry* August 1, 2013;**45**:47–53.

116. Hoppe, Coradini, Frozza, et al. Free and nanoencapsulated curcumin suppress β-amyloid-induced cognitive impairments in rats: involvement of BDNF and Akt/GSK-3β signaling pathway. *Neurobiology of Learning and Memory* November 2013;**106**:134–44.

117. Tiwari, Aggarwal, Seth, et al. Curcumin-loaded nanoparticles potently induce adult neurogenesis and reverse cognitive deficits in Alzheimer's disease model via canonical Wnt/β-catenin pathway. *ACS Nano* January 28, 2014;**8**(1):76–103.

118. Yin, Wang, Li, et al. Effects of curcumin on hippocampal expression of NgR and axonal regeneration in Aβ-induced cognitive disorder rats. *Genetics and Molecular Research* March 24, 2014;**13**(1):2039–47.

119. Zhang, Fang, Xu, et al. Curcumin improves amyloid β-peptide (1-42) induced spatial memory deficits through BDNF-ERK signaling pathway. *Plos One* June 26, 2015;**10**(6):e0131525.

120. Shukitt-Hale, Carey, Simon, et al. Effects of concord grape juice on cognitive and motor deficits in aging. *Nutrition* March 2006;**22**(3):295–302.

121. Rasoolijazi, Joghataie, Roghani, Nobakht. The beneficial effect of (-)-epigallocatechin-3-gallate in an experimental model of Alzheimer's disease in rat: a behavioral analysis. *Iranian Biomedical Journal* October 2007;**11**(4):237–43.

122. Biasibetti, Tramontina, Costa, et al. Green tea (-)epigallocatechin-3-gallate reverses oxidative stress and reduces acetylcholinesterase activity in a streptozotocin-induced model of dementia. *Behavioral Brain Research* January 1, 2013;**236**(1):186–93.

123. Liu, Chen, Sha, et al. (-)-Epigallocatechin-3-gallate ameliorates learning and memory deficits by adjusting the balance of TrkA/p75NTR signaling in APP/PS1 transgenic mice. *Molecular Neurobiology* June 2014;**49**(3):1350–63.

124. Grossi, Rigacci, Ambrosini, et al. The polyphenol oleuropein aglycone protects TgCRND8 mice against Aß plaque pathology. *PLoS One* August 8, 2013;**8**(8):e71702.

125. Sharma, Gupta. Chronic treatment with trans resveratrol prevents intracerebroventricular streptozotocin induced cognitive impairment and oxidative stress in rats. *Life Sciences* October 11, 2002;**71**(21):2489–98.

126. Sharma, Briyal, Gupta. Effect of alpha lipoic acid, melatonin, and trans resveratrol on intracerebroventricular streptozotocin induced spatial memory deficit in rats. *Indian Journal of Physiology and Pharmacology* October–December 2005;**49**(4):395–402.

127. Belkacemi, Doggui, Dao, Ramassamy. Challenges associated with curcumin therapy in Alzheimer disease. *Expert Reviews in Molecular Medicine* November 4, 2011;**13**:e34.

128. Brondino, Re, Boldrini, et al. Curcumin as a therapeutic agent in dementia: a mini systematic review of human studies. *The Scientific World Journal [Electronic Resource]* January 22, 2014;**2014**:174282.

129. Yallapu, Nagesh, Jaggi, Chauhan. Therapeutic applications of curcumin nanoformulations. *The AAPS Journal* November 2015;**17**(6):1341–56.
130. Goozee, Shah, Sohrabi, et al. Examining the potential clinical value of curcumin in the prevention and diagnosis of Alzheimer's disease. *British Journal of Nutrition* February 14, 2016;**115**(3):449–65.
131. Cox, Pipingas, Scholey. Investigation of the effects of solid lipid curcumin on cognition and mood in a healthy older population. *Journal of Psychopharmacology* May 2015;**29**(5):642–51.
132. DiSilvestro, Joseph, Zhao, Bomser. Diverse effects of a low dose supplement of lipidated curcumin in healthy middle aged people. *Nutrition Journal* September 26, 2012;**11**:79.

PART III

Rational Basis Versus Strict Scrutiny

CHAPTER 7

Dose-Adherence and Intent-to-Treat

Contents

A Paradigm Shift to Prevent and Treat Alzheimer's Disease
ISBN 978-0-12-812259-4
http://dx.doi.org/10.1016/B978-0-12-812259-4.00007-2

Ginkgo biloba EXTRACT (EGB 761)

As we reported in the preceding chapters, natural extracts from the leaf of the *Ginkgo biloba* tree, identified in the literature as terpenoids (including bilobalide) and flavonoids (including quercetin), have been shown to enhance the release of the neuroprotective cleavage product sAPPα from amyloid precursor protein (APP)[1] and restore Abeta oligomer-induced synaptic loss and enhance neurogenesis and synaptogenesis.[2] Multiple studies also have shown that a special formula of *Ginkgo biloba* extracts, identified in the scientific literature as *Gingko biloba* extract 761 (EGb 761), protects neurons against oxidative stress.[3] In addition, EGb 761 has been reported to inhibit beta-secretase activity,[4] inhibit production of Abeta peptides,[5] protect hippocampal cells against the toxic effects induced by Abeta peptides,[6] enhance clearance of Abeta from the brain,[7] protect hippocampal neurons against toxicity induced by Abeta fragments "with a maximal and complete protection at the highest concentration tested,"[8] protect neurons against Abeta-induced dysfunction and death "in a concentration-dependent manner,"[9] inhibit the formation of Abeta-derived diffusible neurotoxic ligands (soluble Abeta oligomers) "in a dose-dependent manner,"[10] reduce Abeta oligomers in the hippocampus of mice,[11] increase nearly all aspects of impaired neuroplasticity (long-term potentiation, spine density, neuritogenesis, and neurogenesis),[12] attenuate the loss of synaptic structure proteins, such as postsynaptic density protein 95 (PSD-95), mammalian uncoordinated 18-1 protein (Munc 18-1), and synaptosomal-associated protein 25 (SNAP 25),[4] inhibit the formation of Abeta fibrils,[13] and attenuate zinc-induced tau hyperphosphorylation at Ser262 "in a concentration-dependent manner."[14]

This broad array of therapeutic capabilities, while comparable to the wide-ranging multiple targeting of curcumin and EGCG, is unprecedented among pharmaceutical products for Alzheimer's disease (AD). EGb 761 is also highly bioavailable.[15] Furthermore, not being a single molecule, and standardized to contain 24% flavone glycosides (primarily quercetin, kaempferol, and isorhamnetin) and 6% terpene lactones (2.8%–3.4% ginkgolides A, B, and C, and 2.6%–3.2% bilobalide),[16] EGb 761, is a de facto cocktail of pleiotropic plant compounds that is consistent with the approach presented in this volume toward the prevention and treatment of AD. And while the curcuminoids, epigallocatechin-gallate (EGCG), and resveratrol are probably the most-studied mechanistically with regard to the major pathogenic hallmarks of AD, EGb 761 is the most studied substance of any kind in clinical trials for the prevention and treatment of AD.[17] In this chapter, we review the clinical trials of EGb 761 to prevent and treat AD as a way to evaluate not only the safety and efficacy of EGb 761 itself, but also as a clinical-trials case study for the safety and efficacy of brain-accessible pleiotropic plant compounds in the prevention and treatment of AD.

To our knowledge, the first double-blind, placebo-controlled trial that tested the efficacy of EGb 761 in patients "with presenile and senile primary degenerative dementia of the Alzheimer type" or "multi-infarct dementia" with a dose of 120 mg/day or higher was reported in 1996 by Siegfried Kanowski at the Department of Gerontopsychiatry at the Free University of Berlin in Germany. This study, published in the German medical journal, *Pharmacopsychiatry*, was followed in 1997 by two other double-blind, placebo-controlled studies that were published in the United States in the *Journal of the American Medical Association* and in England in the *Journal of Psychiatric Research*. In 2000, *Dementia and Geriatric Cognitive Disorders*, which is published in Switzerland, was the fourth double-blind, placebo-controlled study using EGb 761 on patients with AD.

All four of these studies led to positive results in AD patients, as indicated below, and initiated two decades of clinical studies involving EGb 761 and AD. Most such studies and several reviews of clinical trials on EGb 761 and AD are presented below. They are tabbed as "Positive" or "Negative" to illustrate the ratio of positive reports to negative ones with respect to EGb 761 and AD and to help identify certain of these studies and reviews for further analysis below.

POSITIVE TRIAL

Ginkgo biloba EXTRACT 761 (↑) MEMORY, COGNITION, BEHAVIOR
Pharmacopsychiatry (Germany) (1996)
Department of Gerontopsychiatry, Free University of Berlin, Germany.
The efficacy of the *Ginkgo biloba* special extract EGb 761 in outpatients with presenile and senile primary degenerative dementia of the Alzheimer type (DAT) and multi-infarct dementia (MID) according to DSM-III-R was investigated in a prospective, randomized, double-blind, placebo-controlled, multi-center study. After a 4-week run-in period, 216 patients were included in the randomized 24-week treatment period. These received either a daily oral dose of **240 mg EGb 761** or placebo. In accordance with the recommended multi-dimensional evaluation approach, three primary variables were chosen: The Clinical Global Impressions (CGI Item 2) for psychopathological assessment, the Syndrom-Kurztest (SKT) for the assessment of the patient's attention and memory, and the Nürnberger Alters-Beobachtungsskala (NAB) for behavioral assessment of activities of daily life. Clinical efficacy was assessed by means of a responder analysis, with therapy response being defined as response in at least two of the three primary variables. The data from the 156 patients who completed the study in accordance with the study protocol were taken into account in the confirmatory analysis of valid cases. **The frequency of therapy responders in the two treatment groups differed significantly in favor of EGb 761, with p < 0.005 in Fisher's Exact Test. The intent-to-treat**

analysis of 205 patients led to similar efficacy results. Thus, the clinical efficacy of the *Ginkgo biloba* special extract EGb 761 in dementia of the Alzheimer type and multi-infarct dementia was confirmed. The investigational drug was found to be well tolerated.[18]

POSITIVE TRIAL

Ginkgo biloba EXTRACT 761 (↑) MEMORY, COGNITION, BEHAVIOR
Journal of the American Medical Association (United States) (1997)
New York Institute for Medical Research, Tarrytown, New York.
EGb 761 is a particular extract of *Ginkgo biloba* used in Europe to alleviate symptoms associated with numerous cognitive disorders. Its use in dementias is based on positive results from only a few controlled clinical trials, most of which did not include standard assessments of cognition and behavior. OBJECTIVE: To assess the efficacy and safety of EGb in Alzheimer disease and multi-infarct dementia. DESIGN: A 52-week, randomized double-blind, placebo-controlled, parallel-group, multicenter study. PATIENTS: Mildly to severely demented outpatients with Alzheimer disease or multi-infarct dementia, without other significant medical conditions. INTERVENTION: Patients assigned randomly to treatment with **EGb (120 mg/d)** or placebo. Safety, compliance, and drug dispensation were monitored every 3 months with complete outcome evaluation at 12, 26, and 52 weeks. PRIMARY OUTCOME MEASURES: Alzheimer's Disease Assessment Scale-Cognitive subscale (ADAS-Cog), Geriatric Evaluation by Relative's Rating Instrument (GERRI), and Clinical Global Impression of Change (CGIC). RESULTS: From 309 patients included in an intent-to-treat analysis, 202 provided evaluable data for the 52-week end point analysis. **In the intent-to-treat analysis, the EGb group had an ADAS-Cog score 1.4 points better than the placebo group (P=.04) and a GERRI score 0.14 points better than the placebo group (P=.004). The same patterns were observed with the evaluable data set in which 27% of patients treated with EGb achieved at least a 4-point improvement on the ADAS-Cog, compared with 14% taking placebo (P=.005); on the GERRI, 37% were considered improved with EGb, compared with 23% taking placebo (P=.003).** No difference was seen in the CGIC. Regarding the safety profile of EGb, no significant differences compared with placebo were observed in the number of patients reporting adverse events or in the incidence and severity of these events. **CONCLUSIONS: EGb was safe and appears capable of stabilizing and, in a substantial number of cases, improving the cognitive performance and the social functioning of demented patients for 6 months to 1 year. Although modest, the changes induced by EGb were objectively measured by the ADAS-Cog and were of sufficient magnitude to be recognized by the caregivers in the GERRI.**[19]

POSITIVE TRIAL

Ginkgo biloba Extract 761 (↑) Memory, Cognition, Behavior
Journal of Psychiatric Research (England) (1997)
Klinik für Psychiatrie I, Johann-Wolfgang-Goethe Universität, Frankfurt, Germany.
In the present study the influence of oral treatment with **240 mg/day** of *Ginkgo biloba* special extract EGb 761 (Tebonin® forte, manufactured by Dr. Willmar Schwabe, Karlsruhe) on the clinical course of DAT [Dementia of the Alzheimer's Type] was investigated in a double-blind, randomized, placebo-controlled parallel-group design in 20 outpatients… **Although the active-treatment group, with a mean sum score of 19.67 points in the S.K.T., had a poorer baseline level than the placebo group (18.11 points), it experienced an improvement to 16.78 points under treatment with EGb 761 whereas the placebo group deteriorated to 18.89 points.** The differences between the baseline and final values formed the basis for a statistical group comparison, which gave a result favourable to EGb 761, at a significance level of p <.013. **In addition to this psychometric confirmation of efficacy, certain descriptive trends were found at the psychopathological (Clinical Global Impression) and dynamic functional (EEG findings) levels, which can be interpreted as evidence of effectiveness of *Ginkgo biloba* special extract EGb 761 in mild to moderate dementia and of local effects in the central nervous system.** Inter-group differences in the ADAS cognitive and non-cognitive subscales did not reach statistical significance, probably because of the small sample size.[20]

POSITIVE TRIAL

Ginkgo biloba Extract 761 (↑) Memory, Cognition, Behavior
Dementia and Geriatric Cognitive Disorders (Switzerland) (2000)
Memory Centers of America Inc., New York.
This intent-to-treat (ITT) analysis was performed to provide a realistic image of the efficacy that could be expected after 26 weeks of treatment with a **120-mg dose** (40 mg, three times a day) of EGb 761 (EGb). The data were collected during a 52-week, double-blind, placebo-controlled, fixed dose, parallel-group, multicenter study. Patients were mildly to severely impaired and diagnosed with uncomplicated Alzheimer's disease or multi-infarct dementia according to ICD-10 and DSM-III-R criteria. The primary outcome measures included the Alzheimer's Disease Assessment Scale-Cognitive Subscale (ADAS-Cog), Geriatric Evaluation by Relative's Rating Instrument (GERRI) and Clinical Global Impression of Change. From 309 patients included in the ITT analysis, 244 patients (76% for placebo and 73% for EGb) actually reached the 26th week visit. **In comparison to the baseline values, the placebo group showed a statistically significant worsening in all domains of assessment,**

while the group receiving EGb was considered slightly improved on the cognitive assessment and the daily living and social behavior... Regarding safety, no differences between EGb and placebo were observed.[21]

These four positive studies, from 1996 to 2000, which involved a total of 854 AD patients, were followed by a see-saw pattern of positive and negative results as follows in a long line of studies on EGb 761 and AD.

NEGATIVE TRIAL

Ginkgo biloba EXTRACT 761 (↓) MEMORY, COGNITION, BEHAVIOR
Journal of the American Geriatrics Society (United States) (2000)

Department of Epidemiology, Maastricht University, The Netherlands.

OBJECTIVES: To evaluate the efficacy, the dose-dependence, and the durability of the effect of the *Ginkgo biloba* special extract EGb 761 (ginkgo) in older people with dementia or age-associated memory impairment. DESIGN: A 24-week, randomized, double-blind, placebo-controlled, parallel-group, multicenter trial. SETTING: Homes for the elderly in the southern part of the Netherlands. PARTICIPANTS: Older persons with dementia (either Alzheimer's dementia or vascular dementia; mild to moderate degree) or age-associated memory impairment (AAMI). 214 Participants were recruited from 39 homes for the elderly. INTERVENTION: The participants were allocated randomly to treatment with EGb 761 (2 tablets per day, total dosage either 240 (high dose) or 160 (usual dose) mg/day) or placebo (0 mg/d). The total intervention period was 24 weeks. After 12 weeks of treatment, the initial ginkgo users were randomized once again to either continued ginkgo treatment or placebo treatment. Initial placebo use was prolonged after 12 weeks. MEASUREMENTS: Outcomes were assessed after 12 and 24 weeks of intervention. Outcome measures included neuropsychological testing (trail-making speed (NAI-ZVT-G), digit memory span (NAI-ZN-G), and verbal learning (NAI-WL)), clinical assessment (presence and severity of geriatric symptoms (SCAG), depressive mood (GDS), self-perceived health and memory status (report marks)), and behavioral assessment (self-reported level of instrumental daily life activities). **RESULTS: An intention-to-treat analysis showed no effect on each of the outcome measures for participants who were assigned to ginkgo (n = 79) compared with placebo (n = 44) for the entire 24-week period.** After 12 weeks of treatment, the combined high dose and usual dose ginkgo groups (n = 166) performed slightly better with regard to self-reported activities of daily life but slightly worse with regard to self-perceived health status compared with the placebo group (n = 48). **No beneficial effects of a higher dose or a prolonged duration of ginkgo treatment were found. We could not detect any subgroup that benefited from ginkgo.** Ginkgo use was also not associated with the occurrence of (serious) adverse events. **CONCLUSIONS: The results of our trial** suggest that ginkgo is not effective as a treatment for older people with mild to moderate dementia or age-associated memory impairment. Our results contrast sharply with those of previous ginkgo trials.[22]

POSITIVE REVIEW

Ginkgo biloba EXTRACT 761 (↑) MEMORY, COGNITION, BEHAVIOR
Public Health Nutrition (England) (2000)
New York University Medical Center, New York.
This review of the literature documents the efficacy of a standard extract of *Ginkgo biloba* (EGb) in managing signs and symptoms associated with memory disorders and dementia. Analysis of the discrepant findings reveals that study outcomes may vary with the type of population studied, the outcome measurements selected, and the dosing tested. Overall, the efficacy of EGb was more frequently reported in trials enrolling dementia patients than healthy volunteers. In contrast to narrow memory tests, broad cognitive assessments were more likely to detect the treatment effect. **Although a dose-response relationship is not yet established, 240 mg/day EGb seems to show a higher rate of treatment response than does 120 mg/day.** Regarding safety, in all trials reviewed the adverse event profile of EGb was not different from that of the placebo.[23]

POSITIVE TRIAL

Ginkgo biloba EXTRACT 761 (↑) MEMORY, COGNITION, BEHAVIOR
Neuropsychobiology (Switzerland) (2002)
New York University Medical Center and Memory Centers of America, New York.
OBJECTIVE: To explore the treatment effect of EGb 761((R)) (EGb) in Alzheimer's disease depending on baseline severity. METHODS: We applied stratification to the intent-to-treat data set collected during a 52-week, randomized, double-blind, placebo-controlled, parallel-group, multicenter study with **120 mg of EGb**, using cutoff points of 23 and 14 for the Mini-Mental State Examination (MMSE) score. Outcome measures used were the cognitive subscale of the Alzheimer's Disease Assessment Scale (ADAS-Cog) and the Geriatric Evaluation by Relative's Rating Instrument (GERRI). **RESULTS: In the severity stratum 1 (MMSE > 23), the placebo group did not show significant changes, while the EGb group improved significantly by 1.7 points on the ADAS-Cog and by 0.09 points on the GERRI. In the severity stratum 2 (MMSE < 24), the placebo group worsened by 4.1 points on the ADAS-Cog and 0.18 points on the GERRI, whereas the EGb group showed 60% less decline on the ADAS-Cog (treatment difference of 2.5 points) and no change on the GERRI (treatment difference of 0.25 points).** The most

severely impaired subgroup (MMSE < 15) showed slightly more pronounced worsening for both treatment groups. However, in comparison to placebo, EGb induced virtually the same magnitude of effect as was observed in the entire stratum 2. **CONCLUSIONS: The results of this retrospective analysis indicated that a treatment effect favorable to EGb could be observed with respect to cognitive performance (p = 0.02) and social functioning (p = 0.02) regardless of the stage of dementia, whether mild or moderately severe.** However, the relative changes from baseline measured at endpoint depended heavily on the severity at baseline. **Improvement was observed in the group of patients with very mild to mild cognitive impairment, while in more severe dementia, the mean EGb effect should be considered more in terms of stabilization or slowing down of worsening, as compared to the greater deterioration observed with placebo.**[24]

POSITIVE TRIAL

Ginkgo biloba EXTRACT 761 (↑) MEMORY, COGNITION, BEHAVIOR
Human Psychopharmacology (England) (2002)
Department of Health Sciences, Liberty University, Lynchburg, Virginia.
The purpose of this research was to conduct the first known, large-scaled clinical trial of the efficacy of *Ginkgo biloba* extract (EGb 761) on the neuropsychological functioning of cognitively intact older adults. 262 community-dwelling volunteers (both male and female) 60 years of age and older, who reported no history of dementia or significant neurocognitive impairments and obtained Mini-Mental State Examination total scores of at least 26, were examined via a 6-week, randomized, double-blind, fixed-dose, placebo-controlled, parallel-group, clinical trial. Participants were randomly assigned to receive either *Ginkgo biloba* extract **EGb 761(n = 131; 180 mg/day)** or placebo (n = 131) for 6 weeks… **Overall, the results from both objective, standardized, neuropsychological tests, and a subjective, follow-up self-report questionnaire provided complementary evidence of the potential efficacy of *Ginkgo biloba* EGb 761 in enhancing certain neuropsychological/memory processes of cognitively intact older adults, 60 years of age and over.**[25]

POSITIVE REVIEW

Ginkgo biloba EXTRACT 761 (↑) MEMORY, COGNITION, BEHAVIOR
The Cochrane Database of Systematic Reviews (England) (2002)
Department of Clinical Gerontology, University of Oxford, United Kingdom.
The aim of the review is to assess the efficacy and safety of *Ginkgo biloba* for the treatment of patients with dementia or cognitive decline. SEARCH STRATEGY: Trials were identified on 26 June 2002 through a search of the CDCIG Specialized Register which contains records from all main medical databases (MEDLINE, EMBASE,

CINAHL, PsycINFO, SIGLE, LILACS), from ongoing trials databases such as Clinicaltrials.gov and Current Controlled Trials and many other sources. The search terms used were ginkgo*, tanakan, EGB-761, EGB761 and "EGB 761". SELECTION CRITERIA: All relevant, unconfounded, randomized, double-blind controlled studies, in which extracts of *Ginkgo biloba* at any strength and over any period were compared with placebo for their effects on people with acquired cognitive impairment, including dementia, of any degree of severity. DATA COLLECTION AND ANALYSIS: Data for the meta-analyses are based on reported summary statistics for each study. **For the intention-to-treat analyses we sought data for each outcome measure on every patient randomized, irrespective of compliance. For the analyses of completers, we sought data on every patient who completed the study on treatment.** For continuous or ordinal variables, such as psychometric test scores, clinical global impression scales, and quality of life scales, there are two possible approaches. If ordinal scale data appear to be approximately normally distributed, or if the analyses reported by the investigators suggest that parametric methods and a normal approximation are appropriate, then the outcome measures will be treated as continuous variables. The second approach, which may not exclude the first, is to concatenate the data into two categories which best represent the contrasting states of interest, and to treat the outcome measure as binary. For binary outcomes, the endpoint itself is of interest and the Peto method of the typical odds ratio is used. MAIN RESULTS: Overall, there are no significant differences between Ginkgo and placebo in the proportion of participants experiencing adverse events. **Most studies report the analyses of data from participants who completed the treatment, there are few attempts at ITT analyses. Therefore, we report completers analyses only.** The CGI scale, measuring clinical global improvement as assessed by the physician, was dichotomized between participants who showed improvement and those who were unchanged or worse. There are benefits associated with Ginkgo (dose less than 200mg/day) compared with placebo at less than 12 weeks (54/63 showed improvement compared with 20/63, OR 15.32, 95% CI 5.90 to 39.80, P=<.0001), and Ginkgo (dose greater than 200mg/day) at 24 weeks (57/79 compared with 42/77, OR 2.16, 95% CI 1.11 to 4.20, P=.02). Cognition shows benefit for Ginkgo (dose less than 200mg/day) compared with placebo at 12 weeks (SMD -0.57, 95% CI -1.09, -0.05, P=0.03, random effects model), Ginkgo (greater than 200mg/day) at 12 weeks (SMD -0.56, 95% CI -1.12 to -0.0, P=0.05), at 12 weeks (Ginkgo any dose) (SMD -0.71, 95% CI -1.23 to -0.19 P=0.008, random effects model) at 24 weeks (Ginkgo any dose) (SMD -0.17, 95% CI -0.32 to -0.02 P=0.03) and at 52 weeks (Ginkgo less than 200 mg/day) (SMD -0.41, 95% CI -0.71 to -0.11, P=<.01). Activities of Daily Living (ADL) shows benefit for Ginkgo (dose less than 200mg/day) compared with placebo at 12 weeks (SMD -1.10, 95% CI -1.79, -0.41, P=0mg/day) compared with placebo at 12 weeks (SMD -1.10, 95% CI -1.79, -0.41, P=<.01), Ginkgo (dose less than 200mg/day) at 24 weeks (SMD -0.25, 95% CI -0.49 to -0.00, P=.05), and at 52 weeks (Ginkgo less than 200mg/day) (SMD

-0.41, 95% CI -0.71 to -0.11, P=<.01). Measures of mood and emotional function show benefit for Ginkgo (dose less than 200 mg/day) compared with placebo at less than 12 weeks (SMD -0.51, 95% CI -0.99 to -0.03, P=.04) and Ginkgo (dose less than 200 mg/day) at 12 weeks (SMD -1.94, 95% CIs -2.73, -1.15 P=<.0001). There are no significant differences between Ginkgo and placebo in the proportion of participants experiencing adverse events. There are no data available on Quality of Life, measures of depression or dependency. REVIEWER'S CONCLUSIONS: *Ginkgo biloba* appears to be safe in use with no excess side effects compared with placebo. Many of the early trials used unsatisfactory methods, were small, and we cannot exclude publication bias. **Overall there is promising evidence of improvement in cognition and function associated with Ginkgo.** However, the three more modern trials show inconsistent results. Our view is that there is need for a large trial using modern methodology and permitting an intention-to-treat analysis to provide robust estimates of the size and mechanism of any treatment effects.[26]

POSITIVE TRIAL

Ginkgo biloba EXTRACT 761 (↑) MEMORY, COGNITION, BEHAVIOR
Pharmacopsychiatry (Germany) (2003)

Abteilung Gerontopsychiatrie, Psychiatrische Klinik und Poliklinik der Freien Universität Berlin.

In 1996, Kanowski et al. reported about the beneficial effects of *Ginkgo biloba* special extract EGb 761 (240 mg/day) in outpatients with pre-senile and senile primary degenerative dementia of the Alzheimer type (DAT) and multi-infarct dementia (MID) of mild to moderate severity. The comparison of the results of this double-blind, placebo-controlled, randomized, multi-center study with other dementia studies is hampered by the fact that only the responder analysis of the per-protocol (PP) population, which was pre-specified in the protocol as confirmatory analysis, has been published in detail so far. Moreover, cognitive functioning was measured using the Syndrom-Kurztest (SKT), whereas results of other studies are based on the Alzheimer's Disease Assessment Scale-Cognitive Subscale (ADAS-cog). Therefore, the conventional intention-to-treat (ITT) analysis of this study is provided with an estimation of ADAS-cog scores based on measured SKT scores. **After 24 weeks of treatment, the ITT analysis of the SKT and estimated ADAS-cog scores revealed a mean decrease in the total score by -2.1 (95 % CI: -2.7; -1.5) points and -2.7 (95 % CI: -3.5; -1.9) points, respectively, for the EGb 761 group, which indicates an improvement in cognitive function.** On the contrary, the placebo group exhibited only a minimal change of -1.0 (95 % CI: -1.6; -0.3) and -1.3 (95 % CI: -2.0; -0.4) points, respectively. The changes from baseline differed significantly between treatment groups by 1.1 (SKT) and 1.4 (estimated ADAS-cog) points, respectively (P = 0.01). The Clinical Global Impression of Change (CGI, Item 2) favored the EGb

761 group with a mean difference of 0.4 points (P = 0.007). Changes in the rating related to activities of daily living (Nürnberger-Alters-Beobachtungs-Skala, NAB) showed a favorable trend for EGb 761R. A subgroup analysis regarding patients with DAT yielded comparable results. Using a decrease of at least 4 points on the estimated ADAS-cog scores as cutoff criterion for treatment response, 35% of EGb 761-treated patients were considered responders versus only 19% for the placebo group (P = 0.01).. **The results of this ITT analysis substantiate the outcomes previously obtained with a responder analysis of the per-protocol population and confirm that EGb 761 improves cognitive function in a clinically relevant manner in patients suffering from dementia.**[27]

NEGATIVE TRIAL

Ginkgo biloba EXTRACT 761 (↓) MEMORY, COGNITION, BEHAVIOR
Journal of Clinical Epidemiology (United States) (2003)
Department of Epidemiology, Maastricht University, The Netherlands.
Preparations based on special extracts of the *Ginkgo biloba* tree are popular in various European countries. **Previous studies have suggested the clinical efficacy of Ginkgo in patients with dementia, cerebral insufficiency, or related cognitive decline. However, most of these studies did not fulfill the current methodologic requirements.** We assessed the efficacy of the *G. biloba* special extract EGb 761 in patients with dementia and age-associated memory impairment in relation to dose and duration of treatment. Our study was a 24-week, randomized, double-blind, placebo-controlled, parallel-group, multicenter trial. Study participants were elderly patients with dementia (Alzheimer's disease or vascular dementia) or age-associated memory impairment (AAMI). A total of 214 participants, recruited from 39 homes for the elderly in the Netherlands, were randomly allocated to Ginkgo (either 240 mg/d or 160 mg/d) or placebo (0 mg/d). After 12 weeks, the subjects in the two Ginkgo groups were randomized to continued Ginkgo treatment or placebo treatment. Primary outcome measures in this study were the Syndrome Kurz Test (SKT; psychometric functioning), the Clinical Global Impression of change (CGI-2; psychopathology, assessed by nursing staff), and the Nuremberg Gerontopsychological Rating Scale for Activities of Daily Living (NAI-NAA; behavioral functioning). One hundred twenty-three patients received Ginkgo (n=79, 240 and 160 mg/d combined) or placebo (n=44) during the 24-week intervention period. We found no statistically significant differences in mean change of scores between Ginkgo and placebo. The differences were SKT: +0.4 (90% confidence interval [CI] -0.9-1.7); CGI-2: +0.1 (90% CI -0.3-0.4), and NAI-NAA: -0.4 (90% CI -1.9-1.2). A positive difference is in favor of Ginkgo. Neither the dementia subgroup (n=36) nor the AAMI subgroup (n=87) experienced a significant effect of Ginkgo treatment. There was no dose-effect relationship and no effect of prolonged Ginkgo treatment. **The**

trial results do not support the view that Ginkgo is beneficial for patients with dementia or age-associated memory impairment.[28]

INCONCLUSIVE TRIAL

Ginkgo biloba EXTRACT 761 (Inconclusive) MEMORY, COGNITION, BEHAVIOR
Current Alzheimer Research (United Arab Emirates) (2005)
Department of Psychiatry and the Behavioral Sciences, Keck School of Medicine, University of Southern California, Los Angeles.
CONTEXT: Previous studies of *Ginkgo biloba* extract (EGb) in patients with various forms of cognitive impairment or dementia have shown promising results. OBJECTIVE: To determine the clinical efficacy of EGb in mild to moderate dementia of the Alzheimer type. DESIGN: Randomized, placebo-controlled, double-blind, parallel-group, multi-center trial. SETTING: Outpatient clinics of universities and private research centers specialized in dementia. PATIENTS: 513 outpatients with uncomplicated dementia of the Alzheimer's type scoring 10 to 24 on the Mini-Mental State Examination and less than 4 on the modified Hachinski Ischemic Score, free of other serious illnesses and not requiring continuous treatment with any psychoactive drug. INTERVENTION: 26-week treatment with EGb at daily doses of 120 mg or 240 mg or placebo. MAIN OUTCOMES: Cognitive subscale of the Alzheimer's Disease Assessment Scale (ADAS-cog), Alzheimer's Disease Cooperative Study Clinical Global Impression of Change (ADCS-CGIC). **RESULTS: There were no significant between-group differences for the whole sample. There was little cognitive and functional decline of the placebo-treated patients, however. For a subgroup of patients with neuropsychiatric symptoms there was a greater decline of placebo-treated patients and significantly better cognitive performance and global assessment scores for the patients on EGb. CONCLUSION: The trial did not show efficacy of EGb, however, the lack of decline of the placebo patients may have compromised the sensitivity of the trial to detect a treatment effect. Thus, the study remains inconclusive with respect to the efficacy of EGb.**[29]

POSITIVE TRIAL

Ginkgo biloba EXTRACT 761 (↑) MEMORY, COGNITION, BEHAVIOR
Arzneimittel-Forschung (Germany) (2007)
Psychiatry Department, National Medical University, Kiev, Ukraine.
In previous trials of the *Ginkgo biloba* special extract EGb 761 improvements in cognitive functioning and behavioural symptoms were found in patients with aging-associated cognitive impairment or dementia. This trial was undertaken to assess the efficacy of EGb 761 in mild to moderate dementia with neuropsychiatric features.

The study was a double-blind trial including 400 patients aged 50 years or above with Alzheimer's disease or vascular dementia, randomized to receive EGb 761 or placebo for 22 weeks. **There was a mean -3.2-point improvement in the SKT upon EGb 761 treatment and an average deterioration by +1.3 points on placebo. EGb 761 was significantly superior to placebo on all secondary outcome measures, including the NPI and an activities-of-daily-living scale. Treatment results were essentially similar for Alzheimer's disease and vascular dementia subgroups. EGb 761 was well tolerated. Adverse events were no more frequent under drug than under placebo treatment. The data add further evidence on the safety and efficacy of EGb 761 in the treatment of cognitive and non-cognitive symptoms of dementia.**[30]

NEGATIVE REVIEW

Ginkgo biloba EXTRACT 761 (↓) MEMORY, COGNITION, BEHAVIOR
The Cochrane Database of Systematic Reviews (England) (2007)
University of Oxford, Nuffield Department of Clinical Medicine, John Radcliffe Hospital, Oxford, United Kingdom.
BACKGROUND: Extracts of the leaves of the maidenhair tree, *Ginkgo biloba*, have long been used in China as a traditional medicine for various disorders of health. A standardized extract is widely prescribed for the treatment of a range of conditions including memory and concentration problems, confusion, depression, anxiety, dizziness, tinnitus and headache. The mechanisms of action are thought to reflect the action of several components of the extract and include increasing blood supply by dilating blood vessels, reducing blood viscosity, modification of neurotransmitter systems, and reducing the density of oxygen free radicals. OBJECTIVES: To assess the efficacy and safety of *Ginkgo biloba* for dementia or cognitive decline. SEARCH STRATEGY: Trials were identified on 10 October 2006 through a search of the Cochrane Dementia and Cognitive Improvement Group's Specialized Register which contains records from all main medical databases (MEDLINE, EMBASE, CINAHL, PsycINFO, SIGLE, LILACS), from ongoing trials databases such as Clinicaltrials.gov and Current Controlled Trials and many other sources. The search terms used were ginkgo★, tanakan, EGB-761, EGB761, "EGB 761" and gingko★.
SELECTION CRITERIA: Randomized, double-blind studies, in which extracts of *Ginkgo biloba* at any strength and over any period were compared with placebo for their effects on people with acquired cognitive impairment, including dementia, of any degree of severity. DATA COLLECTION AND ANALYSIS: Data were extracted from the published reports of the included studies, pooled where appropriate and the treatment effects or the risks and benefits estimated. MAIN RESULTS: Clinical global improvement as assessed by the physician, was dichotomized

between participants who showed improvement or were unchanged and those who were worse. There are benefits associated with Ginkgo (dose greater than 200 mg/day) at 24 weeks (207/276 compared with 178/273, OR 1.66, 95% CI 1.12 to 2.46, P=.001) (2 studies), but not for the lower dose. Cognition shows benefit for Ginkgo (any dose) at 12 weeks (SMD -0.65, 95% CI -1.22 to -0.09 P=0.02, 5 studies) but not at 24 weeks. Five studies assessed activities of daily living (ADLs), using different scales. Some scales are more comprehensive than just ADLs. The results show benefit for Ginkgo (dose less than 200 mg/day) compared with placebo at 12 weeks (MD -5.0, 95% CI -7.88, -2.12, p=0.0007, one study), and at 24 weeks (SMD -0.16, 95% CI -0.31 to -0.01, p=0.03, 3 studies) but there are no differences at the higher dose. No study assessed mood and function separately, but one study used the ADAS-Noncog, which assesses function over several domains, but not cognitive function. There was no difference between Ginkgo and placebo. There are no significant differences between Ginkgo and placebo in the proportion of participants experiencing adverse events. There are no data available on Quality of Life, measures of depression or dependency. **AUTHORS' CONCLUSIONS: *Ginkgo biloba* appears to be safe in use with no excess side effects compared with placebo. Many of the early trials used unsatisfactory methods, were small, and we cannot exclude publication bias. The evidence that Ginkgo has predictable and clinically significant benefit for people with dementia or cognitive impairment is inconsistent and unconvincing.**[31]

POSITIVE TRIAL

Ginkgo biloba EXTRACT 761 (↑) MEMORY, COGNITION, BEHAVIOR
Wiener Medizinische Wochenschrift (Austria) (2007)

Poltava Regional Psychiatry Hospital, Poltava, Ukraine.

In a randomised, double-blind, 22-week trial 400 patients with dementia associated with neuropsychiatric features were treated with *Ginkgo biloba* extract **EGb 761 (240 mg/ day)** or placebo. Patients with probable Alzheimer's disease, possible Alzheimer's disease with cerebrovascular disease, or vascular dementia were eligible if they scored 9 to 23 on the SKT cognitive test battery and at least 5 on the Neuropsychiatric Inventory (NPI). **EGb 761 was significantly superior to placebo with respect to the primary (SKT test battery) and all secondary outcome variables...** The largest drug-placebo differences in favour of EGb 761 were found for apathy/indifference, anxiety, irritability/lability, depression/dysphoria, and sleep/nighttime behaviour.[32]

POSITIVE TRIAL

Ginkgo biloba EXTRACT 761 (↑) MEMORY, COGNITION, BEHAVIOR
Arzneimittel-Forschung (Germany) (2007)

Psychiatry Department, National Medical University, Kiev, Ukraine.

BACKGROUND: In previous trials of the *Ginkgo biloba* special extract EGb 761 improvements in cognitive functioning and behavioural symptoms were found in patients with aging-associated cognitive impairment or dementia. This trial was undertaken to assess the efficacy of EGb 761 in mild to moderate dementia with neuropsychiatric features. METHODS: Double-blind trial including 400 patients aged 50 years or above with Alzheimer's disease (AD) or vascular dementia (VaD), randomized to receive EGb 761 or placebo for 22 weeks. Patients scored below 36 on the Test for the Early Detection of Dementia with Discrimination from Depression (TE4D), between 9 and 23 on the SKT test battery and at least 5 on the Neuropsychiatric Inventory (NPI). **RESULTS: There was a mean -3.2-point improvement in the SKT upon EGb 761 treatment and an average deterioration by +1.3 points on placebo (p < 0.001, two-sided, ANOVA). EGb 761 was significantly superior to placebo on all secondary outcome measures, including the NPI and an activities-of-daily-living scale. Treatment results were essentially similar for AD and VaD subgroups.** The drug was well tolerated; adverse events were no more frequent under drug than under placebo treatment. **CONCLUSION: The data add further evidence on the safety and efficacy of EGb 761 in the treatment of cognitive and non-cognitive symptoms of dementia.**[30]

NEGATIVE TRIAL

Ginkgo biloba EXTRACT 761 (↓) MEMORY, COGNITION, BEHAVIOR
Neurology (United States) (2008)

Department of Public Health, Oregon State University, Corvallis.

OBJECTIVE: To assess the feasibility, safety, and efficacy of *Ginkgo biloba* extract (GBE) on delaying the progression to cognitive impairment in normal elderly aged 85 and older.

METHODS: Randomized, placebo-controlled, double-blind, 42-month pilot study with 118 cognitively intact subjects randomized to standardized GBE or placebo. Kaplan–Meier estimation, Cox proportional hazard, and random-effects models were used to compare the risk of progression from Clinical Dementia Rating (CDR) = 0 to CDR = 0.5 and decline in episodic memory function between GBE and placebo groups. RESULTS: In the intention-to-treat analysis, there was no reduced risk of progression to CDR = 0.5 (log-rank test, p = 0.06) among the GBE group. There was no less of a decline in memory function among the GBE group (p = 0.05). **In the secondary analysis, where we controlled the medication adherence level, the GBE group had a lower risk of progression from CDR = 0 to CDR = 0.5 (HR = 0.33, p = 0.02), and a smaller decline in memory scores (p = 0.04).** There were more ischemic strokes and TIAs in the GBE group (p = 0.01).

CONCLUSIONS: In unadjusted analyses, *Ginkgo biloba* **extract (GBE) neither altered the risk of progression from normal to Clinical Dementia Rating (CDR)=0.5, nor protected against a decline in memory function. Secondary analysis taking into account medication adherence showed a protective effect of GBE on the progression to CDR=0.5 and memory decline. Results of larger prevention trials taking into account medication adherence may clarify the effectiveness of GBE.** More stroke and TIA cases observed among the GBE group requires further study to confirm.[33]

NEGATIVE TRIAL

Ginkgo biloba EXTRACT 761 (↓) MEMORY, COGNITION, BEHAVIOR
Journal of the American Medical Association (United States) (2008)
University of Pittsburgh, Pennsylvania.
CONTEXT: *Ginkgo biloba* is widely used for its potential effects on memory and cognition. To date, adequately powered clinical trials testing the effect of *G. biloba* on dementia incidence are lacking. OBJECTIVE: To determine effectiveness of *G. biloba* vs placebo in reducing the incidence of all-cause dementia and Alzheimer disease (AD) in elderly individuals with normal cognition and those with mild cognitive impairment (MCI). DESIGN, SETTING, AND PARTICIPANTS: Randomized, double-blind, placebo-controlled clinical trial conducted in 5 academic medical centers in the United States between 2000 and 2008 with a median follow-up of 6.1 years. Three thousand sixty-nine community volunteers aged 75 years or older with normal cognition (n=2587) or MCI (n=482) at study entry were assessed every 6 months for incident dementia. INTERVENTION: Twice-daily dose of 120-mg extract of *G. biloba* (n=1545) or placebo (n=1524). MAIN OUTCOME MEASURES: Incident dementia and AD determined by expert panel consensus. RESULTS: Five hundred twenty-three individuals developed dementia (246 receiving placebo and 277 receiving *G. biloba*) with 92% of the dementia cases classified as possible or probable AD, or AD with evidence of vascular disease of the brain. Rates of dropout and loss to follow-up were low (6.3%), and the adverse effect profiles were similar for both groups. The overall dementia rate was 3.3 per 100 person-years in participants assigned to *G. biloba* and 2.9 per 100 person-years in the placebo group. The hazard ratio (HR) for *G. biloba* compared with placebo for all-cause dementia was 1.12 (95% confidence interval [CI], 0.94-1.33; $P = .21$) and for AD, 1.16 (95% CI, 0.97-1.39; $P = .11$). *G. biloba* also had no effect on the rate of progression to dementia in participants with MCI (HR, 1.13; 95% CI, 0.85-1.50; $P = .39$). **CONCLUSIONS: In this study,** *G. biloba* **at 120mg twice a day was not effective in reducing either the overall incidence rate of dementia or AD incidence in elderly individuals with normal cognition or those with MCI.**[34]

NEGATIVE TRIAL

Ginkgo biloba EXTRACT 761 (↓) MEMORY, COGNITION, BEHAVIOR
International Journal of Geriatric Psychiatry (England) (2008)
Department of Psychological Medicine, Imperial College London, England.
OBJECTIVES: Doubt over the cost-effectiveness of the cholinesterase inhibitors in dementia has renewed interest in alternative treatments such as *Ginkgo biloba*. We aimed to determine the effectiveness and the safety profile of *Ginkgo biloba* for treating early stage dementia in a community setting. METHODS: We conducted a community-based, pragmatic, randomised, double-blind, parallel-group trial where participants were given a standardised extract of *Ginkgo biloba* **(120 mg daily)** or a placebo control for 6 months. Our primary outcomes were cognitive functioning (ADAS-Cog) and participant and carer-rated quality of life (QOL-AD). RESULTS: We recruited 176 participants, mainly through general practices. In the ANCOVA model with baseline score as a co-variate (n = 176), Ginkgo did not have a significant effect on outcome at six months on either the ADAS-Cog score (p = 0.392), the participant-rated QOL-AD score (p = 0.787) nor the carer-rated QOL-AD score (p = 0.222). **CONCLUSION: We found no evidence that a standard dose of high purity *Ginkgo biloba* confers benefit in mild-moderate dementia over 6 months.**[35]

NEGATIVE REVIEW

Ginkgo biloba EXTRACT 761 (↓) MEMORY, COGNITION, BEHAVIOR
The Cochrane Database of Systematic Reviews (England) (2009)
Centre for Statistics in Medicine, University of Oxford, Wolfson College, Oxford, United Kingdom.
OBJECTIVES: To assess the efficacy and safety of *Ginkgo biloba* for dementia or cognitive decline. SEARCH STRATEGY: The Specialized Register of the Cochrane Dementia and Cognitive Improvement Group (CDCIG), The Cochrane Library, MEDLINE, EMBASE, PsycINFO, CINAHL and LILACS were searched on 20 September 2007 using the terms: ginkgo*, tanakan, EGB-761, EGB761, "EGB 761" and gingko. The CDCIG Specialized Register contains records from all major health care databases (The Cochrane Library, MEDLINE, EMBASE, PsycINFO, CINAHL, LILACS) as well as from many trials databases and grey literature sources. **SELECTION CRITERIA: Randomized, double-blind studies, in which extracts of *Ginkgo biloba* at any strength and over any period were compared with placebo** for their effects on people with acquired cognitive impairment, including dementia, of any degree of severity. DATA COLLECTION AND ANALYSIS: Data were extracted from the published reports of the included studies, pooled where appropriate and the treatment effects or the risks and benefits estimated. MAIN RESULTS: 36 trials were included but most were small and of duration less than three months. Nine trials were

of six months duration (2016 patients). These longer trials were the more recent trials and generally were of adequate size, and conducted to a reasonable standard. Most trials tested the same standardised preparation of *Ginkgo biloba*, EGb 761, at different doses, which are classified as high or low. The results from the more recent trials showed inconsistent results for cognition, activities of daily living, mood, depression and carer burden. Of the four most recent trials to report results three found no difference between *Ginkgo biloba* and placebo, and one found very large treatment effects in favour of *Ginkgo biloba*. There are no significant differences between *Ginkgo biloba* and placebo in the proportion of participants experiencing adverse events. A subgroup analysis including only patients diagnosed with Alzheimer's disease (925 patients from nine trials) also showed no consistent pattern of any benefit associated with *Ginkgo biloba*. **AUTHORS' CONCLUSIONS: *Ginkgo biloba* appears to be safe in use with no excess side effects compared with placebo. Many of the early trials used unsatisfactory methods, were small, and publication bias cannot be excluded. The evidence that *Ginkgo biloba* has predictable and clinically significant benefit for people with dementia or cognitive impairment is inconsistent and unreliable.[36]**

NEGATIVE TRIAL

Ginkgo biloba EXTRACT 761 (↓) MEMORY, COGNITION, BEHAVIOR
Journal of the American Medical Association (United States) (2009)
Department of Neurology, University of Pittsburgh, Pennsylvania.
CONTEXT: The herbal product *Ginkgo biloba* is taken frequently with the intention of improving cognitive health in aging. However, evidence from adequately powered clinical trials is lacking regarding its effect on long-term cognitive functioning. OBJECTIVE: To determine whether *G. biloba* slows the rates of global or domain-specific cognitive decline in older adults. DESIGN, SETTING, AND PARTICIPANTS: The Ginkgo Evaluation of Memory (GEM) study, a randomized, double-blind, placebo-controlled clinical trial of 3069 community-dwelling participants aged 72 to 96 years, conducted in 6 academic medical centers in the United States between 2000 and 2008, with a median follow-up of 6.1 years. INTERVENTION: Twice-daily dose of 120-mg extract of *G. biloba* (n = 1545) or identical-appearing placebo (n = 1524). MAIN OUTCOME MEASURES: Rates of change over time in the Modified Mini-Mental State Examination (3MSE), in the cognitive subscale of the Alzheimer Disease Assessment Scale (ADAS-Cog), and in neuropsychological domains of memory, attention, visual-spatial construction, language, and executive functions, based on sums of z scores of individual tests. RESULTS: Annual rates of decline in z scores did not differ between *G. biloba* and placebo groups in any domains, including memory (0.043; 95% confidence interval [CI], 0.034-0.051 vs 0.041; 95% CI, 0.032-0.050), attention (0.043; 95% CI, 0.037-0.050 vs 0.048; 95% CI, 0.041-0.054), visuospatial abilities (0.107; 95%

CI, 0.097-0.117 vs 0.118; 95% CI, 0.108-0.128), language (0.045; 95% CI, 0.037-0.054 vs 0.041; 95% CI, 0.033-0.048), and executive functions (0.092; 95% CI, 0.086-0.099 vs 0.089; 95% CI, 0.082-0.096). For the 3MSE and ADAS-Cog, rates of change varied by baseline cognitive status (mild cognitive impairment), but there were no differences in rates of change between treatment groups (for 3MSE, P = .71; for ADAS-Cog, P = .97). There was no significant effect modification of treatment on rate of decline by age, sex, race, education, APOE★E4 allele, or baseline mild cognitive impairment (P > .05). **CONCLUSION: Compared with placebo, the use of G. *biloba*, 120 mg twice daily, did not result in less cognitive decline in older adults with normal cognition or with mild cognitive impairment.**[37]

POSITIVE REVIEW

Ginkgo biloba EXTRACT 761 (↑) MEMORY, COGNITION, BEHAVIOR
Wiener Medizinische Wochenschrift (Austria) (2010)
Institute for Quality and Efficiency in Health Care (IQWiG), Cologne, Germany.
This systematic review determines the benefit of treatment with *Ginkgo biloba* (Ginkgo) in Alzheimer's disease concerning patient-relevant outcomes. Bibliographic databases, clinical trial, and study result registries were searched for randomized controlled trials (RCTs) in patients with AD (follow-up ≥16 weeks) comparing Ginkgo to placebo or a different treatment option. Manufacturers were asked to provide unpublished data. If feasible, data were pooled by meta-analysis. Six studies were eligible; overall, high heterogeneity was shown for most outcomes, except safety aspects. **Among studies administering high-dose Ginkgo (240 mg), all studies favour treatment though effects remain heterogeneous. In this subgroup, a benefit of Ginkgo exists for activities of daily living. Cognition and accompanying psychopathological symptoms show an indication of a benefit. A harm of Ginkgo is not evident.** An estimation of the effect size was not possible for any outcome. Further evidence is needed which focuses especially on subgroups of AD patients.[38]

POSITIVE REVIEW

Ginkgo biloba EXTRACT 761 (↑) MEMORY, COGNITION, BEHAVIOR
BioMed Central Geriatrics (England) (2010)
Institute for Social Medicine, Epidemiology and Health Economics, Charité University Medicine, Berlin, Germany.
BACKGROUND: The benefit of *Ginkgo biloba* has been discussed controversially. The aim of this review was to assess the effects of *Ginkgo biloba* in Alzheimer's disease as well as vascular and mixed dementia covering a variety of outcome domains. METHODS: We searched MEDLINE, EMBASE, the Cochrane databases, CINAHL and PsycINFO

for controlled trials of ginkgo for Alzheimer's, vascular or mixed dementia. Studies had to be of a minimum of 12 weeks duration with at least ten participants per group. Clinical characteristics and outcomes were extracted. Meta-analysis results were expressed as risk ratios or standardized mean differences (SMD) in scores. RESULTS: Nine trials using the standardized extract EGb761(R) met our inclusion criteria. Trials were of 12 to 52 weeks duration and included 2372 patients in total. In the meta-analysis, the SMDs in change scores for cognition were in favor of ginkgo compared to placebo (-0.58, 95% confidence interval [CI] -1.14; -0.01, p = 0.04), but did not show a statistically significant difference from placebo for activities in daily living (ADLs) (SMD = -0.32, 95% CI -0.66; 0.03, p = 0.08). Heterogeneity among studies was high. For the Alzheimer subgroup, the SMDs for ADLs and cognition outcomes were larger than for the whole group of dementias with statistical superiority for ginkgo also for ADL outcomes (SMD = -0.44, 95% CI -0.77; -0.12, p = 0.008). Drop-out rates and side effects did not differ between ginkgo and placebo. No consistent results were available for quality of life and neuropsychiatric symptoms, possibly due to the heterogeneity of the study populations. **CONCLUSIONS:** *Ginkgo biloba* **appears more effective than placebo. Effect sizes were moderate, while clinical relevance is, similar to other dementia drugs, difficult to determine.**[39]

POSITIVE TRIAL

Ginkgo biloba EXTRACT 761 (↑) MEMORY, COGNITION, BEHAVIOR
International Journal of Geriatric Psychiatry (England) (2011)
Geriatric Psychiatry Centre, Alexian Hospital Maria-Hilf, Krefeld and Department of Psychiatry and Psychotherapy, University of Duesseldorf, Germany.
OBJECTIVE: To test the efficacy and safety of a once-daily formulation of EGb 761 in the treatment of patients with dementia with neuropsychiatric features. METHODS: Multi-centre trial of 410 outpatients with mild to moderate dementia (Alzheimer's disease, vascular dementia or mixed form) scoring between 9 and 23 on the SKT cognitive test battery, at least five on the Neuropsychiatric Inventory (NPI) and three or higher in at least one item of the NPI. Patients were randomly allocated to double-blind treatment with **240 mg of EGb 761 or placebo once daily** for 24 weeks. Primary outcomes were the changes from baseline in the SKT total score and the NPI total score. The Alzheimer's Disease Cooperative Study Clinical Global Impression of Change (ADCS-CGIC), Activities of Daily Living International Scale (ADL-IS), NPI distress score, DEMQOL-Proxy quality-of-life scale and Verbal Fluency Test were secondary outcomes. **RESULTS: At endpoint, patients treated with EGb 761 (n = 202) improved by -1.4 (95% confidence interval -1.8; -1.0) points on the SKT and by -3.2 (-4.0; -2.3) on the NPI total score, whereas those receiving placebo (n = 202) deteriorated by +0.3 (-0.1; 0.7) on the SKT and did not**

change on the NPI total score (-0.9; 0.9). **Both drug-placebo comparisons were significant at p < 0.001. EGb 761 was significantly superior to placebo with respect to all secondary outcome measures.** Adverse event rates were similar for both treatment groups. **CONCLUSIONS: EGb 761, 240 mg once-daily, was found significantly superior to placebo in the treatment of patients with dementia with neuropsychiatric symptoms.**[40]

POSITIVE TRIAL

Ginkgo biloba EXTRACT 761 (↑) MEMORY, COGNITION, BEHAVIOR
Neuropsychiatric Disease and Treatment (New Zealand) (2011)
Institute of Gerontology, Academy of Medical Sciences, Kiev, Ukraine.
PURPOSE: To examine the effects of *Ginkgo biloba* extract EGb 761(®) on neuropsychiatric symptoms of dementia. PATIENTS AND METHODS: Randomized, controlled, double-blind, multicenter clinical trial involving 410 outpatients with mild to moderate dementia (Alzheimer's disease with or without cerebrovascular disease, vascular dementia), scoring at least 5 on the Neuropsychiatric Inventory (NPI), with at least one item score of 3 or more. Total scores on the SKT cognitive test battery (Erzigkeit's short syndrome test) were between 9 and 23. After random allocation, the patients took **240 mg of EGb 761(®) or placebo once daily** for a period of 24 weeks. Changes from baseline to week 24 in the NPI composite and in the SKT total score were the primary outcomes. The NPI distress score was chosen as a secondary outcome measure to evaluate caregivers' distress. **RESULTS: The NPI composite score improved by -3.2 (95% confidence interval -4.0 to -2.3) in patients taking EGb 761(®) (n = 202), but did not change (-0.9; 0.9) in those receiving placebo (n = 202), which resulted in a statistically significant difference in favor of EGb 761(®) (P < 0.001). Treatment with EGb 761(®) was significantly superior to placebo for the symptoms apathy/indifference, sleep/night-time behavior, irritability/lability, depression/dysphoria, and aberrant motor behavior. Caregivers' distress evaluation revealed similar baseline pattern and improvements. CONCLUSION: Treatment with EGb 761(®), at a once-daily dose of 240 mg, was safe, effectively alleviated behavioral and neuropsychiatric symptoms in patients with mild to moderate dementia, and improved the wellbeing of their caregivers.**[41]

POSITIVE TRIAL

Ginkgo biloba EXTRACT 761 (↑) MEMORY, COGNITION, BEHAVIOR
Pharmacopsychiatry (Germany) (2012)
Geriatric Psychiatry Centre, Maria-Hilf Hospital Krefeld, Krefeld, Germany.

INTRODUCTION: A 24-week randomised controlled trial was conducted to assess the efficacy of a **240 mg once-daily** preparation of *Ginkgo biloba* extract EGb 761® in 404 outpatients ≥ 50 years diagnosed with mild to moderate dementia (SKT 9-23), Alzheimer's disease or vascular dementia (VaD), with neuropsychiatric features (NPI total score ≥ 5). METHODS: Separate analyses were performed for diagnostic subgroups (probable or possible AD; VaD). **RESULTS: 333 patients were diagnosed with AD and 71 with VaD. EGb 761® treatment was superior to placebo with respect to the SKT total score** (drug-placebo differences: 1.7 for AD, p<0.001, and 1.4 for VaD, p<0.05) and the NPI total score (drug-placebo differences: 3.1 for AD, p<0.001 and 3.2 for VaD, p<0.05). **Significant drug-placebo differences were found for most secondary outcome variables with no major differences between AD and VaD subgroups.** Rates of adverse events in EGb 761® and placebo groups were essentially similar. **CONCLUSION: EGb 761® improved cognitive functioning, neuropsychiatric symptoms and functional abilities in both types of dementia.**[42]

POSITIVE TRIAL

Ginkgo biloba Extract 761 (↑) Memory, Cognition, Behavior
Journal of Psychiatric Research (England) (2012)
Medical Faculty, University of Cologne, Cologne, Germany.
A multi-centre, double-blind, randomised, placebo-controlled, 24-week trial with 410 outpatients was conducted to demonstrate efficacy and safety of a **240 mg once-daily** formulation of *Ginkgo biloba* extract EGb 761 in patients with mild to moderate dementia (Alzheimer's disease or vascular dementia) associated with neuropsychiatric symptoms... **In conclusion, treatment with EGb 761 at a once-daily dose of 240 mg was safe and resulted in a significant and clinically relevant improvement in cognition, psychopathology, functional measures and quality of life of patients and caregivers.**[43]

POSITIVE REVIEW

Ginkgo biloba Extract 761 (↑) Memory, Cognition, Behavior
International Psychogeriatrics (England) (2012)
Department of Psychiatry and Psychotherapy, Medical University of Vienna, Austria.
Research into early intervention for Alzheimer's disease and dementia has involved cohort data from large epidemiological studies and data from specifically designed intervention trials. Cohort data indicate that use of nootropics and *Ginkgo biloba* extract may be associated with a reduced incidence of dementia and death. **Data from large trials have often been inconclusive due to issues with poor medication adherence. However, such trials do indicate potential benefits**

with *Gingko biloba* extract in terms of reduced incidence of dementia of the Alzheimer's disease type, vascular dementia and mixed pathology, reduced progression in terms of the clinical dementia rating, and improvements in attention and memory. Furthermore, *Gingko biloba* extract EGb 761 is a useful option for long-term intervention on the basis of decades of previous experience and an excellent safety record. However, benefits can be expected only with sufficient medication adherence and treatment duration, so clear evidence of a disease-modifying benefit of this extract is needed from adequately designed trials using modern methods to ensure high levels of adherence.[44]

POSITIVE REVIEW

Ginkgo biloba EXTRACT 761 (↑) MEMORY, COGNITION, BEHAVIOR
International Psychogeriatrics (England) (2012)
Academic Unit for Psychiatry of Old Age, St Vincent's Health, Department of Psychiatry, University of Melbourne, Melbourne, Australia.
In June 2011 a two-day expert meeting "The Ageing Brain" took place in Amsterdam, The Netherlands. The main aim was to discuss the available preclinical and clinical data on *Ginkgo biloba* special extract EGb 761® in the context of current developments in the diagnosis and treatment of age-related cognitive decline and Alzheimer's disease. 19 dementia experts covering the disciplines bio- and neurochemistry, gerontology, neurology, pharmacology, and psychiatry from Australia, Asia, Europe and North America reviewed available preclinical and clinical data for EGb 761® and identified core topics for future research. Based on a wide range of preclinical effects demonstrated for *Ginkgo biloba*, EGb 761® can be conceptualized as a multi-target compound with activity on distinct pathophysiological pathways in Alzheimer's disease and age-related cognitive decline. While symptomatic efficacy in dementia and mild cognitive impairment has been demonstrated, interpretation of data from dementia prevention trials is complicated by important methodological issues. Bridging pre-clinical research and clinical research as well as deciding on suitable study designs for future trials with EGb 761® remain important questions. **The participants of the "Ageing Brain" meeting on *Ginkgo biloba* special extract EGb 761® concluded that there is plenty of promising data, both pre-clinical and clinical, to consider future research with the compound targeting cognitive impairment in old age as a worthwhile activity.**[45]

NEGATIVE TRIAL

Ginkgo biloba EXTRACT 761 (↓) MEMORY, COGNITION, BEHAVIOR
The Lancet Neurology (England) (2012)

GuidAge Study Group, 117 Contributors.
BACKGROUND: Prevention strategies are urgently needed to tackle the growing burden of Alzheimer's disease. We aimed to assess efficacy of long-term use of standardised *Ginkgo biloba* extract for the reduction of incidence of Alzheimer's disease in elderly adults with memory complaints. METHODS: In the randomised, parallel-group, double-blind, placebo-controlled GuidAge clinical trial, we enrolled adults aged 70 years or older who spontaneously reported memory complaints to their primary-care physician in France. We randomly allocated participants in a 1:1 ratio according to a computer-generated sequence to a twice per day dose of 120 mg standardised *Ginkgo biloba* extract (EGb 761) or matched placebo. Participants and study investigators and personnel were masked to study group assignment. Participants were followed-up for 5 years by primary-care physicians and in expert memory centres. The primary outcome was conversion to probable Alzheimer's disease in participants who received at least one dose of study drug or placebo, compared by use of the log-rank test. This study is registered with ClinicalTrials.gov, number NCT00276510. FINDINGS: **Between March, 2002, and November, 2004, we enrolled and randomly allocated 2854 participants, of whom 1406 received at least one dose of *Ginkgo biloba* extract and 1414 received at least one dose of placebo**. By 5 years, 61 participants in the ginkgo group had been diagnosed with probable Alzheimer's disease (1·2 cases per 100 person-years) compared with 73 participants in the placebo group (1·4 cases per 100 person-years; hazard ratio [HR] 0·84, 95% CI 0·60-1·18; p=0·306), but the risk was not proportional over time. Incidence of adverse events was much the same between groups. 76 participants in the ginkgo group died compared with 82 participants in the placebo group (0·94, 0·69-1·28; p=0·68). 65 participants in the ginkgo group had a stroke compared with 60 participants in the placebo group (risk ratio 1·12, 95% CI 0·77-1·63; p=0·57). Incidence of other haemorrhagic or cardiovascular events also did not differ between groups. **INTERPRETATION: Long-term use of standardised *Ginkgo biloba* extract in this trial did not reduce the risk of progression to Alzheimer's disease compared with placebo.**[46]

POSITIVE TRIAL

Ginkgo biloba EXTRACT 761 (↑) MEMORY, COGNITION, BEHAVIOR
Journal of Alzheimer's Disease (The Netherlands) (2012)
Department of Neuropsychiatry, Institute of Clinical Medicine, University of Tsukuba, Japan.
Although nutrients or agents with antioxidant properties were reported to show a preventive effect on cognitive decline in animal studies, epidemiologic data on select antioxidants have shown conflicting results. We investigated whether a combination of antioxidants from supplements is effective for the improvement of cognitive function of elderly. **Forty-one subjects from a community dwelling aged 65 years and older took**

supplements containing n-3 polyunsaturated fatty acids (n-3 PUFA), lycopene, and *Ginkgo biloba* extracts (GE) daily for 3 years. The data of 622 subjects without supplement intake were used as control. We investigated the changes in cognitive function during a 3-year follow-up. We also investigated the influence of apolipoprotein E (APOE) genotype on the effect of antioxidants. **We found that a combination of antioxidants improved cognitive function of aged persons after 3 years. Our present study also indicated this improvement in cognitive function with supplement intake in both APOE4 non-carrier (E4-) and APOE4 carrier (E4+) groups. Especially, in E4+, we found a large effect size of the improvement of cognition. When multiple antioxidants are used in combination, they protect against vulnerability to other agents and synergistically potentiate their antioxidant properties. These synergistically potentiated antioxidant effects of agents contribute to the improvement of cognitive function.**[47]

POSITIVE REVIEW

Ginkgo biloba EXTRACT 761 (↑) MEMORY, COGNITION, BEHAVIOR
Shanghai Archives of Psychiatry (China) (2013)
Shanghai Mental Health Center, Shanghai Jiao Tong University School of Medicine, Shanghai, China; Department of Emergency Medicine, Tongji Hospital of Tongji University.
The researchers conducted a meta-analysis of studies about the effect of EGB on cognition and daily functioning in persons with dementia. The researchers searched various English and Chinese databases and identified reports of placebo controlled, randomized trials of *Ginkgo biloba* treatment (lasting a minimum of 22 weeks) for dementia that were published from January 1982 to September 2012… 9 studies with a total of 2578 patients met the inclusion and exclusion criteria. **Pooled results from the 6 studies that were included in the meta-analysis (total n = 1917) found that *Ginkgo biloba* extract was superior to placebo in preventing deterioration in cognitive functioning and in activities of daily living, but these results were only valid for studies with younger subjects (with a mean age below 75).** There were no significant differences in the dropout rates between groups or in the overall rates of adverse events during treatment. However, there was considerable heterogeneity in the results between the studies (primarily based on the age of the subjects) and there were several potential biases in the reports (most of which were supported by pharmaceutical firms), so the overall evidence was considered of "low quality."[48]

POSITIVE REVIEW

Ginkgo biloba EXTRACT 761 (↑) MEMORY, COGNITION, BEHAVIOR
International Psychogeriatrics (England) (2013)

Clinic of Geriatric Psychiatry and Psychotherapy, Alexian Hospital Maria Hilf, Krefeld, Germany, and Department of Psychiatry and Psychotherapy, University of Düsseldorf, Germany. OBJECTIVE: We review four randomised, controlled trials investigating the efficacy of *Ginkgo biloba* extract EGb 761(®) in elderly patients with Alzheimer or vascular dementia with neuropsychiatric features... Three trials compared 2×120 mg/day or 1×240 mg/day EGb 761(®) to placebo while one used donepezil as an active control. The duration of randomised treatment was 22 or 24 weeks. RESULTS: One thousand, two hundred and ninety-four patients were analysed for efficacy. **Patients treated with EGb 761(®) showed improvements of cognitive performance and behavioural symptoms that were associated with advances in activities of daily living and a reduced burden to caregivers.** Placebo-treated patients, on the other hand, showed only minimal improvements or signs of progression. **In each placebo-controlled trial, EGb 761(®) was significantly superior in all mentioned domains** (p < 0.01). In the actively controlled trial, EGb 761(®) and donezepil as well as a combination of both drugs had similar effects. **CONCLUSIONS: The review supports the efficacy of EGb 761(®) in age-related dementia with neuropsychiatric features. The drug was safe and well-tolerated.**[49]

POSITIVE TRIAL

Ginkgo biloba EXTRACT 761 (↑) MEMORY, COGNITION, BEHAVIOR
International Journal of Geriatric Psychiatry (England) (2014)
Mental Health Research Center of the Russian Academy of Medical Sciences, Moscow, Russia. OBJECTIVE: The study was conducted to explore the effects of EGb 761 (Dr. Willmar Schwabe GmbH & Co. KG, Karlsruhe, Germany) on neuropsychiatric symptoms (NPS) and cognition in patients with mild cognitive impairment (MCI). METHODS: One hundred and sixty patients with MCI who scored at least 6 on the 12-item Neuropsychiatric Inventory (NPI) were enrolled in this double-blind, multi-center trial and randomized to receive 240 mg EGb 761 daily or placebo for a period of 24 weeks... **CONCLUSIONS: EGb 761 improved NPS and cognitive performance in patients with MCI. The drug was safe and well tolerated.**[50]

POSITIVE REVIEW

Ginkgo biloba EXTRACT 761 (↑) MEMORY, COGNITION, BEHAVIOR
Clinical Interventions in Aging (New Zealand) (2014)
Alzheimer Disease Research Unit, Memory Clinic, McGill Centre for Studies in Aging, McGill University, Verdun, QC, Canada; Clinical Research Department, Dr. Willmar Schwabe GmbH & Co. KG, Karlsruhe, Germany. The objective of this systematic review was to evaluate current evidence for the efficacy of *Ginkgo biloba* extract EGb 761(®) in dementia. Seven of 15 randomized,

placebo-controlled trials in patients with dementia identified by database searches met all our selection criteria and were included in the meta-analysis. In these trials, patients were treated with 120 mg or 240 mg per day of the defined extract EGb 761 or placebo. Efficacy was assessed using validated tests and rating scales for the cognitive domain, the functional domain (activities of daily living), and global assessment.... Of 2,684 outpatients randomized to receive treatment for 22-26 weeks, 2,625 represented the full analysis sets (1,396 for EGb 761 and 1,229 for placebo). **Standardized mean differences for change in cognition ... activities of daily living ... and global rating ... significantly favored EGb 761 compared with placebo. Statistically significant superiority of EGb 761 over placebo was confirmed by responder analyses as well as for patients suffering from dementia with neuropsychiatric symptoms... In conclusion, meta-analyses confirmed the efficacy and good tolerability of *Ginkgo biloba* extract EGb 761 in patients with dementia.**[51]

NEGATIVE REVIEW

Ginkgo biloba EXTRACT 761 (↓) MEMORY, COGNITION, BEHAVIOR
Journal of the Medical Association of Thailand (Thailand) (2015)
OBJECTIVE: To determine the efficacy of *Ginkgo biloba* for the prevention of dementia in individuals without dementia. MATERIAL AND METHOD: English databases including Medline, Embase, Cochrane Library and PsycINFO, were searched, and randomized double-blind controlled studies comparing *Ginkgo biloba* with placebo in prevention of dementia were considered. **Two trials met inclusion criteria. Methodological quality was assessed using the Jadad criteria. RESULTS: Meta-analysis of the two trials involving 5,889 participants indicated no significant difference in dementia rate between *Ginkgo biloba* and the placebo (347/2,951 vs. 330/2,938, odds ratio = 1.05, 95% CI 0.89-1.23) and there was no considerable heterogeneity between the trials. The two studies revealed no statistically significant differences in the rate of serious adverse effect between *Ginko* [sic] *biloba* and the placebo. CONCLUSION: There is no convincing evidence from this review that demonstrated *Ginkgo biloba* in late-life can prevent the development of dementia. Using it for this indication is not suggested at present.**[52]

POSITIVE REVIEW

Ginkgo biloba EXTRACT 761 (↑) MEMORY, COGNITION, BEHAVIOR
Journal of Alzheimer's Disease (The Netherlands) (2015)
Geriatric Medicine-Memory Unit and Rare Diseases Centre, University of Bari Aldo Moro, Bari, Italy; Neurodegenerative Disease Unit, Department of Basic Medicine, Neuroscience, and

Sense Organs, University of Bari Aldo Moro; Department of Clinical Research in Neurology of the University of Bari Aldo Moro; Geriatric Unit and Laboratory of Gerontology and Geriatrics, Department of Medical Sciences, IRCCS "Casa Sollievo della Sofferenza", San Giovanni Rotondo, Foggia, Italy.

Among nutraceuticals and nutritional bioactive compounds, the standardized *Ginkgo biloba* extract EGb 761 is the most extensively clinically tested herbal-based substance for cognitive impairment, dementia, and Alzheimer's disease. In the last 3 years, notwithstanding negative meta-analytic findings and the discouraging results of preventive trials against AD, some randomized controlled trials focusing particularly on dementia, AD, and mild cognitive impairment (MCI) subgroups with neuropsychiatric symptoms (NPS) and some recent meta-analyses have suggested a renowned role for EGb 761 for cognitive impairment and dementia. **Meta-analytic findings suggested overall benefits of EGb 761 for stabilizing or slowing decline in cognition of subjects with cognitive impairment and dementia. The safety and tolerability of EGb 761 appeared to be excellent at different doses. Subgroup analyses showed that these clinical benefits of EGb 761 were mainly associated with the 240 mg/day dose, and also confirmed in the AD subgroup. More importantly, one of these meta-analyses showed clinical benefits in cognition, behavior, functional status, and global clinical change of EGb 761 at a dose of 240 mg/ day in the treatment of patients with dementia, AD, and MCI with NPS.** The inclusion of the recent randomized controlled trials focusing on dementia, AD, and MCI subgroups with NPS [neuropsychiatric symptoms] may partly explain the conflicting results of these recent meta-analyses and previous pooled findings.[53]

POSITIVE REVIEW

Ginkgo biloba EXTRACT 761 (↑) MEMORY, COGNITION, BEHAVIOR
Journal of Alzheimer's Disease (The Netherlands) (2015)

Department of Neurology, Qingdao Municipal Hospital, College of Medicine and Pharmaceutics, Ocean University of China; Department of Neurology, Qingdao Municipal Hospital, School of Medicine, Qingdao University; Department of Neurology, Qingdao Municipal Hospital, Nanjing Medical University; Memory and Aging Center, Department of Neurology, University of California at San Francisco; Department of Neurology, Qingdao Municipal Hospital, College of Medicine and Pharmaceutics, Ocean University of China.

The objective was to discuss new evidence on the clinical and adverse effects of standardized *Ginkgo biloba* extract EGb 761 for cognitive impairment and dementia… Nine trials met our inclusion criteria. Trials were of 22-26 weeks duration and included 2,561 patients in total. In the meta-analysis, the weighted mean differences in change scores for cognition were in favor of EGb 761 compared to placebo… [T] he standardized mean differences in change scores for activities in daily living (ADLs) were also in favor of EGb 761 compared to placebo… **All these benefits are** mainly associated with EGb761 at a dose of 240 mg/day. For subgroup analysis in patients with neuropsychiatric symptoms, 240 mg/day EGb761 improved cognitive function, ADLs [activities of daily living], CGIC [clinical global impression of change], and also neuropsychiatric symptoms with statistical superiority than for the whole group… Finally, safety data revealed no important safety concerns with EGb761. EGb761 at 240 mg/day is able to stabilize or slow decline in cognition, function, behavior, and global change at 22–26 weeks in cognitive impairment and dementia, especially for patients with neuropsychiatric symptoms.[54]

POSITIVE REVIEW

Ginkgo biloba EXTRACT 761 (↑) MEMORY, COGNITION, BEHAVIOR
The World Journal of Biological Psychiatry (England) (2015)
Service Universitaire de Psychiatrie de l'Age Avancé (SUPAA), Department of Psychiatry, Centre Hospitalier Universitaire Vaudois, Prilly, Switzerland.
OBJECTIVES: To review current evidence of efficacy of *Ginkgo biloba* extract EGb 761® in dementia with behavioural and psychological symptoms (BPSD). METHODS: Randomized, placebo-controlled trials assessing the effects of EGb 761® in dementia patients with BPSD were included if the diagnosis was made in accordance with internationally accepted criteria, the treatment period was at least 22 weeks, outcome measures covered BPSD and at least two of the following domains of assessment, i.e. cognition, activities of daily living and clinical global assessment, and methodological quality was adequate… RESULTS: Four published trials were identified, involving altogether 1,628 outpatients with mild to moderate dementia. Least-square mean differences for change from baseline in cognition, BPSD (including caregiver distress rating), activities of daily living, clinical global impression, and quality of life favoured EGb 761® (P < 0.001 for all comparisons). **CONCLUSIONS: The pooled analyses provide evidence of efficacy of EGb 761® at a daily dose of 240 mg in the treatment of out-patients suffering from Alzheimer's, vascular or mixed dementia with BPSD.**[55]

There are several ways to assess this compilation of reports involving EGb 761 and AD. For example, we listed above an overall total of 39 such reports; 28 positive clinical trials and reviews of those trials, and 10 negative such trials and reviews (with one clinical trial reported as inconclusive). It was this published record of EGb 761 and AD that likely prompted the Mayo Clinic to grade evidence for the use of *Ginkgo biloba* for "dementia" as a "B," which it defines as "good scientific evidence for this use." The Mayo Clinic commented more fully as follows: "Overall, the scientific literature suggests that ginkgo benefits people with dementia. Ginkgo may improve cognitive performance and protect against Alzheimer's. However, conclusions regarding ginkgo for dementia are often conflicting. Additional research is needed in this area."[56]

Another way to assess the reports is through a closer reading of them. For example, the most positive reviews are the ones most recently published (in 2015), which report that "the pooled analyses provide evidence of efficacy of EGb 761 at a daily dose of 240 mg" (*The World Journal of Biological Psychiatry*, 2015), that "all these benefits are mainly associated with EGb 761 at a dose of 240 mg/day" (*Journal of Alzheimer's Disease*, 2015), and that "subgroup analyses showed that these clinical benefits of EGb 761 were mainly associated with the 240 mg/day dose" (*Journal of Alzheimer's Disease*, 2015).

Sharpening this analysis further, four clinical trials cited above that were published in 2011–12 and that used a once-daily dose of 240 EGb 761 (as opposed to 120 mg twice-daily) on a total of 1634 patients with mild to moderate Alzheimer's dementia (a majority) and vascular dementia (a minority) all reported positive results with respect to both efficacy and safety. (See above, *International Journal of Geriatric Psychiatry*, 2011; *Neuropsychiatric Disease and Treatment*, 2011; *Pharmacopsychiatry*, 2012; and *Journal of Psychiatric Research*, 2012.) Furthermore, given the known pharmacology of EGB 761, including as reported in this volume, these results likely reflect disease-modifying effects and are not merely symptomatic. In short, there is no comparable record to date of efficacy and safety within the pharmaceutical and biotech industries on treating patients with mild to moderate AD.

With regard to prevention, we briefly examine here the three negative clinical trials that are most commonly cited as evidence against EGb 761 for AD: The Dodge study published in *Neurology* in 2008; The 2008 Ginkgo Evaluation of Memory (GEM) study led by Steven DeKosky and published in JAMA, and the GuidAge study published in *The Lancet Neurology* in 2012.

The objective of the first of these studies (*Neurology*, 2008) was "to assess the feasibility, safety, and efficacy of *Ginkgo biloba* extract (GBE) on delaying the progression to cognitive impairment in normal elderly aged 85 and older." This was a 42-month, randomized, double blind, placebo-controlled trial with a dose of 240 mg/day of GBE (80 mg, 3 times a day). The researchers, led by Hiroko Dodge at Oregon Health and Science University in Portland, reported the results as follows: "In the intention-to-treat analysis, there was no reduced risk of progression to CDR [clinical dementia rating] … among the GBE group" and "there was no less of a decline in memory function among the GBE group"; thus, a negative result by way of an intention-to-treat analysis. Dodge's group also reported, "in the secondary analysis, where we controlled the medication adherence level, the GBE group had a lower risk of progression [to clinical dementia] … and a smaller decline in memory scores." The researchers concluded that "taking into account medication adherence showed a protective effect of GBE on the progression to CDR = 0.5 and memory decline" and that "results of larger prevention trials taking into account medication adherence may clarify the effectiveness of GBE." The Dodge group also reported that "under all circumstances methods to facilitate and increase adherence in dementia prevention trials will be critical. One simple goal is, where possible, to use once a day dosing."[33] In short, the treatment subjects who adhered to the test dose—240 mg/day of GBE—had a lower risk of developing mild cognitive impairment due to AD or AD dementia itself.

The findings here thus reflect an ongoing clash between results generated by intention-to-treat analysis (ITT), which is a statistical model that "includes all randomized patients in the groups to which they were randomly assigned, regardless of their adherence with the entry criteria, regardless of the treatment they actually received, and regardless of subsequent withdrawal from treatment or deviation from the protocol" and subgroups that adhere to the prescribed treatment dose.[57]

Although the Dodge study used a standardized *Ginkgo biloba* extract from Thorne Research, Inc., which is an excellent company, it nevertheless did not use EGb 761. However, we include the Dodge study here because it is frequently cited by prominent AD researchers as a failed clinical trial involving *Ginkgo biloba* extract. For example, in the 2008 GEM study, the DeKosky-led researchers reported that the Dodge study "did not find a reduction in dementia incidence after an average follow-up of 3.5 years in an intention-to-treat analysis."[34] Another prominent AD researcher, Lon Schneider at the University of Southern California, commented on the Dodge study in the same issue of *JAMA* as follows: "A recent trial involving 118 participants without mild cognitive impairment or dementia, all older than 85 years, who were randomized to receive *G. biloba* extract (240 mg/d) or placebo and were followed up for 42 months showed a non-significant effect for *G. biloba* to delay progression to mild cognitive impairment."[58] Neither DeKosky nor Schneider noted that Dodge reported that "taking into account medication adherence showed a protective effect of GBE."

The objective of the second prevention trial (*JAMA*, 2008)—titled "*Ginkgo biloba* for Prevention of Dementia: A Randomized Controlled Trial," the authors of which were led by DeKosky—was to test "the primary hypothesis that 240 mg of *G biloba* daily would decrease the incidence of all-cause dementia and specifically reduce the incidence of AD" in "elderly individuals with normal cognition and those with mild cognitive impairment (MCI)." The DeKosky researchers concluded that using "*G. biloba* at 120 mg twice a day was not effective in reducing either the overall incidence rate of dementia or AD incidence in elderly individuals with normal cognition or those with MCI." Thus, the dose used in DeKosky's GEM Study—240 mg/day of EGb 761—was featured in both the main objective of the study and in its conclusion. DeKosky's GEM study group (hereinafter "DeKosky") also observed: "The 240-mg dose of EGb 761 was chosen based on information from prior clinical studies suggesting a dose–response relationship up to 240 mg."

DeKosky's GEM study was thus similar to (but much larger than) the Dodge study, including the selection of the 240 mg/day dose. However, while DeKosky, like Dodge, observed that his data was analyzed "using an intention-to-treat approach" (which includes all randomized treatment subjects regardless of the treatment they actually received), unlike Dodge he identified no subgroup of treatment subjects that adhered to the 240 mg/day dose, and thus provided no dose-related secondary analysis of any such subgroup. DeKosky reported that "retention and adherence in this feasibility trial were excellent," but he provided few details relating specifically to dose adherence amidst the claim of excellent adherence. Although the stated objective and conclusion of DeKosky's study is clearly grounded in a 240 mg/day dose of EGb 761, it is not clear on reading the full study, as was made clear in the Dodge study, whether a subgroup of adherents to the 240 mg dose had fared any better (as they did in the Dodge study) than the larger population of mixed-adherent treatment subjects.

This is not to say that there was no discussion of "adherence" in DeKosky's GEM study. Indeed, the DeKosky group reported (emphasis added): "Treatment *adherence* and participant retention were a particular focus in this trial because of the age of the study population. For treatment *adherence* monitoring, participants returned all blister packs at each 6-month visit, and *adherence* was calculated using the number of pills taken vs number of days since previous visit. Additionally, an *adherence* and retention subcommittee led by a geriatrician met monthly to review and discuss challenges to *adherence* and retention with field center staff and to provide study-specific *adherence* and retention tracking data to each field center. Annual site visits to all clinics and a midstudy retreat for field center staff focused extensively on strategies for participant *adherence* and retention specific to this age group." The group also reported that "retention and adherence in this feasibility trial were excellent, reflecting improved methods and growing experience over the past decade in conducting clinical trials in older populations." They further reported that "at the end of the trial, 60.3% of active participants were taking their assigned study medication, and *adherence* did not differ between the 2 groups," and that one of the strengths of the GEM study was its "high rate of follow-up and treatment *adherence*." These passages account for all mentions of "adherence" in the DeKosky study (except for "Figure 2"), none of which address the issue of adherence with respect to any subgroup of adherents to the prominently featured dose of 240 mg/day. Furthermore, Figure 2 in the DeKosky study, which depicted "Cumulative Adherence to Assigned Study Tablets by Scheduled 6-Month Follow-up Visit (Excluding Death and Incident Dementia)," provided no information about adherence in terms of dose.[59]

The third AD prevention study (*The Lancet Neurology*, 2012) was titled "Long-Term Use of Standardised *Ginkgo biloba* Extract for the Prevention of Alzheimer's Disease (GuidAge): A Randomised Placebo-Controlled Trial," the objective of which was to "assess efficacy of long-term use of standardised *Ginkgo biloba* extract for the reduction of incidence of Alzheimer's disease in elderly adults with memory complaints." The study population "was made up of more than 2500 individuals aged 70 years and older, who were free of dementia at baseline, and who had spontaneously reported subjective memory complaints to their PCP [primary care physician]." The dose of EGb 761 used in the study was 120 mg twice per day. The researchers, led by Bruno Vellas at the Gérontopôle, Toulouse

University Hospital in France, reported "we chose a 240 mg daily dose of EGb 761 on the basis of suggested effectiveness in patients with dementia." They also reported that "participants were followed-up for 5 years by investigators and personnel in expert memory centres." With respect to the trial outcome and the dose actually taken by treatment participants, the researchers reported that "the primary outcome was conversion to probable Alzheimer's disease in participants who received at least one [120 mg] dose of study drug [EGb 761] or placebo." The researchers reported their conclusions as follows: "Long-term use of standardised *Ginkgo biloba* extract in this trial did not reduce the risk of progression to Alzheimer's disease compared with placebo"; that "overall, the GuidAge trial was unable to reach our primary endpoint criterion (decreased conversion to Alzheimer's disease in 5 years of follow-up)"; and that "conclusions from this trial were restricted by the lower than expected incidence of Alzheimer's disease." They also reported: "The main limitation of our study was that the number of dementia events was much lower than expected, leading to a lack of statistical power to detect effects." Furthermore, while the researchers observed that "overall, 2487 (95%) patients adhered to their allocated intervention," it was not clear whether adhering to the allocation of EGb 761 meant taking "at least one [120 mg] dose of *Ginkgo biloba*" or the full daily dose of 240 mg.

In commentary that was published in the same issue of *The Lancet Neurology* in which the GuidAge study was published, Lon Schneider wrote that "the GuidAge trial adds to the substantial evidence from the Ginkgo Evaluation of Memory [GEM] trial that *Ginkgo biloba* does not prevent dementia in elderly individuals with or without memory complaints or cognitive impairment and is not effective for prevention of Alzheimer's disease." Schneider also wrote: "*Ginkgo biloba* extract is the most extensively clinically tested substance for Alzheimer's disease and cognitive impairment. Unfortunately, as with all drugs in development that have putative neuroprotective or anti-amyloid β actions, *Ginkgo biloba* extract has not shown meaningful clinical efficacy in well designed and executed clinical trials."[17] Although the GEM trial reported no dose-related subgroup data, as the Dodge study significantly did, to assess prevention-related results of adherence to the 240 mg/day dose of EGb 761, and although the GuidAge researchers mistakenly recruited healthier and better educated participants "than the general elderly population on which our sample size was based," which led "to a lack of statistical power to detect effects," Schneider views the GEM and GuidAge trials, but not the multitude of others, as "well-designed and executed."

From our perspective, and overall, it seems that the Mayo Clinic's B rating of evidence of the efficacy of EGb 761 for AD prevention and treatment, which it defined as "good scientific evidence for this use," is about right. And any large clinical trials using EGb 761 henceforth for the treatment of mild to moderate AD should feature a once-a-day dose of 240 mg. This approach also would be consistent with the advice of Hiroko Dodge that "to facilitate and increase adherence … one simple goal is, where possible, to use once-a-day dosing."

In United States Supreme Court jurisprudence, "strict scrutiny" is a form of judicial review applied to cases that involve fundamental constitutional rights, whereas "rational-basis scrutiny" is applied to lesser rights. Thus, the highest "strict" standard is applied when the constitutional stakes are likewise highest, whereas a "rational-basis" is sufficient to decide cases wherein risks to the constitution are lower.

Since they are very different agents, pharmaceutical products and plant polyphenols should be judged by a similarly two-tiered standard of approved use: "strict scrutiny" for pharmaceuticals, given the higher individual and societal stakes reflected in higher toxicity, prevalence of drug-related health risks, and higher costs, and "rational-basis" for the low-toxicity, low-cost, food-based polyphenolic compounds that humans have been consuming safely for hundreds or thousands of years in the ancient foods of turmeric, green tea, red wine, grape products, and olive oil. An additional allowance should be given to EGb 761, which enjoys an unmatched twenty-year history of clinical-trial testing with significant positive benefits and a side effect profile comparable to placebo, with a slightly higher incidence of hemorrhagic stroke among an "oldest old" treatment population of 85 years and older.

Benchmark criteria for passing the rational-basis test could include the following:

- *Plant polyphenols as nutritional supplements should be standardized extracts from foods that enjoy a longstanding record of safe and beneficial human consumption.* Given the cameos of ancient history and the evidence from epidemiology and population studies reviewed in Chapter 2, the catechins in green tea, the curcuminoids in turmeric, the numerous polyphenols in red wine and other grape products (including resveratrol), and the polyphenolic compounds found in olive oil (including oleocanthal, oleuropein, and hydroxytyrosol), it seems evident that the plant polyphenols featured in this volume have been consumed safely by humans for hundreds or thousands of years with numerous health benefits.

- *The standardized extracts themselves should be nontoxic with a wide margin of safety between the therapeutic dose and the toxic dose within a range of normal or tested dosing.* The abstracts presented in Chapters 3–6 demonstrate the low toxicity and safety of the polyphenolic compounds featured in this volume. This is the case even more so when they are carefully prepared and marketed by high-end supplement manufacturers, of which there are many in the United States.

- *Evidence from epidemiological and population studies should be aligned with mechanistic and pharmacological studies.* Such studies, when they present consistently concordant and complementary results, as they do in Chapters 2–6, should be viewed as a key rational-basis criterion.

- *The plant polyphenols should possess pleiotropic capabilities.* Such capabilities have been shown in the scientific literature and throughout this volume to beneficially modify the multiple dimensions of AD pathophysiology.

- *The pleiotropic capabilities should have been shown in voluminous and detailed mechanistic studies to accommodate various credible theories of AD pathogenesis.* As extensively documented here, the pleiotropic capabilities of the plant polyphenols featured in this volume have been shown to potentially therapeutically accommodate several credible hypotheses about AD pathogenesis as advanced by numerous credible research scientists and as published in established biomedical journals. This includes the most-studied polyphenolic compounds—the curcuminoids, EGCG, and resveratrol, as reported throughout this volume—and EGb 761, as reported in this chapter.

REFERENCES

1. Shi, Wu, Xu, Zou. Bilobalide regulates soluble amyloid precursor protein release via phosphatidyl inositol 3 kinase-dependent pathway. *Neurochemistry International* August 2011;**59**(1):59–64; Colciaghi, Borroni, Zimmermann, et al. Amyloid precursor protein metabolism is regulated toward alpha-secretase pathway by Ginkgo biloba extracts. *Neurobiology of Disease* July 2004;**16**(2):454–60.
2. Tchantchou, Lacor, Cao, et al. Stimulation of neurogenesis and synaptogenesis by bilobalide and quercetin via common final pathway in hippocampal neurons. *Journal of Alzheimer's Disease* 2009;**18**(4):787–98.
3. Zhou, Zhu. Reactive oxygen species-induced apoptosis in PC_{12} cells and protective effect of bilobalide. *The Journal of Pharmacology and Experimental Therapeutics* June 2000;**293**(3):982–8; Bastianetto, Zheng, Quirion. The *Ginkgo biloba* extract (EGb 761) protects and rescues hippocampal cells against nitric oxide-induced toxicity: involvement of its flavonoid constituents and protein kinase C. *Journal of Neurochemistry* June 2000;**74**(6):2268–77; Bridi, Crossetti, Steffen, Henriques. The antioxidant activity of standardized extract of *Ginkgo biloba* (EGb 761) in rats. *Phytotherapy Research* August 2001;**15**(5):449–51; Smith, Luo. Elevation of oxidative free radicals in Alzheimer's disease models can be attenuated by *Ginkgo biloba* extract EGb 761. *Journal of Alzheimer's Disease* August 2003;**5**(4):287–300; Garcia-Alloza, Dodwell, Meyer-Luehmann, et al. Plaque-derived oxidative stress mediates distorted neurite trajectories in the Alzheimer mouse model. *Journal of Neuropathology and Experimental Neurology* November 2006;**65**(11):1082–9; Rhein, Giese, Baysang, et al. *Ginkgo biloba* extract ameliorates oxidative phosphorylation performance and rescues abeta-induced failure. *PLoS One* August 24, 2010;**5**(8):e12359.
4. Liu, Hao, Qin. Long-term treatment with *Ginkgo biloba* extract EGb 761 improves symptoms and pathology in a transgenic mouse model of Alzheimer's disease. *Brain, Behavior, and Immunity* May 2015;**46**:121–31.
5. Yao, Han, Drieu, Papadopoulos. *Ginkgo biloba* extract (EGb 761) inhibits beta-amyloid production by lowering free cholesterol levels. *The Journal of Nutritional Biochemistry* December 2004;**15**(12):749–56.
6. Bastianetto, Quirion. Natural extracts as possible protective agents of brain aging. *Neurobiology of Aging* September–October 2002;**23**(5):891–7.
7. Yan, Zheng, Zhao. Effects of *Ginkgo biloba* extract EGb761 on expression of RAGE and LRP-1 in cerebral microvascular endothelial cells under chronic hypoxia and hypoglycemia. *Acta Neuropathologica* November 2008;**116**(5):529–35.
8. Bastianetto, Ramassamy, Doré, et al. The *Ginkgo biloba* extract (EGb 761) protects hippocampal neurons against cell death induced by beta-amyloid. *The European Journal of Neuroscience* June 2000;**12**(6):1882–90.
9. Bastianetto, Quirion. EGb 761 is a neuroprotective agent against beta-amyloid toxicity. *Cellular and Molecular Biology* September 2002;**48**(6):693–7.
10. Yao, Drieu, Papadopoulos. The *Ginkgo biloba* extract EGb 761 rescues the PC_{12} neuronal cells from beta-amyloid-induced cell death by inhibiting the formation of beta-amyloid-derived diffusible neurotoxic ligands. *Brain Research* January 19, 2001;**889**(1–2):181–90.
11. Tchantchou, Xu, Wu, et al. EGb 761 enhances adult hippocampal neurogenesis and phosphorylation of CREB in transgenic mouse model of Alzheimer's disease. *FASEB Journal* August 2007;**21**(10):2400–8.

12. Müller, Heiser, Leuner. Effects of the standardized *Ginkgo biloba* extract EGb 761® on neuroplasticity. *International Psychogeriatrics* August 2012;**24**(Suppl. 1):S21–4.

13. Luo, Smith, Paramasivam, et al. Inhibition of amyloid-beta aggregation and caspase-3 activation by the *Ginkgo biloba* extract EGb761. *Proceedings of the National Academy of Sciences* September 17, 2002;**99**(19):12197–202; Longpré, Garneau, Christen, Ramassamy. Protection by EGb 761 against beta-amyloid-induced neurotoxicity: involvement of NF-kappaB, SIRT1, and MAPKs pathways and inhibition of amyloid fibril formation. *Free Radical Biology and Medicine* December 15, 2006;**41**(12):1781–94.

14. Kwon, Lee, Cho, et al. *Ginkgo biloba* extract (EGb 761) attenuates zinc-induced tau phosphorylation at Ser262 by regulating GSK3β activity in rat primary cortical neurons. *Food and Function* June 2015;**6**(6):2058–67.

15. Rangel-Ordóñez, Nöldner, Schubert-Zsilavecz, et al. Plasma levels and distribution of flavonoids in rat brain after single and repeated doses of standardized *Ginkgo biloba* extract EGb 761®. *Planta Medica* October 2010;**76**(15):1683–90; Ude, Paulke, Nöldner, et al. Plasma and brain levels of terpene trilactones in rats after an oral single dose of standardized *Ginkgo biloba* extract EGb 761®. *Planta Medica* February 2011;**77**(3):259–64; Woelkart, Feizlmayr, Dittrich, et al. Pharmacokinetics of bilobalide, ginkgolide A and B after administration of three different *Ginkgo biloba* L. preparations in humans. *Phytotherapy Research* March 2010;**24**(3):445–50; Drago, Floriddia, Cro, Giuffrida. Pharmacokinetics and bioavailability of a *Ginkgo biloba* extract. *Journal of Ocular Pharmacology and Therapeutics* April 2002;**18**(2):197–202; Biber, Koch. Bioavailability of ginkgolides and bilobalide from extracts of *Ginkgo biloba* using GC/MS. *Planta Medica* March 1999;**65**(2):192–3; Fourtillan, Brisson, Girault, et al. Pharmacokinetic properties of bilobalide and ginkgolides A and B in healthy subjects after intravenous and oral administration of *Ginkgo biloba* extract (EGb 761). *Therapie* March–April 1995;**50**(2):137–44.

16. EGb 761: *Ginkgo biloba* extract, ginkor. *Drugs in R&D* 2003;**4**(3):188–93.

17. Schneider. *Ginkgo* and AD: key negatives and lessons from GuidAge. *The Lancet Neurology* October 2012;**11**(10):836–7.

18. Kanowski, Herrmann, Stephan, et al. Proof of efficacy of the *Ginkgo biloba* special extract EGb 761 in outpatients suffering from mild to moderate primary degenerative dementia of the Alzheimer type or multi-infarct dementia. *Pharmacopsychiatry* March 1996;**29**(2):47–56.

19. Le Bars, Katz, Berman, et al. A placebo-controlled, double-blind, randomized trial of an extract of *Ginkgo biloba* for dementia. North American EGb study group. *Journal of the American Medical Association* October 22–29, 1997;**278**(16):1327–32.

20. Maurer, Ihl, Dierks, Frölich. Clinical efficacy of *Ginkgo biloba* special extract EGb 761 in dementia of the Alzheimer type. *Journal of Psychiatric Research* November–December 1997;**31**(6):645–55.

21. Le Bars, Kieser, Itil. A 26-week analysis of a double-blind placebo-controlled trial of the *Ginkgo biloba* extract EGb 761 in dementia. *Dementia and Geriatric Cognitive Disorders* July–August 2000;**11**(4):230–7.

22. van Dongen, van Rossum, Kessels, et al. The efficacy of *Ginkgo* for elderly people with dementia and age-associated memory impairment: new results of a randomized clinical trial. *Journal of the American Geriatrics Society* October 2000;**48**(10):1183–94.

23. Le Bars, Kastelan. Efficacy and safety of a *Ginkgo biloba* extract. *Public Health Nutrition* December 2000;**3**(4A):495–9.

24. Le Bars, Velasco, Ferguson, et al. Influence of the severity of cognitive impairment on the effect of the *Ginkgo biloba* extract EGb 761 in Alzheimer's disease. *Neuropsychobiology* 2002;**45**(1):19–26.

25. Mix, Crews Jr. A double-blind, placebo-controlled, randomized trial of *Ginkgo biloba* extract EGb 761 in a sample of cognitively intact older adults: neuropsychological findings. *Human Psychopharmacology* August 2002;**17**(6):267–77.

26. Birks, Grimley, van Dongen. *Ginkgo biloba* for cognitive impairment and dementia. *The Cochrane Database of Systematic Reviews* 2002;(4):CD003120.

27. Kanowski, Hoerr. *Ginkgo biloba* extract EGb 761 in dementia: intent-to-treat analyses of a 24-week, multi-center, double-blind, placebo-controlled, randomized trial. *Pharmacopsychiatry* November 2003;**36**(6):297–303.

28. van Dongen, van Rossum, Kessels, et al. Ginkgo for elderly people with dementia and age-associated memory impairment: a randomized clinical trial. *Journal of Clinical Epidemiology* April 2003;**56**(4):367–76.

29. Schneider, DeKosky, Farlow, et al. A randomized, double-blind, placebo-controlled trial of two doses of *Ginkgo biloba* extract in dementia of the Alzheimer's type. *Current Alzheimer Research* December 2005;**2**(5):541–51.

30. Napryeyenko, Borzenko, GINDEM-NP Study Group. *Ginkgo biloba* special extract in dementia with neuropsychiatric features. A randomised, placebo-controlled, double-blind clinical trial. *Arzneimittel-Forschung* 2007;**57**(1):4–11.

31. Birks, Grimley Evans. *Ginkgo biloba* for cognitive impairment and dementia. *The Cochrane Database of Systematic Reviews* April 18, 2007;(2):CD003120.

32. Scripnikov, Khomenko, Napryeyenko, et al. Effects of *Ginkgo biloba* extract EGb 761 on neuropsychiatric symptoms of dementia: findings form a randomised controlled trial. *Wiener Medizinische Wochenschrift* 2007;**157**(13–14):295–300.

33. Dodge, Zitzelberger, Oken, et al. A randomized placebo-controlled trial of *Ginkgo biloba* for the prevention of cognitive decline. *Neurology* May 6, 2008;**70**(19 Pt. 2):1809–17.

34. DeKosky, Williamson, Fitzpatrick, et al. *Ginkgo biloba* for prevention of dementia: a randomized controlled trial. *Journal of the American Medical Association* November 19, 2008;**300**(19):2253–62.

35. McCarney, Fisher, Iliffe, et al. *Ginkgo biloba* for mild to moderate dementia in a community setting: a pragmatic, randomised, parallel-group, double-blind, placebo-controlled trial. *International Journal of Geriatric Psychiatry* December 2008;**23**(12):1222–30.

36. Birks, Grimley Evans. *Ginkgo biloba* for cognitive impairment and dementia. *The Cochrane Database of Systematic Reviews* January 21, 2009;(1):CD003120.

37. Snitz, O'Meara, Carlson, et al. *Ginkgo biloba* for preventing cognitive decline in older adults: a randomized trial. *Journal of the American Medical Association* December 23, 2009;**302**(24):2663–70.

38. Janssen, Sturtz, Skipka, et al. *Ginkgo biloba* in Alzheimer's disease: a systematic review. *Wiener Medizinische Wochenschrift* December 2010;**160**(21–22):539–46.

39. Weinmann, Roll, Schwarzbach, et al. Effects of *Ginkgo biloba* in dementia: systematic review and meta-analysis. *BioMed Central Geriatrics* March 17, 2010;**10**:14.

40. Ihl, Bachinskaya, Korczyn, et al. Efficacy and safety of a once-daily formulation of *Ginkgo biloba* extract EGb 761 in dementia with neuropsychiatric features: a randomized controlled trial. *International Journal of Geriatric Psychiatry* November 2011;**26**(11):1186–94.

41. Bachinskaya, Hoerr, Ihl. Alleviating neuropsychiatric symptoms in dementia: the effects of *Ginkgo biloba* extract EGb 761. Findings from a randomized controlled trial. *Neuropsychiatric Disease and Treatment* 2011;**7**:209–15.

42. Ihl, Tribanek, Bachinskaya, GOTADAY Study Group. Efficacy and tolerability of a once daily formulation of *Ginkgo biloba* extract EGb 761® in Alzheimer's disease and vascular dementia: results from a randomised controlled trial. *Pharmacopsychiatry* March 2012;**45**(2):41–6.

43. Herrschaft, Nacu, Likhachev, et al. *Ginkgo biloba* extract EGb 761® in dementia with neuropsychiatric features: a randomised, placebo-controlled trial to confirm the efficacy and safety of a daily dose of 240 mg. *Journal of Psychiatric Research* June 2012;**46**(6):716–23.

44. Kasper. Clinical data in early intervention. *International Psychogeriatrics* August 2012;**24**(Suppl. 1):S41–5.

45. Lautenschlager, Ihl, Müller. *Ginkgo biloba* extract EGb 761® in the context of current developments in the diagnosis and treatment of age-related cognitive decline and Alzheimer's disease: a research perspective. *International Psychogeriatrics* August 2012;**24**(Suppl. 1):S46–50.

46. Vellas, Coley, Ousset, et al. Long-term use of standardised *Ginkgo biloba* extract for the prevention of Alzheimer's disease (GuidAge): a randomised placebo-controlled trial. *The Lancet Neurology* October 2012;**11**(10):851–9.

47. Yasuno, Tanimukai, Sasaki, et al. Combination of antioxidant supplements improved cognitive function in the elderly. *Journal of Alzheimer's Disease* 2012;**32**(4):895–903.

48. Jiang, Su, Cui, et al. *Ginkgo biloba* extract for dementia: a systematic review. *Shanghai Archives of Psychiatry* February 2013;**25**(1):10–21.

49. Ihl. Effects of *Ginkgo biloba* extract EGb 761® in dementia with neuropsychiatric features: review of recently completed randomised, controlled trials. *International Psychogeriatrics* November 2013;**17**(Suppl. 1):8–14.

50. Gavrilova, Preuss, Wong, GIMCIPlus Study Group, et al. Efficacy and safety of *Ginkgo biloba* extract EGb 761 in mild cognitive impairment with neuropsychiatric symptoms: a randomized, placebo-controlled, double-blind, multi-center trial. *International Journal of Geriatric Psychiatry* October 2014; **29**(10):1087–95.

51. Gauthier, Schlaefke. Efficacy and tolerability of *Ginkgo biloba* extract EGb 761® in dementia: a systematic review and meta-analysis of randomized placebo-controlled trials. *Clinical Interventions in Aging* November 28, 2014;**9**:2065–77.

52. Charemboon, Jaisin. *Ginkgo biloba* for prevention of dementia: a systematic review and meta-analysis. *Journal of the Medical Association of Thailand* May 2015;**98**(5):508–13.

53. Solfrizzi, Panza. Plant-based nutraceutical interventions against cognitive impairment and dementia: meta-analytic evidence of efficacy of a standardized *Ginkgo biloba* extract. *Journal of Alzheimer's Disease* 2015;**43**(2):605–11.

54. Tan, Yu, Tan, et al. Efficacy and adverse effects of *Ginkgo biloba* for cognitive impairment and dementia: a systematic review and meta-analysis. *Journal of Alzheimer's Disease* 2015;**43**(2):589–603.

55. von Gunten, Schlaefke, Überla. Efficacy of *Ginkgo biloba* extract EGb 761® in dementia with behavioural and psychological symptoms: a systematic review. *The World Journal of Biological Psychiatry* August 27, 2015:1–12.

56. The Mayo Clinic: Drugs and Supplements: Ginkgo (*Ginkgo biloba*): Evidence. At: http://www.mayo-clinic.org/drugs-supplements/ginkgo/evidence/hrb-20059541.

57. Gupta. Intention-to-treat concept: a review. *Perspectives in Clinical Research* July 2011;**2**(3):109–12.

58. Schneider. *Ginkgo biloba* extract and preventing Alzheimer disease. *JAMA* November 19, 2008;**300**(19):2306–8.

59. See, DeKosky, Williamson, Fitzpatrick, et al. *Ginkgo biloba* for prevention of dementia: a randomized controlled trial. *JAMA* November 19, 2008;**300**(19):2253–62. Article Information about the GEM study group's clinical trial involving EGb 761 was reported as follows: "**Corresponding Author**: Steven T. DeKosky, MD, University of Virginia School of Medicine, PO Box 800793, Charlottesville, VA 22908. **Author Contributions**: Dr. DeKosky had full access to all of the data in the study and takes responsibility for the integrity of the data and the accuracy of the data analysis. Study concept and design: DeKosky, Williamson, Kronmal, Lopez, Burke, Fried, Kuller, Robbins, Tracy, Dunn, Nahin, Furberg. Acquisition of data: DeKosky, Williamson, Ives, Saxton, Lopez, Burke, Carlson, Fried, Kuller, Robbins, Tracy, Woolard, Dunn. Analysis and interpretation of data: DeKosky, Williamson, Fitzpatrick, Kronmal, Saxton, Lopez, Burke, Fried, Kuller, Robbins, Tracy, Dunn, Snitz, Nahin, Furberg. Drafting of the manuscript: DeKosky, Williamson, Fitzpatrick, Kronmal, Ives, Dunn, Furberg. Critical revision of the manuscript for important intellectual content: DeKosky, Williamson, Fitzpatrick, Kronmal, Saxton, Lopez, Burke, Carlson, Fried, Kuller, Robbins, Tracy, Woolard, Snitz, Nahin, Furberg. Statistical analysis: Fitzpatrick, Kronmal. Obtained funding: DeKosky, Kronmal, Burke, Fried, Kuller, Robbins, Furberg. Administrative, technical, or material support: DeKosky, Williamson, Fitzpatrick, Kronmal, Ives, Saxton, Lopez, Burke, Carlson, Fried, Kuller, Robbins, Tracy, Woolard, Dunn, Furberg. Study supervision: DeKosky, Williamson, Kronmal, Ives, Saxton, Nahin, Furberg."

INDEX